W9-ART-012

# GREAT EVENTS

## 1900-2001

# GREAT EVENTS

## 1900-2001
### REVISED EDITION

## Volume 3
### 1939-1957

*From*

**The Editors of Salem Press**

SALEM PRESS, INC.

Pasadena, California    Hackensack, New Jersey

*Editor in Chief:* Dawn P. Dawson

*Managing Editor:* R. Kent Rasmussen  *Research Supervisor:* Jeffry Jensen
*Manuscript Editor:* Rowena Wildin  *Acquisitions Editor:* Mark Rehn
*Production Editor:* Joyce I. Buchea  *Page Design and Graphics:* James Hutson
*Photograph Editor:* Philip Bader  *Layout:* William Zimmerman
*Assistant Editor:* Andrea E. Miller  Eddie Murillo

*Cover Design:* Moritz Design, Los Angeles, Calif.

*Cover photos:* Center image—Corbis, Remaining images—AP/Wide World Photos
*Half title photos:* Library of Congress, Digital Stock, AP/Wide World Photos

© 2002 *Great Events: 1900-2001, Revised Edition*
© 1997 *The Twentieth Century: Great Scientific Achievements, Supplement* (3 volumes)
© 1996 *The Twentieth Century: Great Events, Supplement* (3 volumes)
© 1994 *The Twentieth Century: Great Scientific Achievements* (10 volumes)
© 1992 *The Twentieth Century: Great Events* (10 volumes)

∞ The paper used in these volumes conforms to the American National Standard for Permanence of Paper for Printed Library Materials, Z39.48-1992 (R1997).

**Library of Congress Cataloging-in-Publication Data**

Great events : 1900-2001 / editors of Salem Press.— Rev. ed.
    v.  cm.
Includes index.
    ISBN 1-58765-053-3 (set : alk. paper) — ISBN 1-58765-056-8 (vol. 3 : alk. paper)
    1. History, Modern—20th century—Chronology. 2. Twentieth century. 3. Science—History—20th century—Chronology. 4. Technology—History—20th century—Chronology.
D421 .G627 2002
909.82—dc21

                                                                                    2002002008

Second Printing

PRINTED IN THE UNITED STATES OF AMERICA

# CONTENTS

page

# COMPLETE LIST OF CONTENTS

**VOLUME 1**

## VOLUME 2

COMPLETE LIST OF CONTENTS

**xiii**

# VOLUME 3

**XV**

# VOLUME 4

## xvii

# VOLUME 5

**xix**

**xxi**

## VOLUME 6

# VOLUME 7

**xxvii**

## VOLUME 8

**xxix**

# GREAT EVENTS

## 1900-2001

# Allies Win Battle of Atlantic

---

*The fight for control of the Atlantic Ocean was long and difficult, but the United States and Great Britain eventually prevailed over Nazi Germany.*

---

**What:** War
**When:** September, 1939-May, 1945
**Where:** North Atlantic Ocean
**Who:**
FRANKLIN DELANO ROOSEVELT (1882-1945), president of the United States from 1933 to 1945
SIR WINSTON CHURCHILL (1874-1965), prime minister of Great Britain from 1940 to 1945
ADOLF HITLER (1889-1945), chancellor of Germany from 1933 to 1945
HAROLD RAYNSFORD STARK (1880-1972), chief of U.S. Naval Operations from 1939 to 1941
ERNEST JOSEPH KING (1878-1956), chief of U.S. Naval Operations from 1941 to 1945
KARL DÖNITZ (1891-1980), commander of the German submarine fleet from 1936 to 1945, and commander in chief of the German navy from 1943 to 1945

## The War at Sea

When World War II began in September, 1939, neither Great Britain nor Germany realized how important the struggle for control of the Atlantic Ocean would be. Both of these countries were naval powers. Yet Adolf Hitler's strategy for winning the war ignored the advice of his admirals until it was too late. He was convinced that control of the European Continent was the key to victory and that naval weapons were not a first priority. Consequently, until 1943 Hitler provided little support for the construction of submarines. When war broke out, only forty-three U boats were ready for combat.

Aware of this fact, the British felt protected by asdic (sonar), which had only recently been developed. They did not realize what a great impact the German submarines would have on the movement of British warships and merchant ships.

When France fell to the Germans in 1940, the German navy built submarine pens in the Bay of Biscay. This allowed the U-boat fleet to expand operations. Also, Admiral Karl Dönitz, commander of the German submarine fleet, developed new techniques of attack. Instead of moving out alone, U-boats began to hunt in "wolf packs" which operated on the surface of the ocean at night. Catching their targets by moonlight or silhouetted against lights on the coast, the surfaced submarines moved into torpedo-launching position at high speeds. Then, having fired their torpedoes, they submerged and escaped in the resulting confusion.

In the first seven months of the war, Allied merchant ships lost to German submarines averaged about 110,000 tons per month. In June, 1940, however, that figure jumped to 300,000; by October it had reached almost 450,000 tons. With U-boats sinking Allied ships faster than they could be replaced, it appeared that the Allies might lose the Battle of the Atlantic.

## The United States Enters

In desperation, Winston Churchill, who had become prime minister of Great Britain in May, 1940, turned to the United States for help. Though the United States had tried to remain officially neutral in the war, President Franklin D. Roosevelt found ways to help the Allies. In September, Roosevelt authorized giving fifty destroyers to Great Britain in exchange for leases to British lands in the western Atlantic.

Two months later, Churchill explained that Great Britain could no longer afford to buy American products. Roosevelt responded by proposing the Lend-Lease Bill to Congress. This bill, which was passed in March, 1941, in spite of the objections of isolationists, made American equipment and supplies available to defend any country the president decided was important to the defense of the United States.

In April, 1941, Roosevelt warned all nations at war to stay out of the western half of the Atlantic; when American patrols came upon the ships or submarines of any warring nation, their locations would be transmitted by radio in English. Though the United States was remaining officially neutral, this policy was especially helpful to British convoys trying to avoid German submarines.

Three months later, the United States obtained Iceland's permission to set up a naval base there. Soon American warships were escorting British merchant ships from North America to Iceland.

Though the Americans were clearly committed to the Allied cause, Hitler remained cautious. He continued to order his U-boat commanders to avoid hostile contact with American ships. Yet by accident or on purpose, such contact was bound to occur. In October, 1941, the destroyer *Kearny* was torpedoed, and eleven men were lost; six weeks later, the *Reuben James* was sliced in two, and ninety-six men died. On December 11, 1941, Hitler finally declared war on the United States.

In the first year of American involvement in World War II, Germany continued to make gains. Though most of the German surface fleet was stuck in Baltic and Norwegian ports, U-boats became an even greater threat to Allied ships. Admiral Dönitz continually shifted his submarines to positions where they were least expected. All along the western Atlantic, from Newfoundland to Brazil, unescorted merchant ships were attacked. From January through November, 1942, more than 500,000 tons of Allied merchant shipping were lost each month.

Yet by the end of the year, it became clear that time was on the side of the Allies. In December, the loss of Allied merchant tonnage decreased to 400,000 tons; at the same time, the construction of replacement ships rose to 1,000,000 tons for the first time. After March, 1943, this trend was even stronger.

The main reason for the change in Germany's fortunes was simply that the United States was richer than Germany in raw materials and human resources. By the end of the war, most Allied merchant convoys were being escorted by destroyers, and the use of aircraft carriers had become much more common.

American resources were used skillfully under the leadership of chief of Naval Operations Admiral Harold R. Stark and his successor, Admiral Ernest J. King. Stark and King were both deeply convinced that the convoy was the only way to ferry troops and materials over an ocean crisscrossed by submarines. Their belief was proved correct as the Battle of the Atlantic wore on to its end.

With greater experience, American crews were able to make better use of their equipment. Merchant ships could be guided more precisely within their convoys, while escorts learned the fine arts of sonar detection, extended sweeps, and saturation attacks using depth charges.

Technology was another factor in the American victory. Airplanes were equipped with microwave radar, which submarines could not detect or jam; along with the powerful Leigh searchlight, this air-based radar made it quite dangerous for submarines to surface at night. Both Dönitz and Hitler came to believe that microwave radar was the primary cause of Germany's defeat in the Atlantic.

**Consequences**

Though the war had increasingly turned against them, Admiral Dönitz and most of his officers and crews continued the fight to the bitter end. On May 8, 1945, when Dönitz finally ordered his remaining submarines to surrender, forty-nine were still at sea. The Battle of the Atlantic had proved to be a crucial ingredient in Germany's defeat in World War II.

# Germany Invades Poland

*Having succeeded in dividing the Soviet Union from its Western allies, Adolf Hitler sent tanks and warplanes to conquer Poland.*

**What:** War
**When:** September 1, 1939
**Where:** Germany and Poland
**Who:**
ADOLF HITLER (1889-1945), chancellor of Germany from 1933 to 1945
JÓZEF BECK (1894-1944), foreign minister of Poland from 1932 to 1939
JOSEPH STALIN (1879-1953), dictator of the Soviet Union from 1929 to 1953
ARTHUR NEVILLE CHAMBERLAIN (1869-1940), prime minister of Great Britain from 1937 to 1940
ÉDOUARD DALADIER (1884-1970), premier of France from 1938 to 1940

## Maneuvers and Demands

In 1938, Adolf Hitler began planning to annex Czechoslovakia, a plan that was fulfilled on March 15, 1939, when German troops marched into Prague. With this success, Hitler and his henchmen turned their attention to Poland.

On March 21, 1939, German foreign minister Joachim von Ribbentrop summoned Polish foreign minister Józef Beck to Berlin and gave him a list of German demands: The Poles had certain economic rights in Danzig, a free city governed under the supervision of the League of Nations; Ribbentrop demanded that Danzig come under Germany's control; Germany must be allowed to build a German railway and road across the Polish Corridor, a strip of Polish land separating East Prussia from the rest of Germany; Finally, Poland must join Germany in promising to oppose the Soviet Union. Not surprisingly, Beck said no to these demands.

In the meantime, Great Britain and France had been greatly alarmed by Germany's seizure of Prague. In response, they guaranteed Poland's independence; on March 31, British prime minister Neville Chamberlain announced that if Germany invaded Poland, Great Britain and France would give Poland all the support they could. The French and British could not promise direct help, however, because Germany separated them from Poland. Poland's best help could come only from the Soviet Union, and in the following weeks, Moscow became an especially important center of diplomatic negotiations.

Soviet dictator Joseph Stalin was worried about Hitler's rising power. He decided to make contacts with both the Nazis and the Allies and see where his best chance of protection lay. On April 16, Soviet foreign minister Maksim Maksimovich Litvinov approached the Allies with an offer to write an agreement of mutual assistance. The next day, the Soviet ambassador in Berlin asked the German Foreign Ministry whether relations between Germany and the Soviet Union might be improved.

The Germans responded at once, though they did not promise anything. The Allies made no answer at all until three weeks later, for Chamberlain and his Conservative cabinet disliked Communism and were suspicious of the Soviets. Their answer to Litvinov was not very positive. Soon Litvinov, a Jew who was considered sympathetic to the Allies, was replaced by Vyacheslav Mikhailovich Molotov, who was ready to negotiate with the Germans.

## Ready for the Invasion

On May 22, 1939, Hitler scored another success when he made the "Pact of Steel" with Italy. Under this agreement, Italy and Germany promised to fight as allies if either of them went to war. Though Italian foreign minister Galeazzo Ciano objected, Italy's Fascist dictator Benito Mussolini agreed to bring his country into war on Hitler's terms. This alliance made Hitler's position stron-

**839**

*Adolf Hitler salutes his troops as they march toward a wooden bridge constructed by the Germans across the San River, near Jarolaw, Poland, in September, 1939.*

ger, and on May 23 he told his generals that war with Poland was now inevitable.

Hoping to keep the Soviet Union from helping Poland, the Nazis began serious talks with the Soviets on May 30. Things moved slowly over the next two months. The Soviet and German negotiators hesitated to commit their countries to specific promises. By now Great Britain and France had also sent negotiators to Moscow, but they too acted slowly and were unable to gain Beck's support for an agreement with the Soviets. Most of Poland had been occupied by Russia for more than a century before 1918, so it is understandable that the Poles were unwilling to allow the Soviet Red Army to reenter their country in 1939.

For his part, Stalin did not trust Great Britain and France. He doubted that they were really determined to resist Hitler, and he suspected that

they wanted to trap the Soviet Union in a war with Germany. In the end, Stalin decided that Hitler could provide him with more advantages than could Chamberlain and French premier Édouard Daladier.

In the early hours of August 24, the Russo-German Non-Aggression Pact was signed. In a secret section of the treaty, the two countries agreed on how they would divide Poland between them after it was conquered.

The day before the treaty was signed, Hitler had decided that his pact with Stalin would persuade Chamberlain and Daladier that it was useless to try to help Poland. He had ordered his generals to begin their attack at dawn on August 26. Yet the British intended to honor their guarantee. On August 25, Great Britain signed a formal treaty with Poland. That same day, Musso-

**840**

lini, aware that Italy's army had fewer than ten fully equipped divisions, informed Hitler that Italy would remain neutral if a conflict broke out. At this point Hitler paused, and in the evening he ordered the invasion to be postponed.

Hitler now tried to separate the Poles from their allies, but Daladier and Chamberlain stood firm. When it was clear that his efforts had failed, Hitler unleashed his army and air force on Poland. On September 1, at 4:45 A.M., six panzer (armored) divisions rolled across the Polish border, while the Luftwaffe (air force) struck at the Polish airfields.

### Consequences

The Germans had devised a new type of warfare, combining tanks and mobile artillery with aircraft. They moved rapidly and scored quick successes. On September 3, Great Britain and France kept their promise to Poland and declared war on Germany. British and French leaders had hoped that good luck and bad weather would combine to help Poland resist for six months; instead, they saw the Polish army torn to pieces within two weeks.

On September 17, the Soviets hastily invaded Poland from the east to make sure that they got their share of its territory. By October 2, all organized Polish resistance had come to an end. France and Great Britain had been unable to do anything to protect their ally.

The invasion of Poland marked the beginning of World War II. The war would not end until May 8, 1945, after more than thirty million Europeans had been killed.

*Samuel K. Eddy*

# First Color Television Signals Are Broadcast

> *RCA and CBS both demonstrated color television systems.*

**What:** Communications
**When:** 1940
**Where:** New York, New York
**Who:**
PETER CARL GOLDMARK (1906-1977), the head of the CBS research and development laboratory
WILLIAM S. PALEY (1901-1990), the businessman who took over CBS
DAVID SARNOFF (1891-1971), the founder of RCA

## The Race for Standardization

Although by 1928 color television had already been demonstrated in Scotland, two events in 1940 mark that year as the beginning of color television. First, on February 12, 1940, the Radio Corporation of America (RCA) demonstrated its color television system privately to a group that included members of the Federal Communications Commission (FCC), an administrative body that had the authority to set standards for an electronic color system. The demonstration did not go well; indeed, David Sarnoff, the head of RCA, canceled a planned public demonstration and returned his engineers to the Princeton, New Jersey, headquarters of RCA's laboratories.

Next, on September 1, 1940, the Columbia Broadcasting System (CBS) took the first step to develop a color system that would become the standard for the United States. On that day, CBS demonstrated color television to the public, based on the research of an engineer, Peter Carl Goldmark. Goldmark placed a set of spinning filters in front of the black-and-white television images, breaking them down into three primary colors and producing color television. The audience saw what was called "additive color."

Although Goldmark had been a researcher at CBS since January, 1936, he did not attempt to develop a color television system until March, 1940, after watching the Technicolor motion picture *Gone with the Wind* (1939). Inspired, Goldmark began to tinker in his tiny CBS laboratory in the headquarters building in New York City.

If a decision had been made in 1940, the CBS color standard would have been accepted as the national standard. The FCC was, at that time, more concerned with trying to establish a black-and-white standard for television. Color television seemed decades away. In 1941, the FCC decided to adopt standards for black-and-white television only, leaving the issue of color unresolved—and the doors to the future of color broadcasting wide open. Control of a potentially lucrative market as well as personal rivalry threw William S. Paley, the head of CBS, and Sarnoff into a race for the control of color television. Both companies would pay dearly in terms of money and time, but it would take until the 1960's before the United States would become a nation of color television watchers.

RCA was at the time the acknowledged leader in the development of black-and-white television. CBS engineers soon discovered, however, that their company's color system would not work when combined with RCA black-and-white televisions. In other words, customers would need one set for black-and-white and one for color. Moreover, since the color system of CBS needed more broadcast frequency space than the black-and-white system in use, CBS was forced to ask the FCC to allocate new channel space in the ultrahigh frequency (UHF) band, which was then not being used. In contrast, RCA scientists labored to make a compatible color system that required no additional frequency space.

## No Time to Wait

Following the end of World War II, in 1945, the suburbanites that populated new communities in America's cities wanted television sets right away; they did not want to wait for the government to decide on a color standard and then wait again while manufacturers redesigned assembly lines to make color sets. Rich with savings accumulated during the prosperity of the war years, Americans wanted to spend their money. After the war, the FCC saw no reason to open up proceedings regarding color systems. Black-and-white was operational; customers were waiting in line for the new electronic marvel. To give its engineers time to create a compatible color system, RCA skillfully lobbied the members of the FCC to take no action.

There were other problems with the CBS mechanical color television. It was noisy and large, and its color balance was hard to maintain. CBS claimed that through further engineering work, it would improve the actual sets. Yet RCA was able to convince other manufacturers to support it in preference to CBS principally because of its proven manufacturing track record.

In 1946, RCA demonstrated a new electronic color receiver with three picture tubes, one for each of the primary colors. Color reproduction was fairly true; although any movement on the screen caused color blurring, there was little flicker. It worked, however, and thus ended the invention phase of color television begun in 1940. The race for standardization would require seven more years of corporate struggle before the RCA system would finally win adoption as the national standard in 1953.

## Consequences

Through the 1950's, black-and-white television remained the order of the day. Through the later years of the decade, only the National Broadcasting Company (NBC) television network was regularly airing programs in color. Full production and presentation of shows in color during prime time did not come until the mid-1960's; most industry observers date 1972 as the true arrival of color television.

By 1972, color sets were found in more than half the homes in the United States. At that point, since color was so widespread, *TV Guide* stopped tagging color program listings with a special symbol and instead tagged only black-and-white shows. Gradually, only cheap, portable sets were made for black-and-white viewing, while color sets came in all varieties from tiny hand-held pocket televisions to mammoth projection televisions.

*Douglas Gomery*

# Landsteiner Identifies Rhesus Factor

*Landsteiner's discovery of the Rh blood group led to an understanding of a deadly blood disease and a way of preventing it.*

**What:** Medicine
**When:** 1940
**Where:** New York, New York
**Who:**
KARL LANDSTEINER (1868-1943), an immunologist who discovered the ABO blood group in 1901
PHILIP A. LEVINE (1900-1987), an immunologist who shared with Landsteiner the discovery of the MN blood group in 1927 and who helped to characterize the Rh blood group in 1940
ALEXANDER S. WIENER (1907-1976), an immunologist who shared with Landsteiner the discovery of the Rh blood group in 1940

## Antibodies, Antigens, and the Clumping of Red Blood Cells

The liquid part of blood contains T-shaped molecules called "antibodies." Antibodies may bind with structures called "antigens" on the surfaces of cells and cause the cells to clump. The clumping of red blood cells (RBCs) occurs when the two "arms" of an antibody bind with different cells. The clumping of RBCs by antibodies can be so extensive that it can be seen in a drop of blood without the aid of a microscope.

Landsteiner and his colleagues suggested in 1901, after their discovery of the A and B antigens of the ABO blood group, that RBC clumping was responsible for the many cases of illness and death that occurred following blood transfusions. In the United States, the routine analysis of donor and recipient blood to avoid the mixing of blood that might clump began in 1915. Even when donor and recipient bloods were matched, however, transfusions were not always trouble-free. Apparently, another blood group besides the ABO blood group could cause problems.

Karl Landsteiner and Alexander Wiener reported in 1940 that human RBCs could have on their surfaces a number of different antigens ($R^1$, $R^2$, $R^0$, and $R^z$). An antibody called "anti-$Rh^0$" clumped RBCs that had any of these antigens. For simplicity, these four antigens are referred to as the "Rh-positive" ($Rh^+$) antigen. Other forms of the Rh antigen that are not clumped by anti-$Rh^0$ (r, r", r', $r^y$) are referred to as the "Rh-negative" ($Rh^-$) antigen. The Rh antigens constitute what is known as the Rh blood group.

## The Rh Blood Group and Its Relationship to Disease

About one in every thousand babies born each year suffers from anemia (deficiency of RBCs) and jaundice (yellowing of the skin) caused by the destruction of red blood cells. The destruction of fetal RBCs is caused by antibodies produced by the mother, which pass through the placenta from the mother's blood to the baby's blood. The mother's antibodies clump the baby's RBCs. The disease that results is called "erythroblastosis fetalis." Severe cases of the disease result in mental retardation or death. The cause of erythroblastosis fetalis was suggested by Landsteiner and Wiener after their discovery of the Rh antigens on red blood cells in 1940.

Landsteiner and Wiener discovered that rabbit antibodies made against RBCs from the rhesus monkey would clump not only monkey RBCs but also most human red blood cells. This indicated that human and rhesus monkey RBCs both had antigens that were very similar or identical. Landsteiner called the human antigen the Rh antigen because of its similarity to the antigen

**844**

found on the cells of rhesus monkeys. Blood that was clumped by anti-Rh$^0$ was said to be Rh positive (Rh$^+$). It was discovered that blood from approximately 15 percent of Caucasians was not clumped by antibodies against the Rh antigen. These individuals lacked the Rh antigen, and their blood was labeled Rh negative (Rh$^-$). Wiener suggested that some of the destruction of RBCs during blood transfusions was caused by Rh incompatibilities. For example, if an Rh$^-$ person were transfused with Rh$^+$ blood, any antibodies in the recipient against the Rh antigen would clump the transfused blood.

Philip Levine, one of Landsteiner's coworkers, demonstrated the relationship between erythroblastosis fetalis and the Rh antigens. He described a mother who was Rh$^-$ and her Rh$^+$ newborn who was suffering from erythroblastosis fetalis. Levine proposed that the mother had antibodies against Rh antigens and that the antibodies passed from her blood through the placenta to the baby's blood, where they clumped the baby's RBCs. Clumped RBCs cause problems when they block capillaries. In addition, clumped red blood cells are destroyed in the spleen. This leads to anemia and jaundice.

Levine's evidence for antibodies against the Rh antigens was the severe reaction the Rh$^-$ mother had when she was transfused with her Rh$^+$ husband's blood. The destruction of the husband's blood was not caused by ABO blood type incompatibility; both the wife and the husband had the same ABO blood type. They differed only in their Rh blood type, as demonstrated by antibodies against the Rh antigen that clumped the husband's but not the wife's blood.

When an Rh$^-$ mother gives birth to an Rh$^+$ child, the child's RBCs occasionally stimulate the mother to produce antibodies against the Rh antigen. There is no danger to the child, because it takes three to four weeks for the antibodies to be synthesized. Subsequent Rh$^+$ babies, however,

*Karl Landsteiner.*

Library of Congress

may be attacked before birth by the mother's antibodies. These antibodies pass from the mother through the placenta into the fetus. The clumping and destruction of the fetus's RBCs results in anemia and jaundice. Newborn babies suffering from erythroblastosis fetalis often can be saved by a blood transfusion.

By inoculating Rh$^-$ mothers who have given birth to Rh$^+$ babies with antibodies against the Rh antigen, erythroblastosis fetalis can be prevented in subsequent Rh$^+$ babies. It is believed that the development of antibodies against the Rh$^+$ antigens leads to the destruction of the Rh$^+$ RBCs in the Rh$^-$ mother before the Rh$^+$ antigens can stimulate the synthesis of antibodies.

## Consequences

The discovery of the Rh antigen and how it is involved in transfusion reactions and erythroblastosis fetalis has resulted in the saving of thousands of lives each year. Blood used for transfusions is checked not only for its A and B antigens

**845**

but also for its Rh antigens. Red blood cells used in transfusions must be compatible with the recipient's blood. For example, the safest blood to give to a person who has Rh⁻ blood would be Rh⁻ blood. The recipient would have no antibodies that would agglutinate the transfused blood. The use of Rh⁺ blood would be dangerous if the recipient had been exposed to Rh⁺ antigens previously—for example, through a prior transfusion or a pregnancy with an Rh⁺ baby.

The discovery of the Rh antigens and antibodies against them led to an explanation of how erythroblastosis fetalis occurred. Once the mechanism for the hemolytic disease was known, a way of preventing the disease was devised. Presently, Rh⁻ women who give birth to Rh⁺ children are given antibodies against the Rh antigen that may enter their blood. This has been shown to prevent the development of antibodies against the Rh antigens. Thousands of cases of erythroblastosis fetalis have been avoided by this procedure, and the lives of many babies have been saved.

*Jaime S. Colome*

# Royal Air Force Wins Battle of Britain

*The Battle of Britain, fought in the air, was the first major defeat of German armed forces in World War II.*

**What:** War
**When:** 1940-1941
**Where:** England and the English Channel
**Who:**
AIR MARSHAL SIR HUGH DOWDING (1882-1970), commander in chief of the Fighter Command, Royal Air Force, from 1936 to 1940
AIR VICE MARSHAL KEITH R. PARK (1892-1975), commander of Group 11 in the Fighter Command, Royal Air Force
HERMANN GÖRING (1893-1946), German interior minister and minister of aviation from 1933 to 1945

## The Preparations

In June, 1940, after the Germans had conquered Belgium, Holland, and France, Great Britain stood alone against the forces of German chancellor Adolf Hitler. British prime minister Winston Churchill spoke to the British people: "Hitler knows he will have to break us on this Island or lose the war." Most British military leaders believed that the Germans would soon try to invade England.

In that summer, Great Britain did not stand a very good chance of turning back an invasion. The Royal Navy did control the seas around the British island, but many of its ships were needed to protect the Atlantic supply routes from the United States against German U-boat attacks. The army had been weakened, and large numbers of tanks and heavy weapons had been abandoned in France when the British Expeditionary Force was evacuated from Dunkirk.

So it was that the defense of Great Britain was mostly left to the Royal Air Force (RAF), especially the Fighter Command planes. Since 1936,

Sir Hugh Dowding, head of the Fighter Command, had worked to persuade the cabinet and the Air Council to make fighter planes a priority over bombers, for he believed that the coming war would put Great Britain in a defensive position. He also stressed the need for improved detection and early warning of approaching enemy aircraft. In spite of considerable opposition, Dowding had been successful. In 1939 the Air Council had ordered that production of fighter planes be stepped up and that an early warning system be constructed.

British designers created two types of improved fighter planes: the Hurricanes and the Spitfires. Both flew at speeds of up to three hundred miles per hour, which was fast for the time. They had heavy armor, constant-speed propellers, and self-sealing fuel tanks. Each was armed with eight machine guns. These airplanes, together with the pilots who flew them, would win the Battle of Britain.

During 1937, British physicists had worked on aircraft detection through radio-wave signals, and what later would be known as radar was developed. The British began to build a linked system of radar stations, ground observation units, and control bases for the Fighter Command.

Meanwhile, Hitler had been unsuccessful in his efforts to persuade the British to negotiate a settlement and bring an end to the war. Reluctantly, he approved Operation Sea Lion, a plan for the invasion of England. September 21 was set as the day for the first German crossings of the English Channel. Ninety thousand German troops would make up the first assault force.

The Germans prepared rapidly for Operation Sea Lion. More than twelve hundred boats and barges were brought into ports across the Channel from England. Troops were trained in landing procedures, and bases were built for the aircraft that would escort the ships.

The British continued their preparations as well. Under General Alan Brooke, the Home Guard was increased to 500,000 men. Mobile field guns, antitank weapons, and small arms were provided, and more than two million bomb shelters were built.

Field Marshal Hermann Göring, Commander of the Luftwaffe (German air force), was sure that his forces could gain control of the air over southern England. In fact, he believed that his bombers would pulverize Great Britain so badly within a month's time that the British would have to surrender and an invasion across the Channel would be unnecessary. Indeed, with more than thirteen hundred bombers and twelve hundred fighter planes, the Luftwaffe in Western Europe was considerably larger than the RAF.

Dowding's Fighter Command had only about 700 front-line fighters, with another 350 in reserve. It was necessary that these fighters operate all over Great Britain. Group 11, commanded by Air Vice Marshal Keith Park, would have the largest share of fighters, but they would probably be outnumbered ten to one by the German attacking planes. Moreover, the British had only about 1,400 fully trained fighter pilots to fly the planes.

## The Battle

During July and early August, the Luftwaffe attacked from time to time, mostly striking against British ships in the Channel. Dover was bombed as well. About 150 British civilians were killed and twenty ships sunk, but the Luftwaffe also lost dozens of its planes to RAF fighters and to antiaircraft fire.

Next the Germans began bombing the radar stations and airfields in southeastern England. The British lost many men and considerable amounts of equipment, but by mid-August the airfields had been quickly patched up and the radar stations rebuilt. Fighter planes were being produced more rapidly than ever, and more British pilots were being trained.

*Smoke billows beyond the Tower Bridge over the Thames during the first mass air raid on London in September, 1940.*

Operation Eagle, a saturation bombing of the ports and airfields of southern England, was the next German offensive. It began on August 13. The attacking aircraft came in several waves, and Park's fighters rose to meet them. Some of the German bombers got through and inflicted damage, but the Hurricanes and Spitfires shot down forty-seven of the enemy planes. Thirteen British aircraft were lost.

After two days of bad flying weather, the German attacks began anew. Göring declared that the goal was to obliterate the RAF planes and facilities. On August 15 and 16, German bombers managed to destroy four aircraft factories and five airfields near London. Yet in those two days the Luftwaffe lost seventy-six planes—the worst loss it ever suffered in such a short time.

During the last week of August and the first week of September, 1940, there were more than thirty major German attacks, averaging more than one thousand planes each time. Most of the bombs fell on Group 11's airfields and sector stations. In those two weeks, 103 RAF pilots were killed or reported missing. The Fighter Command was on the verge of defeat.

Then, in early September, Göring made a crucial mistake. Believing that the Fighter Command had been practically wiped out, he ordered the Luftwaffe to stop bombing RAF facilities and to make terror attacks on London and other cities instead.

On September 7, the British government sent out the code signal "Cromwell": The expected invasion was at hand. That day, nearly four hundred German bombers hit East London, killing more than one thousand civilians and damaging many houses, docks, and warehouses. That night, another 250 bombers did more damage to the British capital. Park sent up his fighters to intercept them, and in the air battles thirty-eight German planes and twenty-eight British planes went down.

Park had proved that the Fighter Command was still functioning. London was bombed again on September 9, but the attack was weaker because only about half of the attacking planes got through to their targets. Meanwhile, the Royal Navy and the RAF Bomber Command increased their shelling and bombing of the German invasion ports across the Channel. Hitler postponed Operation Sea Lion to September 27.

Göring's last effort came on September 15. One hundred twenty-three bombers, each escorted by five fighters, were sent out, and they were met by Fighter Command squadrons. The air battle began about noon and lasted until evening. At the end of the day, sixty German planes had been destroyed, while the British had lost only twenty-six. September 15, 1940, would later be identified as the day the British won the Battle of Britain.

**Consequences**

On September 17, aware that the Luftwaffe could not gain control of the British skies and that the weather was worsening, Hitler postponed Operation Sea Lion indefinitely. Instead, he began to plan for the invasion of the Soviet Union.

In the "Blitz" of 1940-1941, Great Britain had to endure considerable German bombing. Yet because of the RAF successes in the summer of 1940, there was no longer a threat of invasion. The victory raised the spirits of the British people, and it also convinced others—especially the Americans—that the final defeat of Germany was possible.

*James W. Pringle*

# Spencer Discovers Principle of Microwave Cooking

Percy L. Spencer discovered the use of microwave energy for cooking and helped to develop the microwave oven.

**What:** Food science
**When:** 1940-1955
**Where:** Newton, Massachusetts
**Who:**
PERCY L. SPENCER (1894-1970), an American engineer
HEINRICH HERTZ (1857-1894), a German physicist
JAMES CLERK MAXWELL (1831-1879), a Scottish physicist

## The Nature of Microwaves

Microwaves are electromagnetic waves, as are radio waves, X rays, and visible light. Water waves and sound waves are wave-shaped disturbances of particles in the media—water in the case of water waves and air or water in the case of sound waves—through which they travel. Electromagnetic waves, however, are wavelike variations of intensity in electric and magnetic fields.

Electromagnetic waves were first studied in 1864 by James Clerk Maxwell, who explained mathematically their behavior and velocity. Electromagnetic waves are described in terms of their wavelength and frequency. The wavelength is the length of one cycle, which is the distance from the highest point of one wave to the highest point of the next wave, and the frequency is the number of cycles that occur in one second. Frequency is measured in units called "hertz," named for the German physicist Heinrich Hertz. The frequencies of microwaves run from 300 to 3,000 megahertz (1 megahertz equals 1 million hertz, or 1 million cycles per second), corresponding to wavelengths of 100 to 10 centimeters.

Microwaves travel in the same way that light waves do; they are reflected by metallic objects, absorbed by some materials, and transmitted by other materials. When food is subjected to microwaves, it heats up because the microwaves make the water molecules in foods (water is the most common compound in foods) vibrate. Water is a "dipole molecule," which means that it contains both positive and negative charges. When the food is subjected to microwaves, the dipole water molecules try to align themselves with the alternating electromagnetic field of the microwaves. This causes the water molecules to collide with one another and with other molecules in the food. Consequently, heat is produced as a result of friction.

## Development of the Microwave Oven

Percy L. Spencer apparently discovered the principle of microwave cooking while he was experimenting with a radar device at the Raytheon Company. A candy bar in his pocket melted after being exposed to microwaves. After realizing what had happened, Spencer made the first microwave oven from a milk can and applied for two patents, "Method of Treating Foodstuffs" and "Means for Treating Foodstuffs," on October 8, 1945, giving birth to microwave-oven technology.

Spencer wrote that his invention "relates to the treatment of foodstuffs and, more particularly, to the cooking thereof through the use of electromagnetic energy." Though the use of electromagnetic energy for heating was recognized at that time, the frequencies that were used were lower than 50 megahertz. Spencer discovered that heating at such low frequencies takes a long time. He eliminated the time disadvantage by using shorter wavelengths in the microwave region. Wavelengths of 10 centimeters or shorter were comparable to the average dimensions of foods. When these wavelengths were used, the heat that was generated became intense, the energy that was required was minimal, and the process became efficient enough to be exploited commercially.

Although Spencer's patents refer to the cooking of foods with microwave energy, neither deals directly with a microwave oven. The actual basis for a microwave oven may be patents filed by other researchers at Raytheon. A patent by Karl Stiefel in 1949 may be the forerunner of the microwave oven, and in 1950, Fritz Gross received a patent titled "Cooking Apparatus," which specifically describes an oven that is very similar to modern microwave ovens.

Perhaps the first mention of a commercial microwave oven was made in the November, 1946, issue of *Electronics* magazine. This article described the newly developed Radarange as a device that could bake biscuits in 29 seconds, cook hamburgers in 35 seconds, and grill a hot dog in 8 to 10 seconds. Another article that appeared a month later mentioned a unit that had been developed specifically for airline use. The frequency used in this oven was 3,000 megahertz. Within a year, a practical model 13 inches wide, 14 inches deep, and 15 inches high appeared, and several new models were operating in and around Boston. In June, 1947, *Electronics* magazine reported the installation of a Radarange in a restaurant, signaling the commercial use of microwave cooking. It was reported that this method more than tripled the speed of service. The Radarange became an important addition to a number of restaurants, and in 1948, Bernard Proctor and Samuel Goldblith used it for the first time to conduct research into microwave cooking.

In the United States, the radio frequencies that can be used for heating are allocated by the Federal Communications Commission (FCC). The two most popular frequencies for microwave cooking are 915 and 2,450 megahertz, and the 2,450 frequency is used in home microwave ovens. It is interesting that patents filed by Spencer in 1947 mention a frequency on the order of 2,450 megahertz. This fact is another example of Spencer's vision in the development of microwave cooking principles. The Raytheon Company concentrated on using 2,450 megahertz, and in 1955, the first domestic microwave oven was introduced. It was not until the late 1960's, however, that the price of the microwave oven decreased sufficiently for the device to become popular. The first patent describing a microwave heating system being used in conjunction with a conveyor was issued to Spencer in 1952. Later, based on this development, continuous industrial applications of microwaves were developed.

## Consequences

Initially, microwaves were viewed as simply an efficient means of rapidly converting electric energy to heat. Since that time, however, they have become an integral part of many applications. Because of the pioneering efforts of Percy L. Spencer, microwave applications in the food industry for cooking and for other processing operations have flourished. In the early 1970's, there were eleven microwave oven companies worldwide, two of which specialized in food processing operations, but the growth of the microwave oven industry has paralleled the growth in the radio and television industries. In 1984, microwave ovens accounted for more shipments than had ever been achieved by any appliance—9.1 million units.

By 1989, more than 75 percent of the homes in the United States had microwave ovens. By 2001, this figure exceeded 90 percent. During the 1990's, microwavable foods were among the fastest-growing products in the food industry. Microwave energy facilitates reductions in operating costs and required energy, higher-quality and more reliable products, and positive environmental effects. To some degree, the use of industrial microwave energy is still in its infancy. New and improved applications of microwaves will continue to appear.

# Germany Invades Norway

*Together, German sea, land, and air forces moved into Norway, taking its people by surprise and easily occupying the country.*

**What:** War
**When:** April 9, 1940
**Where:** Norway
**Who:**
ADOLF HITLER (1889-1942), chancellor of Germany from 1933 to 1945
ERICH RAEDER (1876-1960), commander in chief of the German navy from 1928 to 1943
VIDKUN QUISLING (1887-1945), a Norwegian politician
SIR CHARLES MORTON FORBES (1880-1960), commander in chief of the British Home Fleet from 1938 to 1940

## Germany Makes Its Move

During World War I, the Allies blockaded Germany with ships patrolling from the Shetland Islands to the coast of Norway. As a result, the German people suffered hardships, and there were shortages of supplies for the German armed forces. The blockade also kept most German surface raiders from moving out of harbor to attack Allied ships.

After that war, planners in the German navy realized that if Germany had seized Norway, the blockade would not have been as effective. German naval forces could have made their way to the Atlantic Ocean more easily from Norway than they could from Germany. Grand Admiral Erich Raeder, who became commander in chief of the German navy in 1928, gave this matter serious attention.

After World War II broke out, Raeder explained to Adolf Hitler that Germany would benefit from capturing Norway. Hitler did not show much interest in the idea, however, until December, 1939. At that time Vidkun Quisling, a former Norwegian officer and cabinet minister, came to Berlin to offer to help the Nazis overthrow his country. A staff was appointed to study the possibilities.

In February, 1940, British ships entered Norwegian territorial waters to rescue British prisoners from a German boat. This act infuriated Hitler, who began to fear that the British had plans of their own for taking Norway. The Germans began to step up their plans, and by April 7 six reinforced divisions of troops under the command of General Nikolaus von Falkenhorst, twelve hundred aircraft, and most of the German navy were ready to attack.

Actually, the British had been considering occupying Norway, at the urging of First Lord of the Admiralty Sir Winston Churchill. The capture of the port of Narvik would prevent the Germans from importing Swedish iron ore during the winter months. Also, if the British occupied northern Norway, they would stand in the way of the Germans' seizing the ore fields in Sweden.

Yet the British cabinet would allow only the laying of mines in Norwegian territorial waters, in order to force German ships carrying ore to take to the high seas, where they could be legally seized by British ships. On April 5, the British minelayers took off toward Norway.

On April 7, the Germans, unaware of the British movement, also sent warships toward Norway. Before dawn on April 9, German naval forces rushed upon all Norway's main ports. The Norwegians were taken by surprise, and all the ports except Oslo fell. At Oslo, however, the airport was taken by troops landed from the air, and the city was soon in German hands.

By the evening of April 9, the Germans held all the main ports and airfields of Norway and were working quickly to get them ready for the air groups flying in from northern Germany. By the next day, strong squadrons of fighters and bombers had already begun to operate from Norway's airfields.

## The British Respond

The British had realized that something was going on. On the evening of April 7, the Home Fleet, under Admiral Sir Charles Forbes, set out from port. On the morning of April 9, Forbes was prepared to sail to Bergen, a southern Norwegian port, in pursuit of the Germans; the naval force he commanded was far superior to theirs.

German air attacks on Forbes's fleet during the early afternoon of April 9, however, prevented the British from attacking. The destroyer HMS *Ghurka* was sunk, and Forbes's flagship, HMS *Rodney*, was lightly damaged by a heavy bomb. So it was that the Germans used their air force to make up for the inferiority of their navy. Forbes decided that his ships were in serious danger, and that any boats he might send up the fjords of Norway would be in even greater danger. He abandoned his plan to attack Bergen.

Having given up the idea of attacking only at sea, by April 17 the British had landed troops at three different places in Norway. Yet these soldiers had arrived too late, there were too few of them, and they were too widely scattered to be able to force the Germans out of the country. The British lacked armored vehicles, field artillery, and aircraft, so that they were no match for the well-equipped, fast-moving Germans. The British troops were soon forced to evacuate Norway.

## Consequences

In some ways, the Germans' campaign in Norway represented a new type of warfare. It was the first time that a fleet was successfully bombed at sea; for the British, it was the last time that a fleet operated off enemy-held shores in the expectation that it could beat off air attacks with antiaircraft fire alone. The battle for Norway showed that it was now absolutely necessary for air, land, and sea forces to work together.

*Members of the German army make their way up a snowy slope during the invasion of Norway.*

National Archives

The capture of Norway was helpful for the Germans as they continued fighting the war. Trondheim, Norway, became an important submarine base, and other air and naval bases in the far north were used for attacks on Allied convoys going to Russia between 1941 and 1945.

Quisling was proclaimed Norway's political leader. Throughout the war he tried hard to convert Norway to National Socialism (Nazism), but many Norwegian people resisted. After Norway was liberated by the Allies, Quisling was found guilty of treason and was shot.

Germany's capture of Norway also had political consequences in Great Britain. It was an important factor in the fall of Neville Chamberlain's administration, which was replaced by a coalition government headed by the aggressive, determined Sir Winston Churchill.

*Samuel K. Eddy*

**853**

# Antibiotic Penicillin Is Developed

*The antibiotic penicillin was concentrated and clinically tested, leading to its development as the century's greatest "wonder drug."*

**What:** Medicine
**When:** May, 1940
**Where:** Oxford, England
**Who:**
BARON FLOREY (1898-1968), an Australian pathologist, cowinner of the 1945 Nobel Prize in Physiology or Medicine
ERNST BORIS CHAIN (1906-1979), an émigré German biochemist, cowinner of the 1945 Nobel Prize in Physiology or Medicine
SIR ALEXANDER FLEMING (1881-1955), a Scottish bacteriologist, cowinner of the 1945 Nobel Prize in Physiology or Medicine

## The Search for the Perfect Antibiotic

During the early twentieth century, scientists were aware of antibacterial substances but did not know how to make full use of them in the treatment of diseases. Sir Alexander Fleming discovered penicillin in 1928, but he was unable to duplicate his laboratory results of its antibiotic properties in clinical tests; as a result, he did not recognize the medical potential of penicillin. Between 1935 and 1940, penicillin was purified, concentrated, and clinically tested by pathologist Baron Florey, biochemist Ernst Boris Chain, and members of their Oxford research group. Their achievement has since been regarded as one of the greatest medical discoveries of the twentieth century.

Florey was a professor at the University of Oxford in charge of the Sir William Dunn School of Pathology. Chain had worked for two years at the University of Cambridge in the laboratory of Frederick Gowland Hopkins, an eminent chemist and discoverer of vitamins. Hopkins recommended Chain to Florey, who was searching for a candidate to lead a new biochemical unit in the Dunn School of Pathology.

In 1938, Florey and Chain formed a research group to investigate the phenomenon of antibiosis, or the antagonistic association between different forms of life. The union of Florey's medical knowledge and Chain's biochemical expertise proved to be an ideal combination for exploring the antibiosis potential of penicillin. Florey and Chain began their investigation with a literature search in which Chain came across Fleming's work and added penicillin to their list of potential antibiotics.

Their first task was to isolate pure penicillin from a crude liquid extract. A culture of Fleming's original *Penicillium notatum* was maintained at Oxford and was used by the Oxford group for penicillin production. Extracting large quantities of penicillin from the medium was a painstaking task, as the solution contained only one part of the antibiotic in ten million. When enough of the raw juice was collected, the Oxford group focused on eliminating impurities and concentrating the penicillin. The concentrated liquid was then freeze-dried, leaving a soluble brown powder.

## Spectacular Results

In May, 1940, Florey's clinical tests of the crude penicillin proved its value as an antibiotic. Following extensive controlled experiments with mice, the Oxford group concluded that they had discovered an antibiotic that was nontoxic and far more effective against pathogenic bacteria than any of the known sulfa drugs. Furthermore, penicillin was not inactivated after injection into the bloodstream but was excreted unchanged in the urine. Continued tests showed that penicillin did not interfere with white blood cells and had

**854**

no adverse effect on living cells. Bacteria suscep-tible to the antibiotic included those responsible for gas gangrene, pneumonia, meningitis, diph-theria, and gonorrhea. American researchers later proved that penicillin was also effective against syphilis.

In January, 1941, Florey injected a volunteer with penicillin and found that there were no side effects to treatment with the antibiotic. In Febru-ary, the group began treatment of Albert Alexan-der, a forty-three-year-old policeman with a seri-ous staphylococci and streptococci infection that was resisting massive doses of sulfa drugs. Alexan-der had been hospitalized for two months after an infection in the corner of his mouth had spread to his face, shoulder, and lungs. After re-ceiving an injection of 200 milli-grams of penicillin, Alexander showed remarkable progress, and for the next ten days his condition improved. Unfortunately, the Ox-ford production facility was unable to generate enough penicillin to overcome Alexander's advanced infection completely, and he died on March 15. A later case involving a fourteen-year-old boy with staph-ylococcal septicemia and osteomy-elitis had a more spectacular re-sult: The patient made a complete recovery in two months. In all the early clinical treatments, patients showed vast improvement, and most recovered completely from infections that resisted all other treatment.

## Consequences

Penicillin is among the greatest medical discoveries of the twentieth century. Florey and Chain's chem-ical and clinical research brought about a revolution in the treatment of infectious disease. Almost every organ in the body is vulnerable to bacteria. Before penicillin, the only antimicrobial drugs available were quinine, arsenic, and sulfa drugs. Of these, only the sulfa drugs were useful for treatment of bacterial

infection, but their high toxicity often limited their use. With this small arsenal, doctors were helpless to treat thousands of patients with bacte-rial infections.

The work of Florey and Chain achieved partic-ular attention because of World War II and the need for treatments of such scourges as gas gan-grene, which had infected the wounds of numer-ous World War I soldiers. With the help of Florey and Chain's Oxford group, scientists at the U.S. Department of Agriculture's Northern Regional Research Laboratory developed a highly effi-cient method for producing penicillin using fer-mentation. After an extended search, scientists were also able to isolate a more productive peni-cillin strain, *Penicillium chrysogenum*. By 1945, a

*Sir Alexander Fleming (right), the discoverer of penicillin, stands before a poster advertising the new drug.*

**855**

strain was developed that produced five hundred times more penicillin than Fleming's original mold had.

Penicillin, the first of the "wonder drugs," is still the most powerful antibiotic in existence. Diseases such as pneumonia, meningitis, and syphilis are now treated with penicillin. Penicillin and other antibiotics also had a broad impact on other fields of medicine, as major operations such as heart surgery, organ transplants, and management of severe burns became possible once the threat of bacterial infection was minimized.

Florey and Chain received numerous awards for their achievement, the greatest of which was the 1945 Nobel Prize in Physiology or Medicine, which they shared with Fleming for his original discovery. Florey was among the most effective medical scientists of his generation, and Chain earned similar accolades in the science of biochemistry. This combination of outstanding medical and chemical expertise made possible one of the greatest discoveries in human history.

*Peter Neushul*

# Churchill Becomes British Prime Minister

*After Great Britain suffered early defeats in World War II, the leaders of its political parties asked Winston S. Churchill to become prime minister. He was an old warrior with a controversial reputation, but he had been the most resolute foe of Nazi aggression in the 1930's.*

**What:** Government; War
**When:** May 10, 1940
**Where:** London
**Who:**
WINSTON S. CHURCHILL (1874-1965), statesman, author, military leader
LORD HALIFAX (Edward Wood, first earl of Halifax, 1881-1959), foreign secretary
NEVILLE CHAMBERLAIN (1869-1940), prime minister

### Chamberlain and the Nazis

Before the outbreak of war in 1939, Neville Chamberlain, Britain's Conservative prime minister had fostered a policy of appeasement, which meant giving in to Adolf Hitler, the fascist dictator of Germany. Winston S. Churchill had resolutely opposed appeasement, but he had only a very small following in the House of Commons. He had alienated fellow Conservatives and members of the rival Labour Party through engaging in various political controversies over the years. He had called for rearmament and an all-out stand against Hitler's aggression in Europe.

Late in 1938, Chamberlain realized that the Nazis were utterly untrustworthy and had to be stopped. He inaugurated rapid British rearmament and guaranteed support for some of the smaller states in Hitler's path. When Poland was invaded in September, 1939, Britain and France declared war on Germany. To strengthen the government, Churchill became first lord of the Admiralty, in charge of the British navy. He had held this post during part of World War I.

Hitler's forces were unleashed against Norway in the spring of 1940, and British forces sailed north to defend that nation. The campaign was a disaster, and many from all parties in the House of Commons called for Chamberlain's resignation. Ironically, Churchill at the Admiralty was more responsible for failure of the campaign, and he took full responsibility for it in the House of Commons. Even so, the Norwegian campaign was what brought down the Chamberlain government.

Chamberlain's possible successors were Lord Halifax, the foreign secretary, or Churchill. After Norway's occupation, Hitler's armies suddenly attacked Belgium, the Netherlands, and France on May 10.

### The Events of May 9

One of the most important and dramatic meetings in British political history took place the day before Churchill became prime minister. At that meeting were Chamberlain, Halifax, and Churchill. Chamberlain and many Conservative leaders preferred Halifax to be the new prime minister. Churchill's candidacy was hindered by his reputation for being erratic and intemperate. Also, his best years were thought to be behind him as he was sixty-five years old. Yet by this time, Churchill had a growing number of supporters in the House of Commons, attracted by his strength and resolution in facing danger.

When Chamberlain told Halifax that he was regarded as the more acceptable candidate, Halifax pointed out the difficulty of governing from where he sat, which was in the House of Lords. Churchill said nothing, implying that he agreed and wanted the post for himself. Of course, all three knew that Parliament could have passed an enabling act to allow Halifax to govern from the House of Commons during the emergency. Halifax politely said Churchill would be a better choice, and again, Churchill said nothing. Therefore, because Halifax hesitated, and Churchill said nothing to encourage him, Chamberlain reluctantly had to give approval to Churchill.

**857**

*Winston S. Churchill.*

AP/Wide World Photos

The next day, May 10, Hitler's western on-slaught was unleashed. Chamberlain thought that because of this new crisis, he should continue to lead the nation as the head of a national government of all parties. Word came to him, however, that the Labour Party would not serve under him. He resigned, and Churchill went to Buckingham Palace to be appointed prime minister on one of the darkest days of World War II.

## Consequences

Churchill wrote of his appointment, "I felt as if I were walking with destiny and that all my past life had been but a preparation for this hour and for this trial." Before the war began, he was regarded as a politician whose accomplishments, though many, had never matched his great talents and had been marred by some notable failures. When the war was over, in 1945, he was widely hailed as a great politician.

Churchill took over as prime minister when Britain stood all alone. Germany would soon crush France, the United States was resolutely neutral, and the Soviet Union was collaborating with Nazi Germany. By the summer of 1940, all Europe lay at Hitler's feet except for Britain, which fought on and would not surrender. If Britain had been conquered or induced to accept peace on the basis of German dominance, the whole history of the world would have changed dramatically. Without British resistance in the west, Hitler would have been able, in all probability, to win a close victory over the Soviet Union after he invaded Russia in June, 1941. A victory over the Soviets would have enabled Hitler and his fascist dictatorship to gain dominance over Europe.

Britain's survival depended on Churchill's leadership and the courage of the pilots who fought the Battle of Britain in 1940. Churchill was a great war leader whose speeches inspired the soldiers and the rest of the country during those desperate days. He had a sense of history that enabled him to portray the struggle of freedom against tyranny. His vision accounts for his dogged determination to inspire his people never to surrender. He had been a brave man all his life, facing death on the battlefield many times, and he continued to give encouraging examples of his fearlessness throughout the war. He had an American mother and a great appreciation for the United States, which helped him play a key role in forming a strong alliance with the United States. Churchill's becoming prime minister on May 10, 1940, was therefore one of the most significant events in the history of the twentieth century.

*Henry G. Weisser*

# British Evacuate from Dunkirk

*As German tanks advanced toward the French port of Dunkirk, the British Royal Navy, along with hundreds of British civilians, carried out a heroic evacuation of the British Expeditionary Force.*

**What:** War
**When:** June 4, 1940
**Where:** Dunkirk, France
**Who:**

ADOLF HITLER (1889-1945), chancellor of Germany from 1933 to 1945

KARL RUDOLF GERD VON RUNDSTEDT (1875-1953), commander of the German forces in France

HERMANN GÖRING (1893-1946), German interior minister and minister of aviation from 1933 to 1945

SIR WINSTON CHURCHILL (1874-1965), prime minister of Great Britain from 1940 to 1945

SIR BERTRAM H. RAMSAY (1883-1945), British flag officer at Dover

## The German Advance

When the Germans invaded Poland in 1939, the British sent an expeditionary force to France. By May, 1940, it had increased to ten infantry divisions.

In the meantime, Chancellor Adolf Hitler of Germany had planned a general attack on France, Belgium, and Holland. At dawn on May 10, the German army and air force struck with shattering power. Two French armies and the British Expeditionary Force hurried into Belgium.

On May 13, six German panzer (armored) divisions, commanded by General Gerd von Rundstedt, broke through the French defenses along the Meuse River at Dinant and Sedan to the south of Belgium. The Germans then moved northwestward toward the English Channel, hoping to divide the Allied forces entering Bel-

gium from the main body of the French army to the south. The first German tanks reached the English Channel at Abbéville on May 20.

The amazing progress of the Germans toward the coast had seriously alarmed the commander in chief of the British Expeditionary Force, General John Gort. On May 19, he had informed the British government that he was considering sending nine divisions of his army to the English Channel for evacuation by sea. The British authorities responded quickly; on May 20, the Admiralty, the War Office, and Admiral Bertram Ramsay, who commanded British naval forces at Dover, began to make plans and to bring ships together for the evacuation. The operation was given the code name "Dynamo."

Meanwhile, the fighting in France went on. The Germans advanced northeastward from Abbéville along the coast. The Allies rushed troops into Boulogne and Calais, two French ports north of Abbéville, where they succeeded in slowing down the German armored vehicles. By this time, Dunkirk was the only port that could be used for a British evacuation.

Rundstedt's leading tanks reached Gravelines, between Calais and Dunkirk, on May 24. There, upon his order, they halted to rest and regroup. Three of Rundstedt's panzer divisions had already lost 50 percent of their tanks, and he feared that if they lost many more, his army group would be unable to move southward with other German forces into the heart of France.

At this point, Field Marshal Hermann Göring, Commander of the German Luftwaffe (air force), told Hitler that if the Allies left Dunkirk by sea, the Luftwaffe could stop them by dropping bombs from the air. Hitler gave him permission to try and supported Rundstedt's decision to stop the tanks. For two days the advance of

National Archives

*British soldiers taken prisoner at Dunkirk, June, 1940.*

Eden and the war cabinet, on the afternoon of May 26. Shortly before 7:00 P.M. that day, the order was flashed from the Admiralty in London: "Operation Dynamo is to commence."

In fact, Ramsay had already sailed the first set of ships from Dover Harbor. The British hoped to rescue about forty-five thousand men. That same day, the Germans renewed their attack near Gravelines, mostly with infantry. Yet the Allies, fighting with stubborn courage, made it difficult for the Germans to advance toward Dunkirk.

Evacuation of the British began during the night of May 26. The French continued to consider defending Dunkirk as a fortress; it was not until May 28 that they decided to withdraw their troops. Admiral Jean Abrial, who commanded French naval forces at Dunkirk, received the order to cooperate with Ramsay.

Exhausted British soldiers were picked up by ships from the quays and breakwaters of the port and from the beaches east of it. Evacuation from the shore was difficult; at first there were not enough small boats to take the men out to the ships. Then the British government called for the help of civilians who were working fishing boats and yachts. Large numbers of boat owners responded, and a fleet of hundreds of small craft put out to sea from England to rescue the British soldiers and see them safely home.

The fighting around Dunkirk was heavy. When the weather was favorable, the Luftwaffe sent in heavy attacks; the Royal Air Force kept up a defense. Fierce air combats raged around the port. The center of the town of Dunkirk and the harbor area were bombed to rubble, but the Royal Air Force and antiaircraft fire persisted against the Germans.

the German panzers was held up. This delay allowed the British to move to the coast and to organize a defense, along with French forces, around Dunkirk.

**The Evacuation**

Meanwhile, it had become clear that French plans to attack the Germans from the south had come to nothing. The decision to evacuate was made by Prime Minister Winston Churchill, along with secretary of state for war Anthony

In the early morning hours of June 4, the last British ship left Dunkirk. The evacuation was over.

## Consequences

The evacuation of Dunkirk was a feat of heroism and great organizational skill. Altogether, the British, with some French help, rescued 338,000 men, many more than expected.

The cost was high. The British used 760 ships; of these, 228 (mostly small craft) were sunk by air attack. Of France's 300 ships, 60 were lost. One hundred seventy-seven aircraft of the Royal Air Force were shot down, and the Germans lost about the same number.

The evacuation of the British Expeditionary Force from Dunkirk made it possible for the British army to continue in the war. Although the British had been forced to leave behind all of their heavy equipment, tanks, artillery, and transport vehicles, their trained men had been rescued. Reequipped, these men formed the core of a new army that, with its American ally, eventually liberated Western Europe from Nazi aggression.

# France Surrenders to Germany

*In one month, Adolf Hitler's forces were able to accomplish what the Germans had not achieved in four years of World War I: the conquest of France.*

**What:** War
**When:** June 22, 1940
**Where:** France
**Who:**
ADOLF HITLER (1889-1945), chancellor of Germany from 1933 to 1945
KARL RUDOLF GERD VON RUNDSTEDT (1875-1953), commander of German forces in France
ERICH VON MANSTEIN (1887-1973), Rundstedt's chief of staff
PAUL REYNAUD (1878-1966), premier of France from March to June, 1940
MAXIME WEYGAND (1867-1965), inspector general of the French army in 1940
HENRI-PHILIPPE PÉTAIN (1856-1951), premier of unoccupied France from 1940 to 1944

## The Germans Prepare

When World War II broke out in September, 1939, France was weak and demoralized. Unable to stand up to Germany on its own, the French government had followed the British in appeasing German chancellor Adolf Hitler. France's internal political problems had become worse in the years since World War I, and many French people felt disillusioned with the politicians of the Third French Republic. The French military had tried to prepare for war with Germany, but its strategies were badly out of date.

Hidden behind the Maginot Line—a set of fortifications running along France's border with Germany—the French army waited for the Germans to attack. French forces did not go to the aid of Poland when the Germans invaded that country. Between September, 1939, and May, 1940, the border between France and Germany was a quiet zone; the French were reluctant to move out of their defensive position. All their mobile forces were concentrated on the French-Belgian border, ready to help if the Germans invaded Belgium.

Meanwhile, after Nazi armies had moved into Czechoslovakia and Poland, Hitler made plans to invade France. In the winter of 1939-1940 he appointed General Gerd von Rundstedt as commander of Army Group A. Hitler also approved a plan for the Germans to sweep through Belgium and attack northern France.

Because the French were prepared for such an attack, however, Rundstedt's chief of staff, General Erich von Manstein, proposed instead that the German armies try to push through the Ardennes Forest. This would allow them to avoid both the Maginot Line and the French mobile forces. The French believed that the Ardennes could not be penetrated by armored vehicles, and they had posted few forces in the area.

Hitler approved of Manstein's plan. Following the Germans' conquest of Norway and Denmark in April, 1940, they prepared to invade France.

## The Attack

On May 10, the German armies invaded Belgium. Army Group A moved into the Ardennes and was rewarded with a quick breakthrough.

Leaders of France were shocked and dismayed. General Maurice Gamelin, Inspector General of the French army, was criticized for having committed all of his armored forces to the fighting in Belgium, holding none in reserve. The French cabinet, led by Premier Paul Reynaud, fired Gamelin from his post and replaced him with General Maxime Weygand. Reynaud took over the Ministry of War himself and appointed an aging World War I hero, Marshall Philippe Pétain, as his deputy.

Weygand's forces could not contain the Germans. German troops swept close to Paris, encircling the French forces and trapping them against the back of the Maginot Line. Reynaud urgently requested help from Great Britain, especially fighter planes. Though British prime minister Winston Churchill was deeply sympathetic, he did not send fighters, for he realized that he would need them to protect England. He did send other typ to Tours as the German armies approached Paris. Churchill begged the French not to surrender but to retreat to North Africa and carry on the fight from there. Reynaud agreed with this strategy, but the French military saw no reason to carry on the fight—especially after Italy's dictator Benito Mussolini declared war and sent troops into southern France.

On June 16, Reynaud realized that France was on the brink of surrender; strongly opposed to this move, he resigned his cabinet positions. Fearing that if France resisted it would be bombed and torn apart the way Warsaw, Poland, and Rotterdam, the Netherlands, had been, most French leaders wanted to sign an armistice with the Germans. Some of them believed that the Germans' brutality would be softened as the invaders came into contact with France's superior civilization.

Pétain was named premier in Reynaud's place. He appealed to Hitler for the opportunity to surrender, and on June 22, 1940, the French surrendered to Germany.

*Adolf Hitler (center) poses in front of the Eiffel Tower in June, 1940, after taking France.*

## Consequences

Under the terms of the surrender, France was divided into two parts. Northern France was put under German occupation, while southern France remained independent in name. The French fleet was confined to its home ports and was not to be used during the remainder of the war.

As head of southern France, Pétain formed a government at the small southern town of Vichy. In July, 1940, he formally abolished the constitution of the Third French Republic and established himself as a sort of dictator. It was a move that he and other authoritarian politicians had been wanting to make for a number of years.

The Vichy government cooperated fully with the Nazis and assisted in rounding up Jews in France. Though these were shameful years for France, there were also pockets of heroic resistance. In the village of Le Chambon-sur-Lignon, for example, the townspeople sheltered many Jews and helped them escape to safety.

*José M. Sánchez*

**863**

# Japan Occupies Indochina

*During World War II, Japan "swallowed" the French colony of Indochina in two "bites": the northern half in September, 1940, and the southern half in July, 1941.*

**What:** War
**When:** September, 1940-July, 1941
**Where:** French Indochina
**Who:**
YOSUKE MATSUOKA (1880-1946), foreign minister of Japan from 1940 to 1941
CORDELL HULL (1871-1955), United States secretary of state from 1933 to 1944
FRANKLIN DELANO ROOSEVELT (1882-1945), president of the United States from 1933 to 1945

## Japanese Expansion

Indochina, a French colony, was under the protection of the French Republic. Unfortunately for Indochina, however, in 1940 France was not in a position to defend its colonies. By June, 1940, the efficient German armies had conquered France. The southern part of France, which Germany did not occupy, was organized into a government that collaborated with the Nazis. Though more than fifty thousand French troops remained in Indochina, they could expect little help from the mother country.

During these years, Japan was working to bring all Asia under its political and economic influence. Japan called it the "Great East Asia Co-prosperity Sphere." In 1940, however, Japan was having a hard time persuading China to cooperate. Japanese armies had been waging war in China since 1937, with quite a bit of military success but no political success. China would not surrender.

Guerrilla warfare in China kept large numbers of Japanese troops busy. It was an expensive war for Japan. The war effort would be helped if Japan could occupy Indochina, for then it could cut off the Chinese supply line that ran from Haiphong in Indochina to Chungking in China. In any case, before Japan could move into richer portions of Southeast Asia, Indochina would have to be taken.

The United States disapproved of Japan's actions. Since the start of the Sino-Japanese War in 1937, American officials had protested. At first these protests were mild; in the spring of 1940, however, Germany was winning many military victories in Europe, and it became more likely that Japan would move into Southeast Asia. The United States stepped up its objections to Japanese expansion.

The Americans also took a concrete action to oppose Japan: The U.S. Fleet was transferred from San Diego, California, to Pearl Harbor in the Hawaiian Islands. By placing the fleet near Japan, President Franklin D. Roosevelt and Secretary of State Cordell Hull were warning Japan that the United States would not stand by quietly while Japan pushed into Southeast Asia and the oil-rich Netherlands East Indies.

## The Conquest

Soon, however, Japan decided that the time had come to move into the northern half of French Indochina. Diplomatic negotiations to prepare for the occupation began in June, 1940. French officials at the talks stalled, hoping that the United States would come to the aid of their colony. The French troops in Indochina were willing to fight the Japanese if they had help.

Yet the United States was not ready to confront Japan over northern Indochina, and by themselves the French forces were no match for the Japanese army. So it was that when the Japanese declared that they were coming in regardless of whether the French authorities agreed, France gave in.

The United States responded by placing an embargo on iron and steel-scrap exports to Japan. This was a rather mild punishment, designed to persuade Japan that any further expansion would not be tolerated. The embargo was a nuisance to Japan, but it did not cause serious problems.

Had Secretary of State Hull intended to force a real confrontation, he could have used his major weapon, an oil embargo. Oil was so important to the Japanese, however, that Hull feared an oil embargo would send Japan off to take over the oil reserves in the Netherlands East Indies.

In any case, it seemed possible that the Japanese move into northern Indochina had to do mostly with the war against China, not with an attempt to expand farther to the south. Hull had no desire to confront Japan over its war with China, so he avoided serious economic penalties. He did, however, warn Japan not to move into the Netherlands East Indies.

Actually, the American scrap embargo encouraged Japan to expand farther south. The Japanese had complained that they needed to establish their "Co-prosperity Sphere" because the Western Powers, especially the United States and Great Britain, had control over necessary raw materials around the world. Without a secure supply of those raw materials, Japan would always be dependent on the West. Moreover, Japanese officials began to fear that with Germany's recent conquest of the Netherlands, the United States or Great Britain might move into the Netherlands East Indies. The scrap embargo and Hull's warning seemed to prove the point.

Japanese foreign minister Yosuke Matsuoka pressed his government to make an alliance with Germany. He believed that even after Japan moved into southern Indochina, the United States would be intimidated by the Japanese-German alliance and would stay neutral. Japan and Germany did sign a treaty, and Japan decided to push southward.

The opportunity came in June, 1941, when Germany turned its tanks and armies toward the Soviet Union. With the Soviet Union tied up in that conflict, Japan had to decide whether to join in the fight against its long-time enemy or to take advantage of the situation to move southward. The decision was to go south, and on July 27 Japan occupied the rest of French Indochina.

## Consequences

At this point the United States no longer had doubts about Japan's intentions. Clearly, Japan was moving to conquer the Indies and would probably also attack the American-held Philippine Islands and the British naval base of Singapore. The United States and Great Britain responded with full economic sanctions—which meant an embargo on oil.

Again Japan was faced with a choice: to give up or to push forward toward the oil in the Indies. Japanese leaders decided to strike out for the oil. Before Japan could do so, however, it needed to remove the threat of the American fleet in Hawaii. So it was that Japanese warplanes bombed Pearl Harbor on Sunday morning, December 7, 1941. This Japanese action finally brought the United States into World War II.

*Jonathan G. Utley*

# Boys Discover Prehistoric Cave Paintings at Lascaux

*The discovery of cave paintings at Lascaux helped archaeologists to explain how prehistoric art evolved and how early it began.*

**What:** Archaeology
**When:** September 12, 1940
**Where:** Near Montignac, in southwestern France
**Who:**
HENRI-ÉDOUARD-PROSPER BREUIL (1877-1961), a French archaeologist
MONSIEUR RAVIDAT, the seventeen-year-old youth who was the first to enter the cave

## A Mysterious Cave

On September 12, 1940, shortly after the beginning of World War II, three local boys and two refugees from the German-occupied region of France were roaming through a field in southwestern France when they heard the faint sound of barking. Their dog had fallen into a small hole at the base of a fallen tree. One of the local youths, Ravidat, went in after it and slid about 20 feet to a sandy floor. Striking matches to light up a large underground hall measuring approximately 60 feet by 30 feet, he became the first person in fifteen thousand years to see the remarkable multicolored cave paintings of Lascaux (named for the ruins of a place called Château Lascaux that was on the same property as the cave).

Although they had no way of knowing it, what the five boys had discovered looked much like the famous Altamira paintings found in northern Spain seventy years before. The Lascaux cavern walls were covered with paintings of wild beasts: horses and stags, oxlike creatures with strange long bodies, and bulls with strange spotted patterns covering parts of their bodies. Scattered among the animals in this prehistoric scene was a series of checkerboard symbols and leaflike designs. One of the refugee boys, Estréguil, made quick sketches of the paintings.

The boys brought news of their discovery to their schoolmaster, who passed on the news to experienced archaeologists. One of these was Abbé Bouyssonie, the archaeologist who had discovered the famous Neanderthal man skeleton in 1908. Henri-Édouard-Prosper Breuil was another expert for whom this discovery was especially important. Breuil was interested in devising a general theory of how prehistoric art had evolved. The cave paintings at Lascaux looked as if they might contain a key.

Breuil and the other archaeologists explored the cave more fully, finding eighty paintings on their first try, both in the main hall and in a side cave. Most of the paintings were found on blocks of stone that had fallen from the cavern ceiling. Although it would take some time before any real theory could be developed, Breuil sent an article about the find to the distinguished English scientific journal *Nature*.

## The Paintings

The paintings ranged from about a foot to more than 15 feet in length. Many techniques had clearly been used for artistic effect, and many paintings had been touched up since the time they were first made. One of the most useful paintings for suggesting how people lived during that time showed a man lying beside his hunting tools—a javelin and a throwing stick. Looking at a bison that had been gored by his spear, the hunter seemed to be fatally wounded. Also notable was the outline of a child's hand and forearm, perhaps a "signature" for the group of artists. Strangely, among all the paintings in the cavern, this was the only reminder of the community of artists.

Most of the paintings were of many different animals, but some seemed to be telling practical stories. One of these showed a number of horses, some upside down, showing how primitive hunters had driven such animals off cliffs. This painting was later called "Falling Horses." Other paintings were too strange to be readily interpreted. One of these came to be known as "The Apocalyptic Beast." The beast did not look like any animal known to archaeologists, though it may have been an ox or a prehistoric rhinoceros. The body was massive and sagging, as if in a late stage of pregnancy, but the head was much too small for the rest of it. It was spotted with oval-shaped rings and had "horns" that looked like straight sticks and were covered with tufts. Perhaps the most important discovery regarding the Lascaux cavern was suggested by Breuil as soon as he saw it. Based on paintings he had studied in the nearby museum of prehistoric cultures at Les Eyzies, he was sure that the Lascaux paintings were older than the Magdalenian archaeological period (15,000 B.C.E. to 10,000 B.C.E.). Breuil and other archaeologists came to believe that prehistoric art had first developed in an earlier age, which he called "Perigordian."

## Consequences

Although archaeologists had studied prehistoric artifacts throughout Western Europe since the first half of the nineteenth century, the discovery in 1868 of cave paintings at Altamira had offered the first evidence that prehistoric humans had practiced painting.

The paintings at Altamira, like the ones at Lascaux, were very colorful and were mostly of bison and other animals. There were also some designs—such as checkers, squares, and dots—as well as some engraved (not painted) "semihuman" figures. Some archaeologists have suggested that these pictures may show early rituals.

Archaeologists' greatest disappointment about the paintings at Altamira had been that there was no way to figure out how to date the paintings in order. Generally, they agreed that prehistoric art had first appeared in the Aurignacian period (about 25,000 B.C.E. to 15,000 B.C.E.), but they were unable to tell when the paintings first began to be as sophisticated as those found at Altamira. Breuil's work at Lascaux, based on his earlier work at many different sites, helped to show that complex painting techniques had developed much earlier than anyone had thought.

*Byron D. Cannon*

**867**

# Polyester Is Patented

*The development of synthetic polyester polymers ultimately led to their wide use in fabrics and other industrial materials.*

**What:** Materials
**When:** 1941
**Where:** Great Britain
**Who:**
WALLACE H. CAROTHERS (1896-1937), an American polymer chemist
HILAIRE DE CHARDONNET (1839-1924), a French polymer chemist
JOHN R. WHINFIELD, a British polymer chemist

## A Story About Threads

Human beings have worn clothing since prehistoric times. At first, clothing consisted of animal skins sewed together. Later, people learned to spin threads from the fibers in plant or animal materials and to weave fabrics from the threads (for example, wool, silk, and cotton). By the end of the nineteenth century, efforts were begun to produce synthetic fibers for use in fabrics. These efforts were motivated by two concerns. First, it seemed likely that natural materials would become too scarce to meet the needs of a rapidly increasing world population. Second, a series of natural disasters—affecting the silk industry in particular—had demonstrated the problems of relying solely on natural fibers for fabrics.

The first efforts to develop synthetic fabric focused on artificial silk, because of the high cost of silk, its beauty, and the fact that silk production had been interrupted by natural disasters more often than the production of any other material. The first synthetic silk was rayon, which was originally patented by a French count, Hilaire de Chardonnet, and was later much improved by other polymer chemists. Rayon is a semisynthetic material that is made from wood pulp or cotton.

Because there was a need for synthetic fabrics whose manufacture did not require natural materials, other avenues were explored. One of these avenues led to the development of totally synthetic polyester fibers. In the United States, the best-known of these is Dacron, which is manufactured by E. I. du Pont de Nemours. Easily made into threads, Dacron is widely used in clothing. It is also used to make audiotapes and videotapes and in automobile and boat bodies.

## From Polymers to Polyester

Dacron belongs to a group of chemicals known as "synthetic polymers." All polymers are made of giant molecules, each of which is composed of a large number of simpler molecules ("monomers") that have been linked, chemically, to form long strings. Efforts by industrial chemists to prepare synthetic polymers developed in the twentieth century after it was discovered that many natural building materials and fabrics (such as rubber, wood, wool, silk, and cotton) were polymers, and as the ways in which monomers could be joined to make polymers became better understood. One group of chemists who studied polymers sought to make inexpensive synthetic fibers to replace expensive silk and wool. Their efforts led to the development of well-known synthetic fibers such as nylon and Dacron.

Wallace H. Carothers of Du Pont pioneered the development of polyamide polymers, collectively called "nylon," and was the first researcher to attempt to make polyester. It was British polymer chemists John R. Whinfield and J. T. Dickson of Calico Printers Association (CPA) Limited, however, who in 1941 perfected and patented polyester that could be used to manufacture clothing. The first polyester fiber products were produced in 1950 in Great Britain by London's British Imperial Chemical Industries, which had secured the British patent rights from CPA. This polyester, which was made of two monomers, terphthalic acid and ethylene glycol, was called Terylene. In 1951, Du Pont, which had acquired Terylene patent rights for the Western Hemisphere, began to market its own version of this polyester, which was called Dacron. Soon, other companies around the world were selling polyester materials of similar composition.

At the beginning of the twenty-first century, Dacron and other polyesters were being used in many items in the United States. Made into fibers and woven, Dacron becomes cloth. When pressed into thin sheets, it becomes Mylar, which is used in videotapes and audiotapes. Dacron polyester, mixed with other materials, is also used in many industrial items, including motor vehicle and boat bodies. Terylene and similar polyes-

ter preparations serve the same purposes in other countries.

The production of polyester begins when monomers are mixed in huge reactor tanks and heated, which causes them to form giant polymer chains composed of thousands of alternating monomer units. If T represents terphthalic acid and E represents ethylene glycol, a small part of a necklace-like polymer can be shown in the following way: (TETETETETE). Once each batch of polyester polymer has the desired composition, it is processed for storage until it is needed. In this procedure, the material, in liquid form in the high-temperature reactor, is passed through a device that cools it and forms solid strips. These strips are then diced, dried, and stored.

*A worker tests Dacron polyester for strength in a North Carolina plant in 1955.*

When polyester fiber is desired, the diced polyester is melted and then forced through tiny holes in a "spinneret" device; this process is called "extruding." The extruded polyester cools again, while passing through the spinneret holes, and becomes fine fibers called "filaments." The filaments are immediately wound into threads that are collected in rolls. These rolls of thread are then dyed and used to weave various fabrics. If polyester sheets or other forms of polyester are desired, the melted, diced polyester is processed in other ways. Polyester preparations are often mixed with cotton, glass fibers, or other synthetic polymers to produce various products.

## Consequences

The development of polyester was a natural consequence of the search for synthetic fibers that developed from work on rayon. Once polyester had been developed, its great utility led to its widespread use in industry. In addition, the profitability of the material spurred efforts to produce better synthetic fibers for specific uses. One example is that of stretchy polymers such as Helance, which is a form of nylon. In addition, new chemical types of polymer fibers were developed, including the polyurethane materials known collectively as "spandex" (for example, Lycra and Vyrenet).

The wide variety of uses for polyester is amazing. Mixed with cotton, it becomes wash-and-wear clothing; mixed with glass, it is used to make boat and motor vehicle bodies; combined with other materials, it is used to make roofing materials, conveyor belts, hoses, and tire cords. In Europe, polyester has become the main packaging material for consumer goods, and the United States does not lag far behind in this area.

The future is sure to hold more uses for polyester and the invention of new polymers. These spinoffs of polyester will be essential in the development of high technology.

*Sanford S. Singer*

# Touch-Tone Telephone Dialing Is Introduced

*Push-button dialing began in Baltimore in 1941 but did not become cost effective until 1963.*

**What:** Communications
**When:** 1941
**Where:** Baltimore, Maryland
**Who:**
BELL LABS, the research and development arm of the American Telephone and Telegraph Company

## Dialing Systems

A person who wishes to make a telephone call must inform the telephone switching office which number he or she wishes to reach. A telephone call begins with the customer picking up the receiver and listening for a dial tone. The action of picking up the telephone causes a switch in the telephone to close, allowing electric current to flow between the telephone and the switching office. This signals the telephone office that the user is preparing to dial a number. To acknowledge its readiness to receive the digits of the desired number, the telephone office sends a dial tone to the user. Two methods have been used to send telephone numbers to the telephone office: dial pulsing and touch-tone dialing.

"Dial pulsing" is the method used by telephones that have rotary dials. In this method, the dial is turned until it stops, after which it is released and allowed to return to its resting position. When the dial is returning to its resting position, the telephone breaks the current between the telephone and the switching office. The switching office counts the number of times that current flow is interrupted, which indicates the number that had been dialed.

## Introduction of Touch-tone Dialing

The dial-pulsing technique was particularly appropriate for use in the first electromechanical telephone switching offices, because the dial pulses actually moved mechanical switches in the switching office to set up the telephone connection. The introduction of touch-tone dialing into electromechanical systems was made possible by a special device that converted the touch-tones into rotary dial pulses that controlled the switches. At the American Telephone and Telegraph Company's Bell Labs, experimental studies were pursued that explored the use of "multifrequency key pulsing" (in other words, using keys that emitted tones of various frequencies) by both operators and customers. Initially, plucked tuned reeds were proposed. These were, however, replaced with "electronic transistor oscillators," which produced the required signals electronically.

The introduction of "crossbar switching" made dial pulse signaling of the desired number obsolete. The dial pulses of the telephone were no longer needed to control the mechanical switching process at the switching office. When electronic control was introduced into switching offices, telephone numbers could be assigned by computer rather than set up mechanically. This meant that a single touch-tone receiver at the switching office could be shared by a large number of telephone customers.

Before 1963, telephone switching offices relied upon rotary dial pulses to move electromechanical switching elements. Touch-tone dialing was difficult to use in systems that were not computer controlled, such as the electromechanical step-by-step method. In about 1963, however, it became economically feasible to implement centralized computer control and touch-tone dialing in switching offices. Computerized switching offices use a central touch-tone receiver to detect dialed numbers, after which the receiver sends the number to a call processor so that a voice connection can be established.

Touch-tone dialing transmits two tones simultaneously to represent a digit. The tones that are

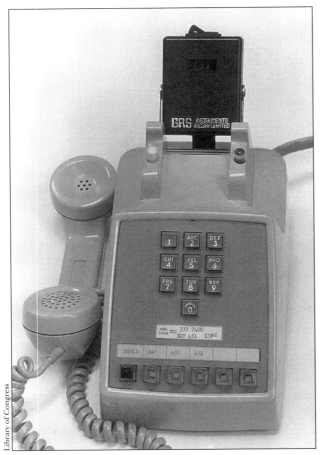

Library of Congress

*Touch-tone telephones like this have replaced most rotary dial telephones.*

transmitted are divided into two groups: a high-band group and a low-band group. For each digit that is dialed, one tone from the low-frequency (low-band) group and one tone from the high-frequency (high-band) group are transmitted. The two frequencies of a tone are selected so that they are not too closely related harmonically. In addition, touch-tone receivers must be designed so that false digits cannot be generated when people are speaking into the telephone.

For a call to be completed, the first digit dialed must be detected in the presence of a dial tone, and the receiver must not interpret background noise or speech as valid digits. In order to avoid such misinterpretation, the touch-tone receiver uses both the relative and the absolute strength of the two simultaneous tones of the first digit dialed to determine what that digit is.

A system similar to the touch-tone system is used to send telephone numbers between telephone switching offices. This system, which is called "multifrequency signaling," also uses two tones to indicate a single digit, but the frequencies used are not the same frequencies that are used in the touch-tone system. Multifrequency signaling is currently being phased out; new computer-based systems are being introduced to replace it.

## Consequences

Touch-tone dialing has made new caller features available. The touch-tone system can be used not only to signal the desired number to the switching office but also to interact with voice-response systems. This means that touch-tone dialing can be used in conjunction with such devices as bank teller machines. A customer can also dial many more digits per second with a touch-tone telephone than with a rotary dial telephone.

Touch-tone dialing has not been implemented in Europe, and one reason may be that the economics of touch-tone dialing change as a function of technology. In the most modern electronic switching offices, rotary signaling can be performed at no additional cost, whereas the addition of touch-tone dialing requires a centralized touch-tone receiver at the switching office. Touch-tone signaling was developed in an era of analog telephone switching offices, and since that time, switching offices have become overwhelmingly digital. When the switching network becomes entirely digital, as will be the case when the integrated services digital network (ISDN) is implemented, touch-tone dialing will become unnecessary. In the future, ISDN telephone lines will use digital signaling methods exclusively.

*Timothy Hanson*

# Seaborg and McMillan Create Plutonium

*Edwin Mattison McMillan and Glenn Theodore Seaborg discovered the first elements heavier than uranium, the most important of which is plutonium.*

**What:** Physics
**When:** February 23, 1941
**Where:** Berkeley, California
**Who:**
EDWIN MATTISON MCMILLAN (1907-1991), an American nuclear physicist, cowinner of the 1951 Nobel Prize in Chemistry
GLENN THEODORE SEABORG (1912-1999), an American chemist, cowinner of the 1951 Nobel Prize in Chemistry
PHILIP ABELSON (1913-          ), an American physical chemist

## Elements Beyond Uranium

In nuclear physics, fission is defined as the splitting of an atomic nucleus into two parts, especially after that nucleus has been bombarded by a neutron. When the nuclei of the heavier atomic elements, such as uranium and plutonium, are split, tremendous amounts of energy are released. Plutonium's story has fission at its beginning and at its end. The discovery of fission in 1938 was the stimulus for scientists such as Edwin Mattison McMillan and Glenn Theodore Seaborg to discover the transuranium elements neptunium and plutonium, and the discovery of a fissionable isotope of plutonium led to the atomic bomb that was dropped on Nagasaki, Japan, in 1945.

Ironically, the discovery of nuclear fission by the German chemists Otto Hahn and Fritz Strassmann was secondary to their search for a transuranium element. They had not intended to split the uranium nucleus; they were simply trying to add a neutron to this element to make a heavier element. The splitting of the nucleus into such smaller nuclei to form the elements barium and lanthanum came as a surprise, as did the produc-

tion of large amounts of energy. The news of the discovery of fission in January, 1939, excited Edwin Mattison McMillan tremendously. He had worked with the American physicist Ernest Orlando Lawrence on the development of the cyclotron (a machine that accelerates atomic particles), and this new discovery stimulated him to think of various experiments that could be done with the cyclotron to investigate this new phenomenon.

After the discovery of fission, McMillan quickly began to investigate this new phenomenon. He

*Glenn Theodore Seaborg.*

was interested in the range that the fission fragments would have. To study this, he put a thin coating of uranium oxide on a cigarette paper and then exposed it to a neutron beam formed by a cyclotron in order to produce fissions in the uranium. He noticed, however, that after the neutron bombardment, the original uranium oxide had a new radioactivity. McMillan suspected that this activity might be from a new element, element 93. According to then-current chemical theories, the new element should have had properties similar to those of the element rhenium. Tests, however, showed that the new element did not behave like rhenium.

McMillan remained puzzled by this enigma and a year later decided to try the investigation once more with a colleague, Philip Abelson. The hypothesis this time was that the new observed radioactivity came from element 93, despite its different chemical properties. This proved correct after exhaustive chemical tests. Before this discovery, there were only 92 known elements. With the discovery of element 93, in 1940, McMillan had opened a new field of transuranium elements.

McMillan named the new element "neptunium," after the planet Neptune, because the new element followed from uranium, which was named after the planet Uranus.

## Element 94

McMillan immediately started experiments directed at finding element 94, using deuterons from a Berkeley, California, cyclotron. He found new patterns of radioactivity, but before he could complete the work he was called away to the East Coast in 1940 to help develop radar for the War Department. With McMillan's permission and notes, Seaborg and a team of colleagues took up the research and obtained definitive proof that a second new element had indeed been made in the cyclotron.

In February, 1941, through bombarding uranium with deuterons in a cyclotron, they discovered element 94, plutonium, in the form of plutonium 238. (This element was named "plutonium" for the next planet in the solar system after Neptune, which is Pluto.) After more experimentation, they discovered plutonium 239, which proved to be a fissionable isotope that might serve as the explosive ingredient in a nuclear bomb and as a nuclear fuel. In 1942, Seaborg and another team of scientists created and identified a second major source of nuclear energy, the isotope uranium 233, which is the key to the use of the abundant element thorium as a nuclear fuel.

## Consequences

In the spring of 1942, Seaborg went to join the operation to make material for an atomic bomb. He moved to the University of Chicago to continue research on plutonium 239. He led a team whose goal was to develop chemical techniques that could be used to manufacture massive quantities of plutonium from uranium.

In the course of its work, Seaborg's team developed new techniques for handling minuscule amounts of radioactive material, transforming such common apparatus as test tubes, flasks, and balances into devices that could handle adeptly tiny quantities of material. These techniques enabled his group to work out the chemistry of plutonium. In an important early experiment, they succeeded, on September 10, 1942, in weighing the first visible amount of plutonium 239 (about one-ten-millionth of an ounce).

The successful solutions to the problems of the chemical separation of plutonium led to the construction, in Hanford, Washington, of large plutonium-producing nuclear reactors and a massive plant designed for the chemical separation of plutonium. As is well known, the labors of these and many other scientists and technicians eventually produced enough pure plutonium for use in two bombs, one that was successfully tested at Alamogordo, New Mexico, on July 16, 1945 (the world's first detonation of an atomic bomb), and another that was dropped on Nagasaki on August 9, 1945.

*Robert J. Paradowski*

# Germany Invades Balkan Peninsula

---

*Though Adolf Hitler had intended to keep the small Balkan states neutral or friendly toward Germany as war raged on in Europe, pressure from Italy and the Soviet Union brought German troops into the Balkan Peninsula.*

---

**What:** War
**When:** April 6, 1941
**Where:** Balkan Peninsula
**Who:**
ADOLF HITLER (1889-1945), chancellor of
    Germany from 1933 to 1945
CAROL II OF HOHENZOLLERN-SIGMARINGEN
    (1893-1953), king of Romania from
    1930 to 1940
BENITO MUSSOLINI (1883-1945), dictator of
    Italy from 1922 to 1943
BORIS III OF SAXE-COBURG (1894-1943),
    king of Bulgaria from 1918 to 1943

## Germany Versus the Soviet Union

By the summer of 1940, the Axis Powers—Germany and its allies—dominated the European Continent. France lay defeated, and Great Britain had been isolated. The Soviet Union in the East was still Germany's ally, though a shaky one, by the terms of the 1939 pact of mutual assistance. The smaller countries of Europe—including the nations of Southeast Europe, known as the Balkan states—were either occupied by the Axis or neutral powers.

Even before the war began, these countries had come into the German economic sphere, for most of their foreign trade was with Germany. The parliamentary governments that had been set up in the Balkans in the 1920's became royal or military dictatorships in the 1930's. All these governments copied parts of the Fascist and Nazi programs and policies of Italy and Germany.

Through an agreement, German chancellor Adolf Hitler recognized the Balkans as part of the Italian sphere of influence. After the Germans occupied Czechoslovakia in the spring of 1939, Italian dictator Benito Mussolini soon fol-

lowed their lead and incorporated Albania into the Italian kingdom.

Events of the first year of the war, however, changed the situation somewhat. Germany's victories were so rapid that they dismayed not only its foes but also its friends—especially the Soviet Union. Shortly after the fall of France, the Soviet Union demanded the Romanian province of Bessarabia, which had belonged to the Russian Empire before the 1917 revolution. Romania's King Carol II turned to Hitler for advice; Hitler, however, was unwilling to go to war against the Soviets at this time, and he told Carol to give in.

The 1939 agreement between Germany and the Soviet Union had not divided the Balkans into German and Soviet spheres of influence. Thus, the Soviets' growing involvement in the area could pose a threat to German interests. Germany was particularly concerned to keep friendly relations with Bulgaria.

After occupying Bessarabia, in the summer of 1940 the Soviets suggested to the Bulgarian government that they could help Bulgaria regain the Dobrudja region, which it had lost to Romania in 1913. At this time Hitler decided to invade the Soviet Union the following spring. To keep the Balkans in a state of friendly neutrality while Germany was at war in the East, Hitler was anxious to settle Bulgaria's and Hungary's territorial disputes with Romania using diplomatic means. By September, he had succeeded in bringing these three countries into agreement—and thus had gained an edge over the Soviet Union.

## Invading the Balkans

Yet there was another obstacle to Hitler's plans. Mussolini had joined the war during Germany's attack on France, partly because he thought that Hitler's military successes were lessening his own prestige. Now, with the Romanian

**875**

settlements, Germany had become active in Southeastern Europe, which was part of the Italian sphere. Mussolini resented this interference and decided to begin a military campaign in the Balkans without Germany's involvement.

It would have been most natural for Italy to attack Yugoslavia, but Hitler objected to any offensive against that country. Mussolini planned an invasion of Greece; though he informed Hitler of his goal, he was determined to make this a glorious Italian conquest. On October 28, 1940, the Italian army crossed the Albanian border into Greece.

The Italians enjoyed some early successes in Greece, but soon the Greeks, stoutly defending their country, drove their enemy back. Mussolini was humiliated. He had to ask an angry Hitler for help in his Balkan campaign.

*Two German soldiers patrol a small town on the Drava River.*

The German army, which had moved into Romania in preparation for the planned invasion of the Soviet Union, was now to be used to bail out the Italians. On December 13, Hitler ordered "Operation Marita," aimed at the quick defeat of Greece. The plan required that German troops invade Greece through Bulgaria while Yugoslavia remained neutral.

Seeking cooperation with this plan, Hitler requested both Bulgaria and Yugoslavia to join the recently formed Three-Power Pact (Germany, Italy, and Japan). Though King Boris III of Bulgaria preferred to remain neutral, he agreed, and at the beginning of March, 1941, Bulgarian prime minister Bogdan Filov traveled to Vienna to sign the pact. Soon afterward, German troops entered Bulgaria in force.

On March 26, the prime minister and foreign minister of Yugoslavia also signed the Three-Power Pact. The next day, however, a military coup seized power in Yugoslavia; the pro-German regent, Prince Paul, was dismissed. His young nephew King Peter II was elevated to the throne. Cheering crowds in Belgrade, Yugoslavia's capital, expressed their opposition to Nazism.

Leaders of the new Yugoslav government assured Hitler that they would still remain neutral in the upcoming campaign against Greece. Nevertheless, Hitler was outraged at the behavior of the Yugoslavs and decided that their country, along with Greece, would be invaded and destroyed.

On the morning of April 6, 1941, German warplanes and troops attacked both Yugoslavia and Greece. Following *Blitzkrieg* strategies, the Germans defeated both nations in a matter of days.

### Consequences

Along with Italy, Bulgaria, and Hungary, Germany continued to occupy Greece and Yugoslavia until it was driven out in 1944. Yugoslavia as a nation disappeared during these years. Two small states—an occupied Serbia and an independent Croatia ruled by an Italian monarch and a Fascist (Ustashi) government—replaced the central part of the Yugoslav kingdom. The rest of the nation was divided among Germany's allies.

# Ho Chi Minh Organizes Viet Minh

*To help the Vietnamese people gain their independence from the French, Ho Chi Minh formed a nationalist organization known as the Viet Minh.*

**What:** Political independence
**When:** May, 1941
**Where:** Northern Vietnam
**Who:**
Ho Chi Minh (1890-1969), Vietnamese leader trained in the Soviet Union
Truong Chinh (1908-1988), secretary general of the Indochinese Communist Party from 1941 to 1956, and in 1986
Le Hong Phong (born 1902), a Soviet-trained Vietnamese agent of the Comintern
Vo Nguyen Giap (1911-    ), a history teacher who became responsible for the Viet Minh's military planning
Pham Van Dong (1906-2000), a Vietnamese Communist who supported the decision to found the Viet Minh

## Unifying the Vietnamese

At a meeting in November, 1939, the Indochinese Communist Party (ICP) approved a new policy: Japanese aggression and French colonial power were to be seen as equally dangerous to the Vietnamese people. Instead of emphasizing class struggle within Vietnamese society, the ICP would call all the people of Vietnam—whatever their class—to oppose France and Japan.

World War II had begun in Europe, and Japan was pushing farther into Chinese territory. It seemed to ICP leaders that the party was in a good position to pull the Vietnamese people together in the name of nationalism and to move Vietnam toward political independence.

Many Vietnamese people agreed that foreign rule had to be ended. Hundreds of thousands of Vietnamese peasants were working under terri-ble conditions on French rubber, coffee, and tea plantations in southern Vietnam. They were treated badly by both French and Vietnamese overseers, they did not eat well, and they were given almost no medical care. Many of them were dying, so that the French were constantly bringing in more workers from central and northern Vietnam.

At the same time, many northerners had to work in French-owned industries; men worked in coal and phosphate mines, while women toiled in spinning and textile factories. Pay was pitifully low, hours were long (up to fifteen hours per day), and conditions were unsafe.

Nguyen Ai Quoc, who later became known as Ho Chi Minh, believed that the Vietnamese people were ready to rise up against their oppressors. Ho had helped to found the French Communist Party and the Communist parties of Thailand and Malaya; he also had helped reorganize the ICP in 1930. He was not very active in the mid-1930's, but by 1939 he had gained new influence within the ICP.

## Time of Crisis

A new leader was badly needed. Anti-French protests in September and October, 1939, failed, and so did uprisings in southern Vietnam in November and December, 1940. More than two thousand ICP members were arrested in late 1939, and large numbers of arrests continued in 1940. Many people suspected of being Communists were executed, while others were imprisoned.

By the middle of 1940, the French colonial government in Indochina had been put under the authority of the pro-German government in Vichy, France. The Indochina administration began cooperating with the Japanese, who wanted to use the airfields and rice supplies in the area.

Ho Chi Minh called the ICP's eighth general conference in northern Vietnam in May, 1941. He wanted to find ways to get more Vietnamese people involved in the struggle for independence.

The conference reached three very important decisions. First, new leaders were chosen. Truong Chinh, who had studied the ideas of Chinese Communist leader Mao Zedong, was named secretary general. Mao was recruiting thousands of Chinese peasants for the Communist cause, and the ICP wanted to follow his example.

Second, the ICP decided to create a Communist military force. Pham Van Dong and Vo Nguyen Giap were both present at the meeting, and it is thought that they had received military training from the Communists in China. They were given special responsibility for organizing secret military training for Communists in Hanoi and in the Tonkin Delta area.

Finally, the ICP formed a new organization called the Vietnam Independence League, or Viet Minh. It was separate from the ICP, but ICP members controlled it. Subgroups in the Viet Minh were formed for peasants, women, youth, students, and Catholics. The ICP created the Viet Minh as a way to unify the Vietnamese people under Communist Party control, but most ordinary people who joined the Viet Minh did so because they wanted to help liberate their land from foreign rule.

**Consequences**

The Viet Minh was not able to recruit many members immediately, for French and Japanese security forces were active in Indochina. Also, a number of rival political parties were being formed in Vietnam—some of them sponsored by the French or Japanese. Arrests and assassinations were all too frequent as various forces and groups competed for influence.

Over time, however, the Viet Minh began to succeed. Traditional Vietnamese politics kept women out, but the Viet Minh included women as members and established the Women's Association for National Salvation. Female textile workers were encouraged to take part in strikes, and some women joined military forces. Getting women involved helped the Viet Minh to gain broader support.

Beginning in 1943, the Viet Minh began to grow quickly. At a secret ICP conference in February, 1943, Communist leaders agreed on new steps to expand the organization. They were helped by the cruel policies of the French government: Though many people in northern and central Vietnam were going hungry, more and more rice was being sold to Japanese forces or stored in government warehouses. The rice shortage became worse in 1944, but the French began to use stored rice to make fuel for vehicles. It is believed that more than one million Vietnamese people died in 1944 and 1945 because of food shortages.

The Viet Minh sometimes sent armed bands to raid rice storehouses. In its propaganda, it blamed the French for the famine. As a result, the Viet Minh became more and more popular among the Vietnamese people.

By the middle of 1945, the Viet Minh's military forces numbered around 5,000 active combatants, with 150,000-200,000 "village defense" forces and support persons. After Japan surrendered to the Allies in August, 1945, Ho Chi Minh sent these forces to seize power in many Vietnamese cities. On September 2, 1945, Ho Chi Minh proclaimed the founding of the Democratic Republic of Vietnam. Like the Viet Minh, the new government was led by the Indochinese Communist Party.

*Laura M. Calkins*

# Whittle's Jet Engine Is Tested

*Sir Frank Whittle developed one of the earliest turbojet engines.*

**What:** Space and aviation
**When:** May 15, 1941
**Where:** Cranwell, Lincolnshire, England
**Who:**

HENRY HARLEY ARNOLD (1886-1950), a chief of staff of the U.S. Army Air Corps

GERRY SAYER, a chief test pilot for Gloster Aircraft Limited

HANS PABST VON OHAIN (1911-     ), a German engineer

SIR FRANK WHITTLE (1907-1996), an English Royal Air Force officer and engineer

## Future Developments in Aircraft Design

On the morning of May 15, 1941, some eleven months after France had fallen to Adolf Hitler's advancing German army, an experimental jet-propelled aircraft was successfully tested by pilot Gerry Sayer. The airplane had been developed in a little more than two years by the English company Gloster Aircraft under the supervision of Sir Frank Whittle, the inventor of England's first jet engine.

Like the jet engine that powered it, the plane had a number of predecessors. In fact, the May, 1941, flight was not the first jet-powered test flight: That flight occurred on August 27, 1939, when a Heinkel aircraft powered by a jet engine developed by Hans Pabst von Ohain completed a successful test flight in Germany. During this period, Italian airplane builders were also engaged in jet aircraft testing, with lesser degrees of success.

Without the knowledge that had been gained from Whittle's experience in experimental aviation, the test flight at the Royal Air Force's Cranwell airfield might never have been possible. It was Whittle's repeated efforts to develop turbojet propulsion engines begun in 1928, when as a twenty-one-year-old Royal Air Force (RAF) flight cadet at Cranwell Academy, he wrote a thesis titled "Future Developments in Aircraft Design." One of the principles of Whittle's earliest research was that if aircraft were eventually to achieve very high speeds over long distances, they would have to fly at very high altitudes, benefiting from the reduced wind resistance encountered at such heights.

Whittle later stated that the speed he had in mind at that time was about 805 kilometers per hour—close to that of the first jet-powered aircraft. His earliest idea of the engines that would be necessary for such planes focused on rocket propulsion (that is, "jets" in which the fuel and oxygen required to produce the explosion needed to propel an air vehicle are entirely contained in the engine, or, alternatively, in gas turbines driving propellers at very high speeds). Later, it occurred to him that gas turbines could be used to provide forward thrust by what would become "ordinary" jet propulsion (that is, "thermal air" engines that take from the surrounding atmosphere the oxygen they need to ignite their fuel). Eventually, such ordinary jet engines would function according to one of four possible systems: the so-called athodyd, or continuous-firing duct; the pulsejet, or intermittent-firing duct; the turbojet, or gas-turbine jet; or the propjet, which uses a gas turbine jet to rotate a conventional propeller at very high speeds.

## Passing the Test

The aircraft that was to be used to test the flight performance was completed by April, 1941. On April 7, tests were conducted on the ground at Gloster Aircraft's landing strip at Brockworth by chief test pilot Sayer. At this point,

**879**

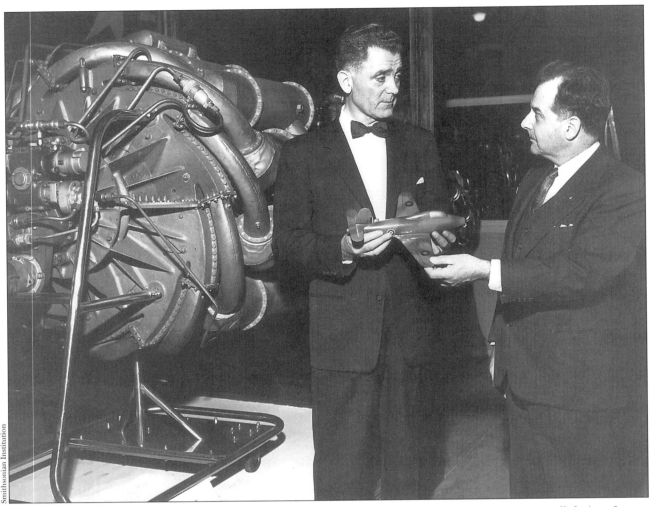

Smithsonian Institution

*Sir Frank Whittle (left) stands by a turbine engine he designed for use in Great Britain's first jet-propelled aircraft.*

all parties concerned tried to determine whether the jet engine's capacity would be sufficient to push the aircraft forward with enough speed to make it airborne. Sayer dared to take the plane off the ground for a limited distance of between 183 meters and 273 meters, despite the technical staff's warnings against trying to fly in the first test flights.

On May 15, the first real test was conducted at Cranwell. During that test, Sayer flew the plane, now called the Pioneer, for seventeen minutes at altitudes exceeding 300 meters and at a conservative test speed exceeding 595 kilometers per hour, which was equivalent to the top speed then possible in the RAF's most versatile fighter plane, the Spitfire.

Once it was clear that the tests undertaken at Cranwell were not only successful but also highly

promising in terms of even better performance, a second, more extensive test was set for May 21, 1941. It was this later demonstration that caused the Ministry of Air Production (MAP) to initiate the first steps to produce the Meteor jet fighter aircraft on a full industrial scale barely more than a year after the Cranwell test flight.

**Consequences**

Since July, 1936, the Junkers engine and aircraft companies in Hitler's Germany had been a part of a new secret branch dedicated to the development, of a turbojet-driven aircraft. In the same period, Junkers's rival in the German aircraft industry, Heinkel, Inc., approached von Ohain, who was far enough along in his work on the turbojet principle to have patented a device very similar to Whittle's in 1935. A later model of this jet engine would power a test aircraft in August, 1939.

In the meantime, the wider impact of the flight was the result of decisions made by General Henry Harley Arnold, chief of staff of the U.S. Army Air Corps. Even before learning of the successful flight in May, he made arrangements to have one of Whittle's engines shipped to the United States to be used by General Electric Corporation as a model for U.S. production. The engine arrived in October, 1941, and within one year, a General Electric-built engine powered a Bell Aircraft plane, the XP-59 A Airacomet, in its maiden flight.

The jet airplane was not perfected in time to have any significant impact on the outcome of World War II, but all of the wartime experimental jet aircraft developments that were either sparked by the flight in 1941 or preceded it prepared the way for the research and development projects that would leave a permanent revolutionary mark on aviation history in the early 1950's.

*Byron D. Cannon*

# Germany Invades Soviet Union

*Though the Nazi invasion took Soviet leaders by surprise, the Germans were not able to maintain the battle over the vast Soviet territory and against the fighting spirit of the Red Army.*

**What:** War
**When:** June 22, 1941
**Where:** Soviet Union
**Who:**
ADOLF HITLER (1889-1942), chancellor of Germany from 1933 to 1945
JOSEPH STALIN (1879-1953), dictator of the Soviet Union from 1924 to 1953
WALTHER VON BRAUCHITSCH (1881-1948), commander in chief of the German army
WILHELM VON LEEB (1876-1956), German commander of Army Group North on the Russian Front
FEDOR VON BOCK (1880-1945), German commander of Army Group Center on the Russian Front
KARL RUDOLF GERD VON RUNDSTEDT (1875-1953), German commander of Army Group South on the Russian Front

## Hitler's Anti-Soviet Ambition

Under the terms of the Nazi-Soviet Pact of August, 1939, Germany and the Soviet Union promised not to make war on each other. They also agreed to divide Eastern Europe into German and Soviet spheres of influence. This agreement allowed Adolf Hitler, chancellor of Germany, to be sure of Soviet neutrality when he began World War II by attacking Poland on September 1, 1939. Two days later, after Germany had refused to withdraw from Polish territory, Great Britain and France declared war on Germany. Within a month, a defeated Poland was divided between Nazi Germany and the Soviet Union.

As World War II grew steadily into a conflict ranging across the globe, the alliance between the Nazis and the Soviets grew weaker. During June, 1940, while the German armies were con-

quering France, Soviet dictator Joseph Stalin took the opportunity to annex the Baltic States—Latvia, Lithuania, and Estonia—and to occupy the Romanian territories of Northern Bukovina and Bessarabia.

Hitler was alarmed at these moves. He did not agree that the Nazi-Soviet Pact permitted Stalin to conquer these lands, only to keep them in his sphere of influence. Also, Germany's main source of oil was Romania, and Hitler did not like to see the Soviets moving in on that crucial area.

In late July, 1940, just as the Battle of Britain was about to begin, Hitler informed Field Marshal Walther von Brauchitsch, commander in chief of the German army, and his other military chiefs that he intended to invade and destroy the Soviet Union. At last Hitler would be able to fulfill the dream he had expressed years before in his political autobiography, *Mein Kampf* (1925-1926): to destroy Soviet Communism and take over *Lebensraum* (living space) for Germans in Eastern Europe.

## The Attack

During late 1940 and early 1941, Hitler gave much thought to plans for the invasion of the Soviet Union. Though the Battle of Britain was raging fiercely, Great Britain refused to negotiate an end to the war, and Hitler became convinced that only the defeat of the Soviets would persuade the British to give in.

By early October, Hitler had moved large numbers of German troops into Poland and Romania. On December 18, he named the operation against the Soviet Union "Barbarossa" (for the twelfth century German Emperor Frederick Barbarossa of Hohenstaufen). Hitler ordered his military planners to be ready by May 15, 1941, "to crush Soviet Russia in a quick campaign even before the end of the war against England." He believed that Germany could fight on two fronts at once.

In late 1940 and early 1941, Italy's attempt to conquer Greece was defeated through the combined efforts of the Greek army and British reinforcements. Before invading the Soviet Union, then, German forces had to come to the aid of their ally Italy. In April, the German armed forces launched a great *Blitzkrieg* attack on Yugoslavia and Greece; within a month, both countries had been subdued.

Though this success had been quick, it forced Hitler to put off the invasion of the Soviet Union for five weeks. That invasion began at dawn on Sunday, June 22, 1941, with a force of three million German troops. There were three army groups in the invasion force, each supported by the Luftwaffe (air force). Army Group North, commanded by Field Marshal Wilhelm von Leeb, was given the task of pushing the Soviets out of the Baltic States and then taking Leningrad (St. Petersburg). Army Group Center, un-

der Field Marshal Fedor von Bock, was to take Smolensk and thus open the way to Moscow. Finally, Army Group South, commanded by Field Marshal Gerd von Rundstedt, was to strike the Ukraine.

In the opening months of the campaign, these armies scored many successes against the Soviet Union's Red Army. By the fall of 1941, the Germans had surrounded Leningrad and captured a number of major Soviet cities, including Smolensk and Kiev. Only then did Hitler allow his forces in Army Group Center to push toward Moscow.

The battle for Moscow raged throughout October and November. One of the worst winters in years fell on the Soviet Union, and the Soviets were able to block the Nazi advance less than twenty miles from the Soviet capital.

It was a terrible fight. By early December, more than 800,000 Germans had been killed,

*German troops in the Soviet Union, during the 1941 invasion.*

**883**

captured, or wounded, while the Soviets had lost more than five million. Nevertheless, the Soviets were able to make up for some of their losses by moving reserve forces from the Far East to the Moscow area. The Red Army then began a powerful counterattack on the Germans. This attack reached its peak on December 6, 1941; the following day, the Japanese attack on Pearl Harbor brought the United States into World War II.

## Consequences

Germany did not succeed in conquering the Soviet Union. There were several reasons for that failure. First, the Soviet Union's vast size allowed the Red Army to retreat and regain strength after the first attacks. Second, Hitler's forces could not concentrate on a single goal; forces were switched between army groups moving on Leningrad and Moscow, and neither was captured. Third, the Germans had not understood the strength of the Red Army—or the importance of its reserves in the Far East. Fourth, the German army lacked enough tracked vehicles to move its soldiers and supplies over such a vast country with poor roads. Finally, the time required for the conquest of Yugoslavia and Greece may have been crucial: The five-week delay meant that German forces had to fight in severe winter weather.

With the United States' entry into the war, World War II became a war that stretched all around the globe. Hitler's forces lost in the Soviet Union, and as the war drew to a close Stalin took his place among the Allied leaders to plan a peace settlement. With the end of the war, the Soviet Union gained controlling influence over virtually all of Eastern Europe, including the eastern half of Germany, which became a Communist state.

# Japan Bombs Pearl Harbor

> *Japan's surprise attack on the United States Fleet at Pearl Harbor, Hawaii, brought the United States into World War II.*

**What:** War
**When:** December 7, 1941
**Where:** Pearl Harbor, Oahu, Hawaii
**Who:**
FRANKLIN DELANO ROOSEVELT (1882-1945), president of the United States from 1933 to 1945
CORDELL HULL (1871-1955), U.S. secretary of state from 1933 to 1944
KICHISABURO NOMURA (1887-1964), Japanese ambassador to the United States
ISOROKU YAMAMOTO (1884-1943), creator of the plan to make a surprise attack on Pearl Harbor

## American-Japanese Conflicts

Between 1939 and 1941, the United States focused anxiously on the progress of World War II in Europe. At the same time, events in the Far East caused growing concern among U.S. leaders. Japan took advantage of the European war to begin creating what it called the "Great East Asia Co-prosperity Sphere." Much of China had come under Japanese control by 1939. In September, 1940, Japan officially became an Axis Power, when it signed the Three-Power (Tripartite) Pact—a military alliance with Germany and Italy.

By the summer of 1941, Japan had moved into Indochina and was threatening to move out into Thailand, the Soviet Union's Siberian provinces, the British colony of Singapore, Burma, the Netherlands East Indies, and the Philippines.

To try to stop Japan's program of expansion, the United States made complaints and imposed economic penalties. Though Japan had seized Manchuria in the 1930's and had then gone on to fight China, the United States remained unwilling to use force against the Japanese. President Franklin Roosevelt felt sympathy for the Chinese but was more concerned about the German threat. Hoping to give liberals within Japan time to take the government away from the imperialists, Roosevelt was cautious in dealing with Japan.

Actually, the United States was slow even to impose economic sanctions in response to Japan's conquests. In August, 1940, the United States forbade the export of aviation gasoline to Japan; this was followed by a similar embargo on scrap iron and steel in September, 1940. In July, 1941, Japanese assets in the United States were frozen, and an oil embargo was imposed.

Meanwhile, Japanese leaders were coming to believe that their country was being encircled by the Western Powers. After the oil embargo was imposed, they estimated that Japan might have to go to war within a year, before its oil reserves were used up.

Negotiations continued between Japan and the United States, but neither side wanted to compromise. Roosevelt and Secretary of State Cordell Hull were pessimistic, but they wanted to continue the discussions to give the United States more time to make defense preparations.

At Pearl Harbor, Hawaii, and in Washington, D.C., army and navy intelligence gained evidence that the Japanese might be planning a surprise attack—but there also seemed to be evidence to the contrary. Intercepted messages led American military planners to believe that something would happen if the United States rejected a final Japanese proposal, but most of the evidence pointed to an attack somewhere in Southeast Asia.

On November 10, Japanese ambassador Kichisaburo Nomura presented to Hull what was to be the final Japanese proposal. Hull decided that it was not acceptable, and on November 26 he made a counteroffer that he knew the Japanese would also reject. The negotiation process had led nowhere.

## Bombs Fall

Japan's preparations for an attack on Pearl Harbor, where the U.S. Pacific Fleet was stationed, had begun in the early months of 1941. Japanese military planners realized that if Japan's advance into Southeast Asia were to succeed, the U.S. Fleet would have to be destroyed. Admiral Isoroku Yamamoto's daring plan to bomb the fleet at anchor in Pearl Harbor was at first considered impractical. Yamamoto worked hard to get approval for the plan, however, and it began to look possible.

Pilots began training in September. Because of the shallow waters of Pearl Harbor, wooden torpedoes were designed, together with a new method of propelling them to their target. Careful precautions were taken to maintain secrecy.

A special task force of thirty-one vessels, including six aircraft carriers that carried 432 airplanes (fighters, dive-bombers, high-level attack bombers, and torpedo planes), began to leave Japanese ports in early November. On November 22, this force gathered in the Southern Kuriles. Four days later, it headed out to sea for a run of 3,500 miles, to a meeting place 275 miles north of Pearl Harbor. There it was to await final clearance before attacking.

On December 2, the signal "Climb Mount Niitaka" came through. Early on December 7, the strike force reached position so that the first Japanese planes were flying over Pearl Harbor by 7:55 A.M. local time.

The weather was ideal for an attack, and Pearl Harbor was caught totally unprepared. The bombing had been carefully planned for Sunday morning, when the Pacific Fleet ships were moored in a neat row and their crews were either having breakfast ashore or relaxing on board.

There was no advance warning. An operator at an American radar post observed the oncoming Japanese squadrons at 7:02 A.M. He reported the blips on the screen, but the watch officer did not pass on the information; he thought that the squadron was a group of American bombers due to arrive that morning from the West Coast.

The Japanese planes swooped to the attack. Fighters and dive-bombers strafed and bombed the rows of aircraft at Wheeler Field and the Naval Air Station. Torpedo planes and dive-bombers

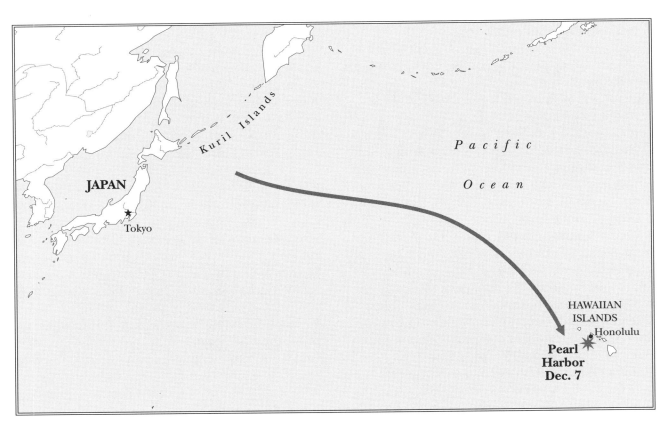

Digital Stock

*The Navy destroyer USS* Shaw *explodes during the Japanese attack on Pearl Harbor.*

also attacked Battleship Row in the attack's first phase, which lasted thirty minutes.

After a lull of fifteen minutes, the Japanese launched bombing attacks on the harbor, airfields, and shore facilities, followed by more attacks by dive-bombers, which descended amid antiaircraft fire. The last planes left at 9:45 A.M., less than two hours after the attack had begun.

## Consequences

The bombing left 2,403 dead and 1,178 wounded. Three battleships—the *West Virginia,* the *Arizona,* and the *California*—were sunk; the *Oklahoma* lay capsized, and the *Tennessee,* the *Nevada,* the *Maryland,* and the *Pennsylvania* had been damaged. Several smaller warships were sunk, and others were severely damaged. Almost all combat aircraft on the islands were harmed or destroyed. For their part, the Japanese had lost twenty-nine airplanes, five midget submarines, and one full-sized submarine.

The Japanese had succeeded beyond their hopes. Yet this great success was actually a failure in two important ways. First, the Japanese missed the most crucial targets: the aircraft carriers *Lexington* and *Enterprise* (both of which were at sea) and the *Saratoga* (which was in dry dock on the West Coast). Second, the Japanese failed to destroy the United States' huge oil storage facilities. If these had been ruined, the Pacific Fleet would have been forced to return to the West Coast.

Until this time, the American people had been rather reluctant to become involved in World War II. The most significant effect of the bombing of Pearl Harbor was to unify Americans to wage war against Japan and its Axis allies. President Roosevelt immediately declared war on Japan, and on December 11, 1941, Germany and Italy declared war on the United States.

*Theodore A. Wilson*

**887**

# Japan Occupies Dutch East Indies, Singapore, and Burma

> *Invading the British colonies of Singapore and Burma, along with the Dutch colony of Netherlands East Indies, Japan revealed that the European imperial powers were too weak to defend their colonies.*

**What:** War
**When:** December 8, 1941-August, 1945
**Where:** Southeast Asia
**Who:**
TOMOYUKI YAMASHITA (1885-1946), commander of the Japanese 25th Army in Malaya
MASANOBU TSUJI (1902-1968), head of operations for the Japanese 25th Army in Malaya
ARTHUR PERCIVAL (1887-1966), the British general in command of Malaya
KAREL W. F. M. DOORMAN (1889-1942), Allied naval commander of Eastern striking force in the Battle of Java Sea
HEIN TER POORTEN (1887-1948), commander of Dutch land forces in Java
SHOJIRO IIDA (born 1888), commander of the Japanese 15th Army in Burma

## Malaya and Singapore

While Great Britain was fighting off Nazi bombing attacks, British colonies in the Far East were coming under a growing threat of Japanese invasion. Malaya produced about 40 percent of the world's rubber and 60 percent of its tin, as well as iron and bauxite. Four hundred miles to the south lay Singapore, an island that was important to the defense of all Great Britain's Asian colonies. Complaining that the Western Powers controlled too many of the world's natural resources, Japanese leaders had become determined to remedy the situation. Japanese armies had already taken over Indochina, and Malaya and Singapore were logical next targets.

British leaders were aware of the danger, but because of the intensity of the war in Europe, they were not able to build up the air forces needed to protect Malayan airfields and to prevent a mass landing there. By 1941, it was too late.

On the night of December 7-8, 1941, an invasion force of 27,000 Japanese troops landed at Singora and Patani in southern Thailand and at Kota Bharu in northern Malaya. The 25th Japanese army was led by Lieutenant General Tomoyuki Yamashita, who had served in North China. He was Japan's most skillful general and was deeply admired by the men who served under him. Most of those men were experienced soldiers.

The 25th Army's head of operations, Colonel Masanobu Tsuji, had studied jungle warfare in Taiwan for nearly a year. He made important adjustments in the army's methods. Bicycles were used instead of cavalry, and about two hundred light tanks were used to support the infantry. Japanese agents had gathered information showing that the British were not well prepared to defend their colonies.

The Japanese troops, soon numbering 125,000, were more than a match for British forces, which included 37,000 Indians, 19,600 British, 15,200 Australians, and 16,000 local recruits—a total of nearly 88,000. Many of the British soldiers had been recruited only recently and had received little training.

The British had few aircraft in Southeast Asia, and Japanese pilots flying the new Zeros soon had command of the sky. The British battleship *Prince of Wales* and the battle cruiser *Repulse* were destroyed by air on December 10, so that the Japanese gained control of the sea as well.

Yamashita continually pushed his troops to attack, moving them southward. They carried only light equipment and rice rations; their bicycles allowed them to cover ground quickly and to pass destroyed bridges. When Japanese tanks rolled toward British fortifications, many Indian troops panicked, for they had not been trained in anti-tank methods. The British had not believed that tanks could be used in jungle warfare.

Major battles were fought at Jitra, Slim River, and Muar River. The British forces became discouraged as they lost many men and large quantities of supplies. On January 31, 1942, the last British forces crossed to Singapore Island and quickly prepared for a final showdown.

For a long time Singapore had been called a fortress, but it was a flat island separated from the mainland by the Johore Straits and was not difficult to attack. On the night of February 8, Japanese troops in small boats repaired the causeway. Artillery and tanks crossed over, and on February 15, Lieutenant General Arthur Percival surrendered.

## The Dutch East Indies and Burma

Japanese troops had already landed on Northeast Borneo and the Celebes on January 11, 1942. The Dutch army in the Netherlands East Indies totaled about 100,000 men, and few of them had adequate equipment. Only two hundred aircraft were available.

On January 23, the Japanese landed at Balikpapan, and on February 14, they seized Palembang on Sumatra. After they had landed on Bali and Timor, the important island of Java was surrounded. General Archibald Wavell, Supreme Commander of Allied forces in the Southwestern Pacific, had established his headquarters in Java, but as the Japanese approached he moved to India on February 25. The defense of Java was left to the Dutch General Hein ter Poorten.

The only hope was to defend the island at sea. Although Rear Admiral Karel W. F. M. Doorman commanded a mixed Allied force that did not have even a common signal code, he coura-

*Silhouette of troops at Pandu Ghat as they travel to Myitkyina, Burma, in October, 1944.*

National Archives

**889**

geously led his forces against Japan in the Battle of Java Sea on February 27-28. Yet the Japanese naval forces, commanded by Rear Admiral Takeo Takagi, won a smashing victory. Several Allied cruisers and destroyers were sunk.

Japanese troops landed on Java on the night of February 28. The defending Dutch and Australian forces were no match for the Japanese, who reached Tjilatjap on Java's southern coast on March 6. General Poorten surrendered on March 8, and a formal surrender followed four days later.

In order to close the 750-mile Burma Road, a supply route for Chinese Nationalist forces in Chungking, the Japanese had also decided to conquer Burma. Burma was defended by an Indian and Burmese division under British lieutenant general Thomas J. Hutton, but it was unable to halt the Japanese 15th Army, commanded by Lieutenant General Shojiro Iida. A small Japanese force entered southern Burma on December 11, 1941, and seized Tavoy eight days later. More troops moved in on January 20, 1942.

On January 30-31, the Japanese took Moulmein at the mouth of the Salween River. The poorly trained defending forces were quickly pushed back, and the Japanese entered Rangoon on March 8, capturing large amounts of supplies intended for China. They then turned north, seizing the endpoint of the Burma Road, Lashio, on April 20 and Mandalay on May 2. In June, monsoons completed the defeat of the British forces.

## Consequences

Before Japan invaded Singapore, British leaders had tended to see the Japanese as militarily weak because of their failure to conquer China. By the time Yamashita's forces reached the Johore Straits, however, Japanese troops were considered supermen.

Great Britain was thoroughly humiliated by its defeats in Southeast Asia. The British recovered their overseas empire in 1945, but by then the Asian people would no longer stand to be ruled by a foreign power that had invested so little in their defense.

# Germany and Italy Declare War on United States

Four days after the bombing of Pearl Harbor brought the United States into war against Japan, Germany and Italy supported their Japanese ally by declaring war on the United States.

**What:** War
**When:** December 11, 1941
**Where:** Berlin, Rome, and Washington, D.C.
**Who:**
ADOLF HITLER (1889-1945), chancellor of Germany from 1933 to 1945
BENITO MUSSOLINI (1883-1945), dictator of Italy from 1922 to 1943
FRANKLIN DELANO ROOSEVELT (1882-1945), president of the United States from 1933 to 1945

## The Crisis Builds

During the years between 1936 and 1940, it seemed quite unlikely that the United States would become involved in a war in Europe. The American government and people preferred a policy of "isolationism"—trying to keep out of other nations' conflicts. For its part, Nazi Germany was mostly interested in European affairs rather than the United States or the larger Western Hemisphere.

Most Americans were opposed to Italian dictator Benito Mussolini's invasion of Ethiopia in 1935; Congress responded with the first of the Neutrality Acts, which prohibited weapons from being exported to nations at war.

By 1936, it became clear that Germany and Italy intended to take over new lands and extend their territory. The Rome-Berlin Axis, a treaty of alliance, was formed in 1936, and Japan joined Germany and Italy in a treaty in 1937. That year, the U.S. government passed a new Neutrality Act.

Though most Americans still wished to stay out of a European war, President Franklin D. Roosevelt and his advisers had begun to worry about the danger of foreign aggression in Europe and the Pacific. In 1938, American opposition to the Nazis' persecution of the Jews led to the replacement of the American ambassador in Berlin by a lower-level diplomat. World War II broke out in 1939, and Roosevelt's concern grew.

In 1940, France surrendered to Germany, and the Nazis began bombing England in what became known as the Battle of Britain. The fall of France made Americans realize how powerful and threatening Germany had become. Admiring the stubborn British resistance to German attacks, American officials increased aid to Great Britain. In the summer and fall of 1940, Congress approved spending billions of dollars for defense. Destroyers were sent to Great Britain, and the United States set up its first peacetime draft. Roosevelt declared the United States to be an "arsenal of democracy," though he and other Americans still hoped to avoid war.

American aid to Great Britain, especially in the Battle of the Atlantic, angered German chancellor Adolf Hitler. In early 1941, the Joint Chiefs of Staff of the American armed forces met with British military leaders to plan how to cooperate if the United States did get drawn into the war. They decided that the defeat of Germany should get top priority. In a speech of May 27, 1941, Roosevelt emphasized that Germany was a great danger to the Western Hemisphere, and he declared a state of national emergency.

In June, 1941, German consulates in the United States were closed. That same month Germany invaded the Soviet Union, and the United States promised to help the Soviets. In August, Roosevelt and British prime minister Winston Churchill wrote the Atlantic Charter, an agreement to resist the Axis Powers. In September and October, the Germans responded to the

United States' growing involvement by torpedoing the American destroyer *Greer* and sinking the *Reuben James.* Congress repealed the Neutrality Act and allowed American merchant ships to be armed.

## Uniting Against America

Hitler had wanted to avoid war with the United States, just as the Americans had wished to avoid war with Germany. Yet his actions were crucial in bringing the United States into the conflict. He despised Americans, especially after the Great Depression began in 1929, for he considered the United States an inferior country for its cultural and racial mix.

Aware that the American people were mostly isolationists, Hitler did not realize how great the American industrial capacity was and how quickly the United States could prepare to fight a war. The Germans did not make plans for how to wage war with the United States; in 1939, Hitler was able to convince the Italians that the United States would not fight.

Though he was upset by U.S. aid to Great Britain and the Soviet Union, Hitler tried to keep German warships and submarines away from American ships in the Atlantic Ocean, for he knew that German submarine warfare had brought the United States into World War I. By September, 1941, he gave in to German navy leaders, who wanted to break American supply lines to Great Britain.

Meanwhile, conflict between Japan and the United States was growing. Germany hoped that Japan would counterbalance the Soviet Union in the Far East and would distract U.S. leaders from their commitment to the Allies. On September 27, 1940, Germany and Japan signed a pact of mutual assistance, and on April 4, 1941, Hitler promised the Japanese that Germany would support them if war broke out between Japan and the United States.

The Japanese attack on Pearl Harbor on December 4, 1941, took the Germans and Italians by surprise. On December 9, President Roosevelt announced that he considered Germany just as guilty as Japan for the surprise attack. Secretary of War Henry L. Stimson wanted the United States to declare war on Germany at once, but he was overruled. Yet he did not have to wait long.

On December 11, the governments of Germany and Italy issued declarations of war against the United States. The declarations did not mention Japan; instead, the Germans claimed that the Americans had committed open acts of war against Germany in the Atlantic. Congress responded by passing a unanimous joint resolution stating that the United States was in a state of war with Germany.

## Consequences

The Nazis were pleased by the Japanese attack on Pearl Harbor, for they believed that Japan would weaken the British, the Soviet, and especially the American war effort. Yet by declaring war on the United States just as his armies were getting bogged down in the Soviet Union and his navy was being defeated in the Atlantic, Hitler made the greatest mistake of his career.

When Hitler said that his declaration of war on the United States would be "decisive not only for the history of Germany, but for the whole of Europe and indeed for the world," he was right. By declaring war, Germany and Italy guaranteed their own final defeat. The U.S. entry into World War II brought about an Allied victory and made the United States one of the world's superpowers.

*Leon Stein*

# Kodak Perfects Color Film for Infrared Photography

> *By solving many complex chemical and physical problems, Kodak researchers made it possible for infrared radiation to be shown in different color intensities in photographs.*

**What:** Photography
**When:** 1942
**Where:** Rochester, New York
**Who:**
Eastman Kodak Company, an American manufacturer of cameras and film
Sir William Herschel (1738-1822), an English astronomer

## Invisible Light

Photography developed rapidly in the nineteenth century when it became possible to record the colors and shades of visible light on sensitive materials. Visible light is a form of radiation that consists of electromagnetic waves, which also make up other forms of radiation such as X rays and radio waves. Visible light occupies the range of wavelengths from about 400 nanometers (1 nanometer is 1 billionth of a meter) to about 700 nanometers in the electromagnetic spectrum.

Infrared radiation occupies the range from about 700 nanometers to about 1,350 nanometers in the electromagnetic spectrum. Infrared rays cannot be seen by the human eye, but they behave in the same way that rays of visible light behave; they can be reflected, diffracted (broken), and refracted (bent).

Sir William Herschel, a British astronomer, discovered infrared rays in 1800 by calculating the temperature of the heat that they produced. The term "infrared," which was probably first used in 1800, was used to indicate rays that had wavelengths that were longer than those on the red end (the high end) of the spectrum of visible light but shorter than those of the microwaves, which appear higher on the electromagnetic spectrum. Infrared film is therefore sensitive to the infrared radiation that the human eye cannot see or record. Dyes that were sensitive to infrared radiation were discovered early in the twentieth century, but they were not widely used until the 1930's. Because these dyes produced only black-and-white images, their usefulness to artists and researchers was limited. After 1930, however, a tidal wave of infrared photographic applications appeared.

## The Development of Color-Sensitive Infrared Film

In the early 1940's, military intelligence used infrared viewers for night operations and for gathering information about the enemy. One device that was commonly used for such purposes was called a "snooper scope." Aerial photography with black-and-white infrared film was used to locate enemy hiding places and equipment. The images that were produced, however, often lacked clear definition.

The development in 1942 of the first color-sensitive infrared film, Ektachrome Aero Film, became possible when researchers at the Eastman Kodak Company's laboratories solved some complex chemical and physical problems that had hampered the development of color infrared film up to that point. Regular color film is sensitive to all visible colors of the spectrum; infrared color film is sensitive to violet, blue, and red light as well as to infrared radiation. Typical color film has three layers of emulsion, which are sensitized to blue, green, and red. Infrared color film, however, has its three emulsion layers sensitized to green, red, and infrared. Infrared wavelengths are recorded as reds of varying densities, depending on the intensity of the infrared radiation. The more infrared radiation there is, the darker the color of the red that is recorded.

In infrared photography, a filter is placed over the camera lens to block the unwanted rays of visible light. The filter blocks visible and ultraviolet rays but allows infrared radiation to pass. All three layers of infrared film are sensitive to blue, so a yellow filter is used. All blue radiation is absorbed by this filter.

In regular photography, color film consists of three basic layers: The top layer is sensitive to blue light, the middle layer is sensitive to green, and the third layer is sensitive to red. Exposing the film to light causes a latent image to be formed in the silver halide crystals that make up each of the three layers. In infrared photography, color film consists of a top layer that is sensitive to infrared radiation, a middle layer sensitive to green, and a bottom layer sensitive to red. "Reversal processing" produces blue in the infrared-sensitive layer, yellow in the green-sensitive layer, and magenta in the red-sensitive layer. The blue, yellow, and magenta layers of the film produce the "false colors" that accentuate the various levels of infrared radiation shown as red in a color transparency, slide, or print. The color of the dye that is formed in a particular layer bears no relationship to the color of light to which the layer is

sensitive. If the relationship is not complementary, the resulting colors will be false. This means that objects whose colors appear to be similar to the human eye will not necessarily be recorded as similar colors on infrared film. A red rose with healthy green leaves will appear on infrared color film as being yellow with red leaves, because the chlorophyll contained in the plant leaf reflects infrared radiation and causes the green leaves to be recorded as red. Infrared radiation from about 700 nanometers to about 900 nanometers on the electromagnetic spectrum can be recorded by infrared color film. Above 900 nanometers, infrared radiation exists as heat patterns that must be recorded by nonphotographic means.

## Consequences

Infrared photography has proved to be valuable in many of the sciences and the arts. It has been used to create artistic images that are often unexpected visual explosions of everyday views. Because infrared radiation penetrates haze easily, infrared films are often used in mapping areas or determining vegetation types. Many cloud-covered tropical areas would be impossible to map without infrared photography. False-color infrared film can differentiate between healthy and unhealthy plants, so it is widely used to study insect and disease problems in plants.

Medical research uses infrared photography to trace blood flow, detect and monitor tumor growth, and to study many other physiological functions that are invisible to the human eye.

Some forms of cancer can be detected by infrared analysis before any other tests are able to perceive them. Infrared film is used in criminology to photograph illegal activities in the dark and to study evidence at crime scenes. Powder burns around a bullet hole, which are often invisible to the eye, show clearly on infrared film. In addition, forgeries in documents and works of art can often be seen clearly when photographed on infrared film. Archaeologists have used infrared film to locate ancient sites that are invisible in daylight. Wildlife biologists also document the behavior of animals at night with infrared equipment.

*David L. Chesemore*

*Library of Congress*

*Sir William Herschel.*

**894**

# Reber Makes First Radio Maps of Universe

> *Grote Reber built the first radio telescope and used it to record the first radio contour maps of the Milky Way, establishing the foundations of a new type of astronomy.*

**What:** Astronomy
**When:** 1942-1947
**Where:** Wheaton, Illinois
**Who:**
GROTE REBER (1911-        ), an American radio engineer and amateur astronomer
KARL JANSKY (1905-1950), an American radio engineer
SIR WILLIAM HERSCHEL (1738-1822), a German English musician and astronomer
HARLOW SHAPLEY (1885-1972), an American astronomer

## Opening an Invisible Window

Grote Reber's recording of the first radio contour maps of the universe was a new and unexpected application of radio technology. Reber's work opened the invisible window of radio frequencies, allowing astronomers to see new features of the universe.

Sir William Herschel was one of the first astronomers to recognize the true nature of the dense band of stars across the sky called the Milky Way. From counting stars in various directions in the Milky Way, he concluded in 1785 that the vast majority of stars are contained within a flattened disk shape, forming an island universe or galaxy in space, with the solar system reduced to a tiny speck in the vast universe of stars. Early in the twentieth century, Harlow Shapley was able to use the 254-centimeter Mount Wilson telescope to show that the Milky Way galaxy is far larger than any previous estimate, and that the sun is far from the galactic center, which he located in the direction of the constellation Sagittarius.

In 1932, Karl Jansky reported his accidental discovery of radio waves from space. Using a rotating array of antennas sensitive to 15-meter radio waves, he detected a steady hiss whose emission corresponded to the daily motion of the stars. He concluded that he was receiving cosmic radio waves from beyond the solar system. He was able to identify the source of the most intense radiation in the direction of Sagittarius, suggesting that it came from the center of the Milky Way galaxy. He also showed that weaker radio waves came from all directions in the Milky Way and suggested that their source was in the stars or in the interstellar matter between the stars.

## The Lone Radio Astronomer

Jansky's work was so unrelated to traditional astronomy that no professional astronomer followed it up. As a young radio engineer at the Stewart-Warner Company in Chicago, Reber read Jansky's papers and began to plan how he could measure the detailed distribution of the radiation intensity throughout the sky at different wavelengths. In 1937, he built a 9.4-meter parabolic reflecting dish in his yard, mounted so that it could be pointed in a north-south direction; scanning west to east would result from the earth's rotation. For ten years, he operated this radio telescope in Wheaton as the only radio astronomer in the world.

As the Milky Way crossed the meridian late at night, Reber measured the increasing intensity of the cosmic radio waves. He published his initial results in the February, 1940, *Proceedings of the Institute of Radio Engineers*, where he noted that the intensity of the radiation was too low to come from stars, as Jansky had proposed, but suggested the possibility of radiation from interstellar gases.

In 1941, Reber began a complete sky survey with an automatic chart recorder and more sensitive receiving equipment. The recording pen would slowly rise and fall as the reflecting dish rotated with the earth. After collecting approximately two hundred chart recordings, he plotted the resulting radio contours on the two hemispheres of the sky. The resulting radio maps, published in the *Astrophysical Journal* in November, 1944, revealed interesting details: The greatest radio intensity was coming from the center of the galaxy, in Sagittarius; less intense radio waves were coming from the constellations Cygnus and Cassiopeia. More important was his recognition that radio waves could penetrate the interstellar dust that obscures much visible light in the Milky Way.

Reber's last observations in Wheaton were made from 1945 to 1947. The resulting radio maps, published in the *Proceedings of the Institute of Radio Engineers* in October, 1948, now revealed two noise peaks in the Cygnus region, later identified as a radio galaxy (Cygnus A) and a source associated with a spiral arm of the Milky Way (Cygnus X). An intensity peak in Taurus was later identified with the eleventh century supernova remnant in the Crab nebula, and another in Cassiopeia matches the position of a seventeenth century supernova explosion. These results were the beginning of many important discoveries in the field of radio astronomy.

## Consequences

Reber's pioneering work and resulting radio maps led to a growing interest in radio astronomy and many unexpected discoveries with radio telescopes of increasing sophistication and size.

In 1945, a graduate student at the University of Leiden, in the Netherlands, Hendrik Christoffel van de Hulst, predicted that neutral hydrogen should emit 21-centimeter radio waves. By 1949, the Harvard physicist Edward Mills Purcell began a search for these radio waves with Harold Irving Ewen, a graduate student who was sent to confer with Reber on techniques in radio astronomy. Ewen and Purcell developed special equipment and by 1951 had succeeded in detecting the predicted 21-centimeter radio waves. The Dutch astronomer Jan Hendrik Oort's group then began a seven-year collaboration with Australian radio astronomers to map the spiral arms of the Milky Way galaxy.

In 1960, two radio sources were identified with what appeared to be stars, but each emitted much more radio energy than Earth's sun or any other known star. Four of these "quasars" (*quasi-stellar* radio sources) had been discovered by 1963. At distances of billions of light-years, these objects would have to be more than one hundred times brighter than entire galaxies and would appear to be some kind of highly energetic stage in the early formation of a galaxy.

Another dramatic event in radio astronomy occurred in 1967, when Jocelyn Bell, a graduate student in radio astronomy at Cambridge, discovered "pulsars." These are believed to be fast-spinning "neutron stars" with high magnetic fields that produce a rotating beam of radio emission. A pulsar in the Crab nebula was later identified with the collapsed core of the supernova remnant that had appeared on Reber's radio maps.

Perhaps the most important discovery in radio astronomy was the 1965 detection of microwave background radiation by radio astronomers Arno Penzias and Robert Woodrow Wilson. Using a 6-meter horn antenna at the Bell Telephone Laboratories in Holmdel, New Jersey, they found an unexpected excess of steady radiation with no directional variation. This matched current predictions of cosmic radiation from a primeval fireball in the "big bang" theory. Thus, radio astronomy provided confirmation of the creation and expansion of the universe.

*Joseph L. Spradley*

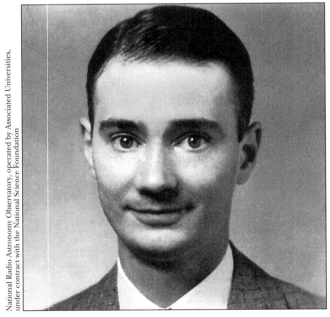

*Grote Reber.*

# Orlon Fibers Are Introduced

*Industrial chemists learned how to make fibers from polyacrylonitrile. These fibers are widely used in textiles and in the preparation of high-strength carbon fibers.*

**What:** Materials
**When:** 1942-1950
**Where:** Wilmington, Delaware, and Germany
**Who:**
HERBERT REIN (1899-1955), a German chemist
RAY C. HOUTZ (1907-     ), an American chemist

## A Difficult Plastic

"Polymers" are large molecules that are made up of chains of many smaller molecules, called "monomers." Materials that are made of polymers are also called polymers, and some polymers, such as proteins, cellulose, and starch, occur in nature. Most polymers, however, are synthetic materials, which means that they were created by scientists.

The twenty-year period beginning in 1930 was the age of great discoveries in polymers by both chemists and engineers. During this time, many of the synthetic polymers, which are also known as plastics, were first made and their uses found. Among these polymers were nylon, polyester, and polyacrylonitrile (PAN). The last of these materials, PAN, was first synthesized by German chemists in the late 1920's. They linked more than one thousand of the small, organic molecules of acrylonitrile to make a polymer. The polymer chains of this material had the properties that were needed to form strong fibers, but there was one problem. Instead of melting when heated to a high temperature, PAN simply decomposed. This made it impossible, with the technology that existed then, to make fibers.

The best method available to industry at that time was the process of melt spinning, in which fibers were made by forcing molten polymer through small holes and allowing it to cool. Researchers realized that if PAN could be put into a solution, the same apparatus could be used to spin PAN fibers. Scientists in Germany and the United States tried to find a solvent or liquid that would dissolve PAN, but they were unsuccessful until World War II began.

## Fibers for War

In 1938, the German chemist Walter Reppe developed a new class of organic solvents called "amides." These new liquids were able to dissolve many materials, including some of the recently discovered polymers. When World War II began in 1940, both the Germans and the Allies needed to develop new materials for the war effort. Materials such as rubber and fibers were in short supply. Thus, there was increased governmental support for chemical and industrial research on both sides of the war. This support was to result in two independent solutions to the PAN problem.

In 1942, Herbert Rein, while working for I. G. Farben in Germany, discovered that PAN fibers could be produced from a solution of polyacrylonitrile dissolved in the newly synthesized solvent dimethylformamide. At the same time Ray C. Houtz, who was working for E. I. du Pont de Nemours in Wilmington, Delaware, found that the related solvent dimethylacetamide would also form excellent PAN fibers. His work was patented, and some fibers were produced for use by the military during the war. In 1950, Du Pont began commercial production of a form of polyacrylonitrile fibers called Orlon. The Monsanto Company followed with a fiber called Acrilon in 1952, and other companies began to make similar products in 1958.

There are two ways to produce PAN fibers. In both methods, polyacrylonitrile is first dissolved in a suitable solvent. The solution is next forced through small holes in a device called a "spin-

*A worker examines a bobbin of Orlon acrylic fiber at the South Carolina plant in 1952.*

neret." The solution emerges from the spinneret as thin streams of a thick, gooey liquid. In the "wet spinning method," the streams then enter another liquid (usually water or alcohol), which extracts the solvent from the solution, leaving behind the pure PAN fiber. After air drying, the fiber can be treated like any other fiber. The "dry spinning method" uses no liquid. Instead, the solvent is evaporated from the emerging streams by means of hot air, and again the PAN fiber is left behind.

In 1944, another discovery was made that is an important part of the polyacrylonitrile fiber story. W. P. Coxe of Du Pont and L. L. Winter at Union Carbide Corporation found that when PAN fibers are heated under certain conditions, the polymer decomposes and changes into graphite (one of the elemental forms of carbon) but still keeps its fiber form. In contrast to most forms of graphite, these fibers were exceptionally strong. These were the first carbon fibers ever made. Originally known as "black Orlon," they were first produced commercially by the Japanese in 1964, but they were too weak to find many uses. After new methods of graphitization were developed jointly by labs in Japan, Great Britain, and the United States, the strength of the carbon fibers was increased, and the fibers began to be used in many fields.

**Consequences**

As had been predicted earlier, PAN fibers were found to have some very useful properties. Their discovery and commercialization helped pave the way for the acceptance and wide use of polymers. The fibers derive their properties from the stiff, rodlike structure of polyacrylonitrile. Known as acrylics, these fibers are more durable than cotton, and they are the best alternative to wool for sweaters. Acrylics are resistant to heat and chemicals, can be dyed easily, resist fading or wrinkling, and are mildew-resistant. Thus, after their introduction, PAN fibers were very quickly made into yarns, blankets, draperies, carpets, rugs, sportswear, and various items of clothing. Often, the fibers contain small amounts of other polymers that give them additional useful properties.

A significant amount of PAN fiber is used in making carbon fibers. These lightweight fibers are stronger for their weight than any known material, and they are used to make high-strength composites for applications in aerospace, the military, and sports. A "fiber composite" is a material made from two parts: a fiber, such as carbon or glass, and something to hold the fibers together, which is usually a plastic called an "epoxy." Fiber composites are used in products that require great strength and light weight. Their applications can be as ordinary as a tennis racket or fishing pole or as exotic as an airplane tail or the body of a spacecraft.

*David MacInnes, Jr.*

# Roosevelt Approves Internment of Japanese Americans

During World War II, about 110,000 Japanese Americans living on the West Coast were forced to move into relocation camps.

**What:** Civil rights and liberties; Ethnic conflict
**When:** February 19, 1942
**Where:** Washington, D.C., California, Oregon, Washington, and Arizona
**Who:**
FRANKLIN DELANO ROOSEVELT (1882-1945), president of the United States from 1933 to 1945
ALLEN W. GULLION (1880-1946), chief law-enforcement officer of the U.S. Army
KARL R. BENDETSEN (1907-1989), an aide to Gullion
JOHN L. DEWITT (1880-1962), commander of the Army's Western Defense Command
EARL WARREN (1891-1974), attorney general of California
JOHN J. McCLOY (1895-1989), U.S. assistant secretary of war
HENRY L. STIMSON (1867-1950), U.S. secretary of war
DILLON S. MYER (1891-1982), director of the War Relocation Administration

## The Call for Removal

At the close of the year 1941, when the Japanese attacked Pearl Harbor, there were approximately 110,000 Japanese Americans living on the Pacific coast of the United States. About one-third of them were "issei"—Japanese who had been born abroad and were not allowed to become U.S. citizens. The rest were the "nisei"—their children, born in the United States. Most of the nisei, who were automatically American citizens through their birth, considered themselves American rather than Japanese.

Immediately after the Pearl Harbor attack, about fifteen hundred Japanese aliens (noncitizens) were rounded up by the U.S. government on suspicion of being disloyal to the United States. Other Japanese Americans were not allowed to travel without permission, barred from areas surrounding military bases, and forbidden to have weapons, shortwave radios, or maps.

But for many other Americans on the Pacific coast, these restrictions were not enough. Patriotic groups, newspapers, and politicians began demanding that all Japanese Americans be removed. California state attorney general Earl Warren warned that all Japanese Americans were suspect because of their race.

Allen W. Gullion, provost marshal general of the U.S. Army, joined the call for a roundup of all Japanese Americans in the West Coast area. His ambitious aide, Major Karl R. Bendetsen, worked eagerly to promote the idea.

At first Lieutenant General John L. DeWitt, commander of the Army's Western Defense Command, opposed total removal of Japanese Americans. By early February, 1942, however, he changed his mind. "The Japanese race is an enemy race," he said.

In Washington, D.C., Assistant Secretary of War John J. McCloy was able to persuade Secretary of War Henry L. Stimson that the Japanese Americans should be moved. U.S. attorney general Francis Biddle, along with others in the Justice Department, did not think a mass evacuation was necessary, but he gave in to the War Department.

Most important, President Franklin D. Roosevelt decided to give his full backing to the evacuation plan. On February 19, 1942, Roosevelt issued Executive Order 9066, which authorized the military to name "military areas" from

**899**

which "any or all persons" could be kept out. In March, Congress passed a law making it a crime for anyone excluded from a military area to stay there.

### The Internment Camps

It seems that no one had thought much about what would be done with the Japanese Americans, once they had been moved out of their homes. At first, the military simply asked Japanese Americans living in the western parts of California, Oregon, and Washington—and in the strip of Arizona along the Mexican border—to leave for the interior of the United States. Yet communities in the interior did not want to welcome the Japanese Americans, and on March 27, 1942, the Army ordered Japanese Americans to remain where they were.

Finally, the Japanese Americans were told to report to assembly centers, where they would stay until more permanent housing could be found. By June, 1942, more than 100,000 Japanese Americans had been evacuated. They were moved from the assembly centers to ten relocation camps in the interior. Each of these camps held between ten and eleven thousand people. A new agency, the War Relocation Authority (WRA), had been formed to oversee the camps.

The camps were surrounded by barbed wire and patrolled by armed guards. Camp residents lived in wooden barracks covered with tar paper. Each barracks contained a number of one-room apartments for families or groups of unrelated persons; each apartment was furnished with army cots, blankets, and a light bulb. The residents shared bathrooms, dining areas, and laundry rooms.

Religious worship was allowed, except for the practice of Shinto, a traditional Japanese religion. Eventually schools were opened for the children and teenagers. Residents grew some of their own food, and some of them began small manufacturing projects. Most of them, however, could not find much productive work to do in the camps. The WRA encouraged them to form camp governments, but these governments did

*In April, 1942, Japanese Americans line up at the assembly center at Santa Anita Park in Southern California before being incarcerated at relocation camps.*

not have any real power and the residents did not respect them.

Conditions were worst at the camp in Tule Lake, California, where "troublemakers" from other camps were sent. The Tule Lake facility came to be dominated by a secret group that wanted to see Japan win the war.

One nisei said later that staying in shacks and lacking material possessions were not the worst parts of living in the camps. "The most devastating effect upon a human soul is not hatred but being considered not human."

At first, Dillon S. Myer, who directed the WRA beginning in June, 1942, saw the relocation camps as "temporary wayside stations." In 1943, the WRA began releasing evacuees who had not shown disloyalty, who had jobs waiting for them away from the Pacific coast, and who could show that the local community would accept them. By the end of 1944, about thirty-five thousand evacuees had left the camps.

By the spring of 1944, the Roosevelt administration was aware that there was no military reason to keep Japanese Americans away from the Pacific coast. Yet Roosevelt's team waited until after the 1944 presidential election to announce that the exclusion order was no longer in effect.

Nearly all of those in the camps were allowed to leave. Many of the evacuees, however, were afraid that they would be treated badly outside, and they stayed in the camps. In June, 1945, the WRA decided to close the camps by the end of the year.

## Consequences

Just over half of the evacuees returned to the Pacific coast, and most found that their homes, businesses, and jobs were gone. Overall, Japanese Americans lost income and property estimated at $350 million.

Some nisei had been eager to show their patriotism and had volunteered for military service during the war. The Japanese American 100th Infantry Battalion and 422d Regimental Combat Team were two of the Army's best units. Other nisei, however, became bitter because of their painful memories of life in the camps. More than five thousand of them gave up their U.S. citizenship.

Demands from Japanese Americans led Congress in 1981 to set up a Commission on Wartime Relocation and Internment to take a new look at internment during World War II. The commission's report concluded that internment was not a "military necessity" but had been caused by "race prejudice, war hysteria and a failure of political leadership." In 1988, Congress formally apologized to Japanese Americans and voted $1.25 billion to be shared among camp residents who still survived.

*George Q. Flynn*

# British Mount First Thousand-Bomber Raid on Germany

*Great Britain's Royal Air Force began a strategy of dropping huge numbers of bombs during night raids over German cities.*

**What:** War
**When:** May 30-31, 1942
**Where:** Cologne, Germany
**Who:**
Sir Winston Churchill (1874-1965), prime minister of Great Britain from 1940 to 1945
Archibald Sinclair (1890-1970), secretary of state for air from 1940 to 1945
Sir Hugh Trenchard (1873-1956), chief of the Air Staff from 1919 to 1929
Sir Charles Portal (1893-1971), chief of the Air Staff from 1940 to 1945
Sir Arthur Travers Harris (1892-1984), commander in chief of Bomber Command from 1942 to 1945

## Trial and Error

During World War I, Germany dropped bombs on London and a few other British cities. The British prepared to retaliate with giant planes that could reach Berlin, but the war ended before such raids could be made. These experiences, however, were not forgotten. During the 1920's, Marshal of the Royal Air Force Sir Hugh Trenchard, chief of the Air Staff, worked to build up an air force designed to attack the cities of any enemy of Great Britain. The goal would be to destroy important plants and factories and thus make the enemy unable to continue its war effort—but also to intimidate the civilian population under a rain of bombs.

Trenchard pursued this goal single-mindedly, with little thought or money spent on developing aircraft that could support the British army in the field of battle. As a result, when World War II began in 1939, Great Britain had no good ground-attack aircraft. The air marshals trusted in their multiengined, heavy bombers designed to attack Germany in daylight, blowing up oil plants and railway centers in the Ruhr Valley and other areas of western Germany. They believed that if the refineries and railroad yards could be blown up, the German war effort would be brought to a standstill. The Allies would have won the war without a bloody struggle with the German army.

Before these ideas could be tried out, however, the British had some experience in daylight attacks on German warships. They quickly learned that their bombers, which were defended only by .303 machine guns, could not survive the attacks of the German fighters, which fired cannons. Instead, they decided to send their crews on precision-bombing night attacks on German industrial targets.

The first such attack took place on the evening of May 15-16, 1940, in response to a German attack on Rotterdam in which thousands of civilians were believed to have been slaughtered. At first the British believed that their raids were effective; in the fall of 1940, the reports of Sir Charles Portal, chief of the air staff, were full of optimism. As photographs of the targets began to accumulate, however, the air marshals realized that almost no damage was being done. The British equipment was not up to precision attacks at night.

## Area Bombing

The Air Staff soon decided to begin "area bombing"—aiming not for certain factories or railroad yards but simply for the German cities themselves. This type of attack was likely to kill many civilians, but British military planners felt

**902**

justified in approving the policy, since London, Coventry, and other British cities had been heavily bombed by the Germans in 1940 and 1941. The German cities would now suffer more than a mere tit for tat. This new policy was approved by Prime Minister Winston Churchill and Secretary of State for Air Archibald Sinclair.

In May, 1942, Sir Arthur Harris was appointed commander in chief of the Bomber Command, and he began the practice of attacking with one thousand bombers at a time. Cologne, Germany, which was important for its war industry and rail facilities, was chosen as the first target.

On the night of May 30-31, 1,046 British bombers were sent to attack Cologne. Some were lost in bad weather over the North Sea, and others turned back because of mechanical problems; only 898 bombers actually reached Cologne. At 12:38 A.M., the first bombers sent down showers of firebombs to start fires, which would light the way for the following bombers.

On the ground, sirens wailed, and the people of Cologne went underground to sit out the raid huddled in fear. Guns positioned within the city flashed into action, firing a heavy barrage, and outside the city area German nightfighters began to try to stop the stream of British bombers. Soon bombers hit by flak were plummeting to earth in flames.

The British continued to roar overhead until 3:10 A.M., when the last aircraft finally turned for home. Altogether, 1,455 tons of bombs—more than half of them firebombs—had been dropped. The crews of the last planes reported that the fires of Cologne were still visible 150 miles away.

Actually, only slightly more than six hundred aircraft had attacked the city; nearly three hundred pilots had mistaken some other town for the target. Even so, the British dropped about thirty-one tons of bombs per square mile of Cologne's built-up area. Nearly five hundred people were killed, and more than five thousand

*Bombs destroyed much of Cologne, Germany.*

National Archives

**903**

were injured. Six hundred acres were gutted by fire and explosion; thirty-three hundred houses were destroyed, and forty-five thousand people were left homeless.

Yet there was fairly little military damage to Cologne. Of the city's 328 large factories, only 36 were forced to stop production for the time being. Much of the damage was repaired within two weeks.

Forty Royal Air Force bombers were shot down, and another twelve were so badly damaged that they had to be scrapped. Just over one hundred other bombers were more lightly hit.

## Consequences

In the summer of 1942, the United States Air Force joined the Royal Air Force in more attacks on cities in Germany and the occupied countries. Before Germany surrendered in 1945, more than 1.5 million tons of bombs had been dropped. About three hundred thousand German civilians had been killed and many more injured.

The British paid a high price for these results. More than eight thousand Royal Air Force bombers were lost, most of them to antiaircraft fire. Bomber Command proved the most dangerous branch of Great Britain's armed forces: 55,888 men were killed, and 9,162 were wounded.

British military historians have calculated that the bombings succeeded in destroying only 3.5 percent of Germany's war production capacity. The German oil plants and railroad system were hard hit, and the Luftwaffe (German air force) was badly hurt by attacks on its factories. If one considers how many people were killed, however, and how much money the British and American governments had invested in their bombers, one cannot call the bombing raids a great success.

# Battle of Midway Turns Tide in Pacific

> *Forewarned of a Japanese attack, the American defenders of Midway Island fought hard and succeeded in damaging or turning back a much larger Japanese fleet.*

**What:** War
**When:** June 3-6, 1942
**Where:** Pacific Ocean, near Midway Island
**Who:**

ISOROKU YAMAMOTO (1884-1943), commander in chief of the Japanese Combined Fleet

CHUICHI NAGUMO (1887-1944), commander of the Japanese First Carrier Striking Force

CHESTER NIMITZ (1885-1966), commander in chief of the U.S. Pacific Fleet

RAYMOND SPRUANCE (1886-1969), commander of U.S. Naval Task Force 16

FRANK JACK FLETCHER (1885-1973), commander of the U.S. Naval Carrier Striking Force

JOSEPH J. ROCHEFORT, JR., commander of U.S. Naval Combat Intelligence at Pearl Harbor

## The Japanese Plan

Admiral Isoroku Yamamoto, commander in chief of Japan's Combined Fleet, was in charge of naval planning for his country. After Japan's bombing of the United States Pacific Fleet at Pearl Harbor, Hawaii, he turned his attention to other targets. The most logical choice seemed to be Midway, an American-owned island about a thousand miles west of Hawaii.

If the Japanese could capture Midway, the island could serve as a key point in the defensive line the Japanese wanted to stretch all the way from the Aleutian Islands in the north to Australia in the south. Some in the Japanese navy believed that Midway would be too expensive to hold as a base. After U.S. lieutenant colonel James Doolittle led a raid on Tokyo and other

Japanese cities on April 18, 1942, however, the doubters were convinced. To protect the nation and the emperor, Midway was clearly necessary.

More than simply capturing Midway, Yamamoto hoped to draw what was left of the U.S. Fleet into battle and destroy it. The United States was working hard to replace ships and aircraft that had been lost in the Pearl Harbor attack. Yamamoto wanted to wipe out the fleet before the Americans could reach their goal.

To launch his attack, Yamamoto put together the largest fleet the Japanese had ever assembled. It included eleven battleships, headed by the *Yamato*, Japan's newest battleship—and the world's largest. There were also four heavy aircraft carriers and four light ones, twenty-one cruisers, sixty-five destroyers, more than fifty smaller boats, and nineteen submarines.

Yet Yamamoto made a crucial mistake: He separated these vessels into a number of groups so far apart that they could not work together effectively. He sent the Northern Force—two light carriers, eight cruisers, thirteen destroyers, and six submarines—to capture Kiska and Attu in the Aleutians, slightly before the main forces were to arrive at Midway. The Northern Force succeeded in taking Kiska and Attu, but these islands were too far away to be of real value in the war, and the American ships stationed there were not especially important in the U.S. Fleet.

Farther south, the forces that were to converge on Midway were badly divided. From the southwest came the Midway Occupation Group and the Second Fleet—two battleships, eight cruisers, a light carrier, and a dozen destroyers. Approaching Midway from the northwest was Yamamoto with the main body and the carrier striking force; his main force was organized around three battleships and a light carrier. Split off to the north to move either to the Aleutians or to Midway (but actually too far from either) was the Guard Force—four battleships along with cruisers and destroyers. Moving ahead of all these forces was the First Carrier Striking Force, commanded by Vice Admiral Chuichi Nagumo; it included four heavy carriers—*Akagi, Kaga, Soryu,* and *Hiryu*—along with support vessels.

The plan was for Nagumo's carriers to attack Midway on June 4; they would destroy the U.S. airfields and planes. When the Americans sent out their fleet from Pearl Harbor, the Main Body would move in and destroy it.

Japan's earlier successes had led Yamamoto and other planners to expect another victory at Midway. They made no plans for what to do if the Americans did not follow the Japanese expectations.

**The Battle**

In fact, the Americans did not cooperate. The intelligence unit at Pearl Harbor, under Commander Joseph J. Rochefort, Jr., had broken the Japanese naval code. They gained enough information to allow them to guess that Midway would be the main target of an upcoming attack.

Admiral Chester W. Nimitz, commander in chief of the Pacific Fleet, called in all available carriers and could come up with only three: *Enterprise* and *Hornet,* commanded by Rear Admiral Raymond Spruance, and *Yorktown,* a damaged vessel under the command of Rear Admiral Frank Jack Fletcher. With a screen of eight cruisers and fourteen destroyers, these made up Nimitz's defending force.

Nimitz ordered that Midway be reinforced with 120 planes, antiaircraft guns, and 3,632 defenders. The three carriers were stationed northeast of Midway, where they lay in wait for the Japanese.

Early on June 3, a scout plane sighted the invasion force six hundred miles to the southwest. Army and Marine pilots moved out from Midway to attack, but scored no major hits.

Not realizing that American ships were anywhere around, Nagumo launched an attack with half his planes (108) before dawn on June 4. The Midway defenders put all of their planes in the air and took heavy punishment, but were not knocked out. Though the Japanese Zero aircraft were superior, the American planes, along with

**SHIPS SUNK AT MIDWAY, JUNE 4, 1942**

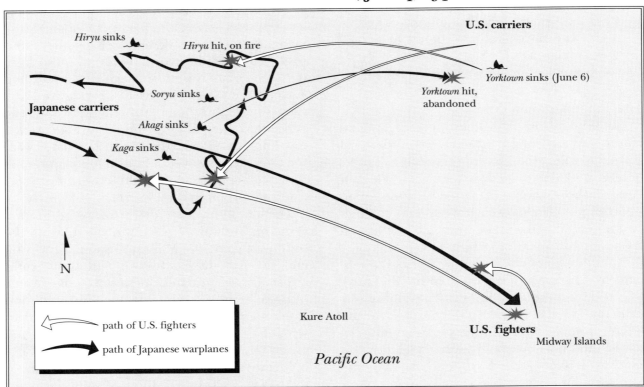

*U.S. Navy fighters fly above a burning Japanese ship during the Battle of Midway.*

National Archives

antiaircraft guns, managed to inflict some losses on the Japanese. By 7:00 A.M. the first raid was over, and the Japanese flight leader radioed Nagumo: "There is need for another attack."

Before the second attack could be launched, Marine and Army pilots zoomed repeatedly over the Japanese carriers. They scored no hits, though nearly all of them died trying. In the midst of these attacks, at 8:20 a Japanese scout plane reported an American carrier within range.

Instead of immediately launching the second wave of planes, which were being rearmed for another attack on Midway, Nagumo decided to prepare these planes for action against the U.S. Fleet. By 9:18 all was ready, although in their haste the Japanese had piled bombs and torpedoes around the carrier decks.

Just then, unprotected American carrier torpedo planes began nearly suicidal attacks. Of the forty-two involved, thirty-eight were lost. None scored hits, but they succeeded in luring the Zeros down low to attack them.

At the end of these attacks, SBD Dauntless di-ve-bombers located and hit the *Kaga, Akagi,* and *Soryu.* The bombs set off explosions on board, so within five minutes these carriers had turned into flaming wrecks.

The Japanese attacked with planes left on the *Hiryu* and damaged the *Yorktown,* while American planes sank *Hiryu.* (*Yorktown* was later sunk on June 6 by Japanese submarine I-168.) Rear Admiral Spruance refused to fight a night battle against Yamamoto's battleships. With Midway still in United States' hands and two American carriers facing him, Yamamoto turned back, leaving behind a heavy cruiser.

## Consequences

Naval battles during World War II ended up relying on airplanes more than they did on battleships. Yamamoto had failed to realize that technology had brought about this change, and as a result his forces lost the crucial battle for Midway. Japan had lost four heavy aircraft carriers, and ultimately it lost the war.

*Charles W. Johnson*

**907**

# Allies Win Battle of Guadalcanal

*After a long, difficult fight that employed air, sea, and land forces, the United States succeeded in taking control of Guadalcanal away from Japan.*

**What:** War
**When:** August, 1942-February, 1943
**Where:** Solomon Islands
**Who:**

CHESTER W. NIMITZ (1885-1966),
   commander in chief of the U.S. Pacific
   Fleet
RICHMOND K. TURNER (1885-1961),
   commander of amphibious force at
   Guadalcanal
FRANK JACK FLETCHER (1885-1973),
   commander of the invasion and naval
   support forces at Guadalcanal
ALEXANDER A. VANDEGRIFT (1887-1973),
   commander of the U.S. Marine
   landing force
ROBERT L. GHORMLEY (1883-1958),
   commander of the South Pacific Area
   until October, 1942
WILLIAM F. HALSEY (1882-1959),
   commander of the South Pacific Area
   from October, 1942, to 1944
GUNICHI MIKAWA (born 1888), commander
   of Japanese navy forces during the
   Battle of Savo Island
KIYOTAKE KAWAGUCHI, commander of
   Japanese Army forces on
   Guadalcanal

## Into Guadalcanal

An important battle between Japanese and American forces took place near Midway Island in early June, 1942. The Japanese lost the battle along with 4 aircraft carriers, 322 planes, and 100 pilots. This major defeat put Japan on the defensive, while American military planners decided to begin taking the offensive in the Pacific.

The logical first goal for the United States was to capture the Solomon Islands. They lay within easy bombing range of the great Japanese air base at Rabaul on New Britain Island, as well as the important Allied base of Port Moresby in southern New Guinea. Also, the Japanese had begun to build a bomber field on Guadalcanal, one of the southern islands in the Solomons. Whoever controlled Guadalcanal and finished the airfield would have a real advantage in the Pacific war.

Under urgent orders from Admiral Ernest J. King, chief of Naval Operations, Admiral Chester W. Nimitz and the other American commanders in the Southwest Pacific began gathering all available forces and equipment for an "amphibious" (sea and land) operation in the Solomon Islands. The targets were Guadalcanal and its neighbor, Tulagi.

The American forces were not familiar with the islands they were to invade and had little time to prepare. The combined force, made up of eighty-two ships carrying the First Marine Divisions, part of the Second Marine Division, and other forces, met near the Fiji Islands in late July.

Early on August 7, a U.S. carrier task force took position south of Guadalcanal. Under its protection, the ships carrying the forces that were to land moved along Guadalcanal's west coast, under the command of Rear Admiral Richmond K. Turner. After the shore was bombarded, Major General Alexander A. Vandegrift's marines waded ashore.

On Guadalcanal, the landings were relatively peaceful (though the forces landing on Tulagi had to fight hard). On August 8, the marines accomplished their main goal—seizing the unfinished airfield, which would soon be named Henderson Field. Japanese forces on Guadalcanal were fewer than twenty-five hundred, and within a few days there were about sixteen thousand U.S. Marines on the islands. Nevertheless, an American victory did not come quickly.

## An Exhausting Fight

The Japanese quickly sent new forces to Guadalcanal. Meanwhile, American Vice Admiral Frank Jack Fletcher, commander of the invasion, withdrew the carrier task force because he feared for its safety. After August 8, the beachhead had little American air protection, and heavy Japanese bombing attacks began.

With the carriers gone, the Japanese decided to send a strong naval force, hoping to isolate the marines by destroying American warships and transports. The Japanese striking force of five heavy cruisers, two light cruisers, and a destroyer slipped past U.S. patrol ships and entered "Iron Bottom Sound" at 1:00 A.M. on August 9. Carefully trained for night action, the Japanese were able to sink four Allied cruisers.

The Japanese force, commanded by Vice Admiral Gunichi Mikawa, made the mistake of not attacking the unprotected support ships and the beachhead. Still, it was a real victory for Japan: Rear Admiral Turner withdrew his amphibious force, leaving behind the sixteen thousand marines, who were not well supplied.

From mid-August, 1942, until early February, 1943, Japanese and Allied forces were locked in bitter conflict on the island. After their early successes, the marines came up against the Japanese force's stubborn resistance, and for several months they made little progress. Each side worked to keep supplies and reinforcements from coming in to the other side.

Large numbers of Japanese were killed. At the Battle of Tenaru River, one thousand Japanese men were wiped out, and at the Battle of Bloody Ridge on September 13-14, a Japanese force of about six thousand men under Major General Kiyotake Kawaguchi was cut to pieces.

Yet final victory could not be won until one side managed to get control of the surrounding ocean. The struggle between Allied and Japanese naval forces continued through the autumn, with the Imperial Navy controlling the waters around Guadalcanal at night and the Allies, because of Henderson Field's aircraft, commanding the area during the day.

On October 15, forty-five hundred Japanese soldiers were landed, bringing their total force on the island to twenty thousand men, and the Imperial Army prepared for a final hard attack. Meanwhile, the marines were suffering from discouragement, malaria and other diseases, and exhaustion. The Japanese had bombed Henderson Field, and more than half of the U.S. planes there could not be flown.

The defeated feeling among American soldiers and commanders began to be lifted on October 16, when Vice Admiral William F. Halsey replaced Vice Admiral Robert L. Ghormley as Commander of the Southwest Pacific. Halsey was determined to gain control of Guadalcanal, and the Joint Chiefs of Staff supported him.

The main Japanese attacks came on October 24 and 25, but they were beaten back. In November, both sides tried to bring in reinforcements. The Americans succeeded, but the Japanese were able to land only about four thousand soldiers, who lacked equipment and supplies.

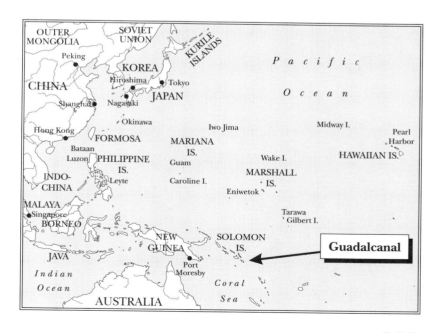

In December, U.S. Army units replaced the exhausted marines, and these fresh forces soon launched a powerful attack, along with air strikes from Henderson Field and from aircraft carriers. The Japanese held on grimly until January 4, 1943, when they received orders from Tokyo to evacuate Guadalcanal within thirty days. With skillful organization, the Imperial Army managed to evacuate more than eleven thousand men via destroyers in early February. Their departure marked an end to the Battle of Guadalcanal.

## Consequences

In the Battle of Guadalcanal, sixteen hundred American men were killed and forty-two hundred wounded. About fourteen thousand Japanese were killed or missing, nine thousand died from disease, and about one thousand were captured. More than six hundred Japanese aircraft had been downed. Having succeeded in their offensive against Japan, the Allies were on their way to victory in the Pacific war of World War II.

*Theodore A. Wilson*

# Battle of Alamein Heralds German Defeat

*At the Battle of Alamein, the British Eighth Army achieved the first major defeat of a German army in World War II.*

**What:** War
**When:** October 23-November 3, 1942
**Where:** Alamein, Egypt
**Who:**
ERWIN ROMMEL (1891-1944), German commander in North Africa
SIR ARCHIBALD WAVELL (1883-1950), commander of British Middle East Forces from 1939 to 1941
SIR CLAUDE AUCHINLECK (1884-1981), commander of British Middle East Forces from 1941 to 1942
SIR HAROLD ALEXANDER (1891-1969), commander of British Middle East forces from 1942 to 1943
BERNARD MONTGOMERY (1887-1976), commander of British Eighth Army from 1942 to 1944

## War in North Africa

After France surrendered to the invading Nazi armies in June, 1940, Great Britain was left to fight alone against Germany. Defending themselves against German bombing raids, the British were in no position to launch an attack of their own on the European Continent.

The same month that France fell, however, Italy had joined the fray by declaring war on the Allies. Italy had several colonies in North Africa, and British military planners saw a chance of success there.

It was very important that Great Britain keep its own colonial holdings in North Africa—especially Gibraltar, Malta, and Suez—so as to maintain control in the whole Mediterranean area. So it was that British prime minister Winston Churchill and General Sir Alan Brooke, chief of the general staff, ordered an offensive against the Italian army in North Africa.

Under General Sir Archibald Wavell, Commander of Middle East forces, the small but well-trained British forces defeated the Italians time and time again during 1940 and 1941. The Italian troops, commanded by Marshal Rodolfo Graziani, were driven out of Libya and Ethiopia, and hundreds of thousands of Italians surrendered to the British.

German chancellor Adolf Hitler began to be perturbed over his ally's losses. Grand Admiral Erich Raeder, Commander of the German navy, was Hitler's best military strategist. His advice was to send large forces to North Africa in order to break Great Britain's hold on its colonies, especially Suez, and gain the oil riches of the Middle East for Germany. In late 1940, however, Hitler had already decided to invade the Soviet Union, a major undertaking that would require a huge commitment of human resources and money. He agreed only to send a small force to North Africa, under the command of General Erwin Rommel.

Rommel's arrival in Africa was well timed. Wavell had been forced to send many of his troops to the defense of Greece, and the British forces were weak when Rommel began to attack in April, 1941. The British were pushed back eastward toward Cairo.

To try to save the situation, Churchill and Brooke appointed General Sir Claude Auchinleck Commander of Middle East Forces. It was not a good choice. Never a forceful soldier, Auchinleck allowed Rommel to take even more ground.

The Germans captured the key town of Tobruk in June, 1942. Though Hitler and Italian dictator Benito Mussolini had originally planned for Rommel's forces to stop there and then attack Malta, they were so delighted with Rommel's successes that they encouraged him to continue his offensive against the British. Hitler hoped that Rommel's army could eventually sweep past Suez and join German forces in the southern Soviet Union.

Meanwhile, the British moved back to a defensive position at Alamein, sixty-five miles from Alexandria and the Nile Delta.

## Britain Strikes Back

In desperation, Churchill and Brooke replaced Auchinleck with General Sir Harold Alexander. To command the main British force, the Eighth Army, they appointed General Bernard Montgomery. Montgomery was able to understand the military situation quickly, and he planned an offensive that would turn back the Axis armies.

Alamein was a good position to defend, for it lay just north of the Qattara Depression, and between Qattara and the Mediterranean Sea there were thirty miles of battle space. When Rommel tried to launch a new attack in September, Montgomery brought him to a standstill at the Battle of Alam el Halfa.

Meanwhile, British troops and supplies, particularly tanks, were arriving in great numbers, while the Germans found themselves low on fuel. The truck caravans that were bringing supplies to Rommel's forces were being attacked by British airplanes. By mid-October, the British had both more men and more tanks than the Germans.

On October 23, 1942, Montgomery led his own attack. At this crucial moment Rommel was on leave in Germany. He hurried back to take command of the Axis forces, but the British had already broken through his lines. In a tank battle that lasted a week, Rommel's forces were defeated. He ordered a general retreat on November 3.

## Consequences

Rommel's forces withdrew to Tunis, and he was recalled to Germany in March, 1943. In 1944 he was appointed to oversee the defense of the French coast, but in that same year he became involved in a plot to assassinate Hitler. Before he could be brought to trial, he committed suicide.

The Battle of Alamein was the first major defeat of a German army in World War II, and both British and German historians have called it the turning point in the war. On November 3, joint American and British forces invaded northwestern Africa; two months later, on the Russian front, the Germans lost Stalingrad. For the remainder of the war, the Germans were on the defensive.

*Italian troops run from falling British bombs during the Battle of Alamein in October, 1942.*

# Allied Forces Invade North Africa

> *Landing in French North Africa, American and British forces were able to avoid fighting the French and succeeded in taking North Africa from the Germans and Italians.*

**What:** War
**When:** November 7-8, 1942
**Where:** French North Africa
**Who:**
SIR WINSTON CHURCHILL (1874-1965), prime minister of Great Britain from 1940 to 1945
FRANKLIN DELANO ROOSEVELT (1882-1945), president of the United States from 1933 to 1945
GEORGE CATLETT MARSHALL (1880-1959), chief of staff of the U.S. Army from 1939 to 1945
DWIGHT D. EISENHOWER (1890-1969), commander of Allied forces in northwestern Africa
JEAN-LOUIS DARLAN (1881-1942), commander in chief of the French forces at Vichy, France

## Choosing a Target

On December 11, 1941, Germany and Italy declared war on the United States. After that time, American military planners gave much thought to defeating their European enemies. Yet officers of the U.S. Army disagreed with British leaders about how this goal should be accomplished.

General George C. Marshall, chief of staff of the U.S. Army, wanted to build up air and ground forces in the British Isles and then cross the English Channel into France, pushing toward Berlin, Germany. He hoped to be ready to invade late in 1942 or, at the latest, in early 1943.

British planners, especially Prime Minister Sir Winston Churchill, preferred a "peripheral" strategy: wearing down Germany and Italy through air attacks and defeating them at the edges of their empires, where their forces were

not strong. Churchill argued that the Allies should cross the Channel and make a head-on attack on the German army only after the Germans were exhausted and spread thin.

President Franklin D. Roosevelt insisted that American troops be in action against the Germans somewhere before the end of 1942. Invading France was not possible in 1942, however, because the Allies lacked landing craft and trained troops. An easier target was needed.

Churchill won the argument when the Allies decided that French North Africa was ideal for invasion. There were no German troops in that western area of North Africa. Defending French forces—which were under the pro-Nazi government that had been set up in Vichy, France, after the French surrender to Germany in June, 1940—would probably not resist the British and American invaders very forcefully.

At the eastern end of North Africa, in Egypt, the British Eighth Army was fighting German forces commanded by Field Marshal Erwin Rommel. An Allied landing in French North Africa would bolster the British forces in Egypt, and perhaps the Germans could be squeezed out of North Africa altogether. If the Allies had control of North Africa, the Mediterranean Sea would be freed for British shipping, so that oil from the Near East and supplies from India could come to the British Isles by the most direct route.

Marshall argued that a landing in North Africa would tie up armies and supplies, delaying an invasion of northern France by two years. Yet Roosevelt overruled him.

## Invading and Attacking

Marshall chose General Dwight D. Eisenhower to lead the invasion force. In the early morning of November 8, 1942, Eisenhower's men landed at Casablanca, Oran, and Algiers.

Marshal Henri-Philippe Pétain, head of the Vichy government, ordered his forces to resist.

Nevertheless, Eisenhower was able to contact Admiral Jean-Louis Darlan, commander in chief of the Vichy troops, and succeeded in striking a deal. Darlan would take control of French North Africa, while Eisenhower ordered his troops to cease fire and prepare to make a later attack on the Germans. The American press criticized Eisenhower for making a bargain with a pro-Nazi, but Churchill and Roosevelt supported him, especially after Eisenhower explained that the deal allowed the Allies to avoid fighting the French and to begin the real job of fighting the Germans.

In November and early December, Eisenhower pushed his forces quickly toward Tunis, hoping to seize that port before the Germans could send large forces into Tunisia. Yet the Allies were stalled by rain, German tanks, and German warplanes.

Meanwhile, the British Eighth Army, commanded by General Bernard L. Montgomery, was driving Rommel back. By February, 1943, Rommel had crossed Libya and reached southern Tunisia. He then turned his forces against Eisenhower's, and at Kasserine Pass the Allied forces suffered a sharp defeat.

Yet the Allies were continuing to build up their force, while the Germans received few reinforcements. By May 13, the last German resistance ended, and Eisenhower captured nearly 300,000 prisoners in Tunisia.

## Consequences

As Marshall had feared, putting large numbers of Allied troops in North Africa made it impossible to launch an invasion across the English Channel in 1943. Since the troops and landing craft were already in the Mediterranean area, they had to be used there. On July 10, 1943, Eisenhower invaded Sicily, a major island off Italy's coast. Under the leadership of two American generals, Omar Bradley and George S. Patton, the Allies soon captured the island.

The fall of Sicily and the Allied bombing of Rome led to the downfall of Italy's dictator, Benito Mussolini, on July 25, 1943. The Italians eventually surrendered to the Allies. Before the negotiations could be worked out, however, the Germans had occupied Italy.

In September, 1943, Eisenhower invaded Italy. The Allies made slow progress up the Italian peninsula and were not able to liberate Rome until June, 1944.

Military planners and historians have remained divided about the value of the North African invasion. Some criticize it, arguing that World War II could have been ended a year earlier if the Allies had attacked Germany directly. Those who defend the invasion point out that between 1943 and 1944 Germany's army and air force were weakened considerably, so that when the cross-Channel Allied invasion finally did take place—in June, 1944—the American and British forces had gained an important advantage.

*Stephen E. Ambrose*

# Germans Besiege Stalingrad

*Through two and a half months of a bitter winter, invading German troops held on to the Soviet city of Stalingrad, but they were finally forced to surrender when they ran out of ammunition.*

**What:** War
**When:** November 19, 1942-January 31, 1943
**Where:** Stalingrad (Volgograd), Soviet Union
**Who:**
ADOLF HITLER (1889-1945), chancellor of Germany from 1933 to 1945
FRIEDRICH PAULUS (1890-1957), commander of the German Sixth Army from 1942 to 1943
ERICH VON MANSTEIN (1887-1973), commander of German army Group Don from 1942 to 1944
JOSEPH STALIN (1879-1953), dictator of the Soviet Union from 1924 to 1953
GEORGI KONSTANTINOVICH ZHUKOV (1896-1974), commander of Soviet forces on the southwestern front from 1942 to 1943

## The Drive for Stalingrad

When Germany invaded the Soviet Union in June, 1941, German chancellor Adolf Hitler was determined to conquer the Soviets and force a surrender within six months. Yet because the invasion was delayed, and because Hitler spread his forces across a broad front instead of sending them directly toward Moscow, his goal was not realized.

Early in December, 1941, the German forces were stopped before they reached Moscow, and Soviet dictator Joseph Stalin ordered a fierce counterattack to try to push the Germans back to their frontier. Hitler insisted that the German troops stand fast, however, and they dug in for the winter. So it was that when the spring of 1942 arrived, the Germans were still deep within the Soviet Union, and Hitler could plan a renewed assault to try to force the Soviets to surrender.

Hitler's plans for 1942 centered on two goals: capturing Leningrad (St. Petersburg) in the north and moving into the Donets industrial area and the Caucasus oil fields in the southern Soviet Union. By midsummer, the Germans had still been unable to take Leningrad, and Hitler took personal responsibility for commanding the German troops in the south. These troops had already driven the Soviets back and had moved well into the oil fields.

Now Hitler turned his attention toward Stalingrad, an industrial city on the Volga River. If it could be captured, important river traffic—especially boats carrying oil to Moscow—could be stopped. Also, such an attack would draw in the Soviety army to battle the Germans. Hitler believed that the Soviets' reserves were few and that the capture of a few more troops would bring down Stalin's government. In September, 1942, then, Hitler ordered General Friedrich Paulus and the German Sixth Army to capture Stalingrad and seize the left bank of the Volga, in order to halt river traffic.

## The Siege

Paulus reached Stalingrad in mid-September. As Hitler had expected, the Red Army hurried to the city's defense. In the process of fighting, many of the city's buildings became heaps of rubble. This made the German task more difficult: The ruins became hideouts for Soviet defenders who continued to launch attacks on the Germans even after the main portion of the city was taken by Paulus. The fighting continued into November.

Meanwhile, Marshal Georgi Zhukov, Soviet commander on the southwestern front, planned a counterattack to take Stalingrad back. Aware

Library of Congress

*Two women make their way through the rubble during the siege of Stalingrad.*

dered Field Marshal Erich von Manstein, Commander of German army Group Don, to send a force to Stalingrad to break the siege and get supplies to Paulus. Hermann Göring, president of the Council for War Economy and Commander of the Luftwaffe (German air force), assured Hitler that he could bring supplies to Paulus by air.

It soon became clear that the Luftwaffe could not carry out this task, however, and Manstein asked Hitler to order Paulus to break out of the city and join the relief force. Otherwise, Manstein said, he would not be able to help Paulus. For Hitler, though, possessing Stalingrad had become a matter of great pride, and he refused Manstein's request.

Manstein's forces came within thirty miles of Stalingrad, but could get no farther, especially after Zhukov began attacking the German supply lines.

By early January, Paulus had lost any chance of breaking out of the encirclement. His troops were tired, cold, hungry, and almost out of ammunition. Hitler still ordered him not to surrender but to fight to the last man if necessary. Yet without ammunition, Paulus's soldiers could not continue resisting. On January 31, 1943, Paulus surrendered the Sixth Army to the Soviets.

that the German troops were stretched thin, Zhukov formed two armies, commanded by General Andrei Yeremenko and General Konstantin Rokossovski, and on November 19, 1942, these two armies began coming together to trap Paulus in Stalingrad. By November 24, Paulus and his 250,000 soldiers were encircled.

Although the Germans could have broken out and retreated at any time until the middle of December, Hitler disregarded his generals' pleas and ordered Paulus to stand fast. He also or-

## Consequences

The defeat of the Germans at Stalingrad was the turning point of World War II on the Eastern Front, for after this time the Germans never again won a major battle in the Soviet Union. The Soviets had gained new confidence, while the Germans and their allies became discouraged.

The Soviet offensive continued to push the Germans back for two years until April, 1945, when the Red Army entered Berlin.

*José M. Sánchez*

# Fire in Boston's Coconut Grove Nightclub Kills 492 People

*A fire at the Coconut Grove nightclub in Boston, Massachusetts, results in the deaths of 492 people, one of the worst disasters in U.S. history. Many of the deaths could have been prevented by strict observance of occupancy limits and better fire safety codes.*

**What:** Disasters
**When:** November 28, 1942
**Where:** Boston, Massachusetts
**Who:**

Barney Welansky, Coconut Grove's owner

Buck Jones (1891-1942), cowboy movie star and celebrity

Stanley Tomaszewski, busboy

## A Disaster Waiting to Happen

The popular Coconut Grove nightclub, established in 1926, was the favored place to go for a festive evening out in Boston in 1942. The one-and-a-half story club, divided into a series of lounges and dining rooms, had a tropical paradise theme, lavishly decorated with artificial palm trees, bamboo and rattan furnishings, and fabric wall hangings and canopies, all highly flammable. Some exits were concealed by the decorations, and some were deliberately locked in an attempt to keep customers from leaving without paying their bills.

The club's owner, Barney Welansky, cut costs by hiring unskilled staff, paying low wages, and providing little or no job training. Lack of preparation by employees was a factor in the extent of the disastrous fire. Seven of eleven available fire extinguishers were never used, and bartenders initially attempted to put out the fire with soda water bottles.

Saturday, November 28, 1942, fell during Thanksgiving weekend. Scores of military personnel were in town, on leave and hoping to enjoy themselves before going overseas. Also, numerous college students and sports fans were out on the town after attending a big football game in which Holy Cross defeated top-ranked home team Boston College. That resulted in the popular nightclub's being packed beyond legal capacity with servicemen and their dates, football fans, local media, and attending celebrities, including cowboy movie star Buck Jones. More than one thousand customers were in the club, which had a rated capacity of less than half that number. Nearly half of these customers were dead by the end of the evening. Had modern fire safety standards been in force, most of these people would have survived. In fact, most would not have been in the nightclub in the first place.

## The Worst Nightclub Fire in U.S. History

The fire started in the downstairs Melody Lounge at approximately 10:05 P.M. Stanley Tomaszewski, a sixteen-year-old busboy, later testified that he struck a match for light while replacing an electric lightbulb. Moments later, flames shot up from an artificial palm tree and spread with incredible swiftness through the flammable decorations. Within fifteen minutes, the entire nightclub was ablaze and filled with toxic smoke. Firefighters were on the scene quickly, and the fire was mostly contained by 10:40 that evening, but the damage was already done.

The occupants of the Melody Lounge headed for the single stairway up toward the main exit. The stairway was quickly clogged with falling people who had succumbed to the flames, smoke, and panic. In the main dining room upstairs, diners were oblivious during the first few minutes of the fire. Then the entire room suddenly "flashed over" with a wall of flame, and the lights went out. Widespread panic ensued. Many of the dead were killed by being trampled underfoot. The main exit from the dining room was a large revolving door, which became a deathtrap. This door was a major cause of casualties because the force exerted by people trying to press on both sides of the door made it inoperable. Efforts to bring in hoses were delayed because firefighters could not get past the piles of bodies at this exit.

Most of the staff of the Coconut Grove escaped the fire through kitchen exits that the patrons did not know about. Had the main doors swung outward or had the patrons been aware of the service exits, many of the 492 victims would have been saved.

## Consequences

The city of Boston was stunned by the disaster as the dead and injured were brought into emergency rooms. Temporary morgues were set up in more than one location, including the nearby Park Square Garage. Survivors of the disaster and hospital workers told harrowing tales of the horror and panic of that evening.

Soon worldwide media attention focused on the sensational aspects of the disaster, including the death of popular cowboy movie star Buck Jones, who died two days after the fire of smoke inhalation and burns to his face, mouth, and throat. There were charges of lax enforcement of building codes and allegations of Mafia connections on the part of the club's management as well as of political corruption in Boston. Welansky

was indicted and later convicted of manslaughter and gross negligence. He was sentenced to twelve years in prison.

The fire left two important legacies. One was advances in the treatment of burns and trauma, and the other was in improved fire safety standards. Survivors of the Coconut Grove fire received what was then cutting-edge burn treatment: intravenous fluids and blood transfusions to prevent shock, morphine for pain, and sulfa drugs and penicillin, the new wonder drug, to prevent infection. This was the first general use of penicillin for burn treatment.

Advances in building and fire safety codes as a result of the lessons of the Coconut Grove fire have had a lasting impact on fire prevention in public buildings. Revolving doors were banned as primary exits unless they were flanked by at least two conventional doors with safety bars. Inward-opening, locked, or barricaded doors as well as flammable decorations and non-flame-retardant furnishings were also banned in public buildings. Backup emergency lighting that activates automatically in the event of a main power outage was required in public buildings. Building codes also required occupant-capacity placards to be posted in a visible location and emergency exits to be identified with lighted signs. Any of these fire safety improvements would have prevented many of the deaths at the Coconut Grove in November, 1942.

*Susan Butterworth*

# Fermi Creates Controlled Nuclear Fission Chain Reaction

> *Enrico Fermi's team demonstrated that nuclear energy could be released in a sustained chain reaction, which led to the development of the atomic bomb and nuclear fission electric power plants.*

**What:** Physics
**When:** December 2, 1942
**Where:** Chicago, Illinois
**Who:**

ENRICO FERMI (1901-1954), an Italian American nuclear physicist who won the 1938 Nobel Prize in Physics

WALTER HENRY ZINN (1906-    ), a Canadian physicist

HERBERT L. ANDERSON (1914-1988), an American physicist

ARTHUR HOLLY COMPTON (1892-1962), an American physicist who won the 1938 Nobel Prize in Physics

## Fleeing Hitler

In December, 1938, Enrico Fermi, a professor of physics in Rome, took advantage of his 1938 Nobel Prize in Physics to leave his native Italy and escape Adolf Hitler's increasing domination of Italy. With his family, Fermi arrived in New York City and settled down to continue his research at Columbia University.

Fermi and his associates in Rome had been studying the new nuclei produced when various chemical elements are bombarded by neutrons. In 1934, experiments on uranium produced a new radioactive isotope. Fermi and his collaborators demonstrated chemically that the new isotope did not belong to any of the elements immediately below uranium on the periodic table. They concluded that they had produced the first element ever found that was heavier than uranium.

The idea of a "transuranic" element caught the imagination of the scientific community and the popular press. When, for example, German chemist Ida Noddack published an article suggesting that Fermi had not ruled out the possibility that his new radioactivity came from a lighter (nontransuranic) chemical element produced when a uranium nucleus split into two parts, she was largely ignored.

Fermi and other scientists, including Irène Joliot-Curie and Paul Savitch in Paris, and Otto Hahn, Lise Meitner, and Fritz Strassmann in Berlin, continued to study the effects of irradiating uranium with neutrons. All the experimenters gradually compiled a list of several different radioactive species that were produced when uranium was bombarded.

In December, 1938, Hahn wrote to Meitner and informed her that he and Strassmann had incontrovertible evidence that the bombardment of uranium with neutrons produced lighter elements, not transuranic elements. Meitner and her nephew, Otto Robert Frisch, a young physicist working with Danish physicist Niels Bohr in Copenhagen, concluded that when a uranium nucleus absorbed a neutron, that nucleus split or fissioned into two lighter nuclei and some extra neutrons, releasing a hundred million times as much energy as was released in a typical chemical reaction between two atoms.

## Inaugurating a New Age

Fermi and Leo Szilard, a Hungarian physicist also driven into exile by Hitler's advance in Europe, realized immediately that if the neutrons from one fission could be used to trigger a second fission, the resulting chain reaction could be used to produce energy. If the multiplication could be made geometric, so that each fission produced at least two fissions, each of which produced at least two more fissions, and so on, the chain reaction would yield a powerful explosion. Szilard feared that Hitler's Germany would construct a superweapon based on these principles. He persuaded his American colleagues, including Fermi, to delay publication of their experimental results on fission.

Meanwhile, the physics community measured the energy released, the number of new nuclei produced, and the number of neutrons released during each fission. In August of 1939, Szilard and fellow Hungarian émigré Eugene Paul Wigner persuaded physicist Albert Einstein to send a letter to U.S. president Franklin D. Roosevelt urging a research program into the possibility of a superweapon. The American government hesitated while the physicists determined that only the rare isotope of uranium—uranium 235—underwent fission, while the isotope uranium 238, which composed 99.3 percent of naturally occurring uranium, did not.

In July, 1941, Fermi and his group received funding to begin experiments in constructing a graphite-uranium "pile" designed to sustain a chain reaction. In December, 1941, Arthur Holly Compton, the American Nobel laureate in physics, was placed in charge of the project. He moved the experiments to the University of Chicago in early 1942. Construction of the pile began in November in a squash court, the only area available that was large enough to hold the 771,000 pounds of graphite, 80,590 pounds of uranium oxide, and 12,400 pounds of uranium metal that were to compose the pile.

Construction crews headed by Walter Henry Zinn and Herbert L. Anderson worked around the clock machining and stacking the graphite and uranium blocks. Control rods that absorbed neutrons were built into the pile and would be withdrawn once it was time to start the chain reaction. Each day, the control rods were withdrawn and measurements were taken to see how close the system was to sustaining a chain reaction.

On the evening of December 1, 1942, Anderson and Zinn decided that the layer of uranium and graphite that the night crew had placed on the pile should be sufficient to sustain a chain reaction. The crew went home for a few hours of sleep and reassembled at 8:30 the following morning. Fermi ordered the main control rods withdrawn, and the final control rod was moved foot by foot out of the pile as the assembled physicists gathered to watch the neutron counters. At about 3:25 P.M., the last foot of the final control rod was removed. The counting rate climbed exponentially. A controlled fission chain reaction had been achieved and was sustained until Fermi ordered the control rods inserted back into the pile at 3:53 P.M. As the group celebrated, they realized that the success of their experiment had inaugurated a new age.

## Consequences

The successful operation of the atomic pile provided physicists with a tool for studying the behavior of nuclear fission chain reactions. These studies were essential for the design and construction of an atomic bomb, since details of critical mass and neutron absorption by materials could be easily measured using atomic piles. Moreover, atomic piles produced a second fissionable isotope, plutonium 239, and the design of large-scale piles for the production of plutonium was soon underway.

Plutonium was to prove more efficient than highly enriched uranium 235 as a fuel for bombs. Finally, the first atomic pile demonstrated that it was possible to produce a sustained energy source from nuclear fission, which makes it possible to construct nuclear electric generating plants.

*Ruth H. Howes*

Gift of Laura Fermi, Courtesy AIP Emilio Segrè Visual Archives

*Enrico Fermi.*

# DNA Is Identified as Genetic Key

*Oswald Avery and coworkers demonstrated that the genetic transformation of bacteria is caused by DNA, providing direct evidence about the chemical nature of hereditary information.*

**What:** Biology; Genetics
**When:** 1943-1944
**Where:** New York, New York
**Who:**

OSWALD AVERY (1877-1955), an American bacteriologist

FREDERICK GRIFFITH (1881-1941), an English public health officer and microbiologist

COLIN MUNRO MACLEOD (1909-1972), an American microbiologist

MACLYN MCCARTY (1911-     ), an American microbiologist

## Genetic Transformation

In the 1920's, the field of genetics had progressed to the point of locating hereditary information within the cell. Genes, which were uncharacterized elements responsible for the inheritable traits of organisms, had been localized to the chromosomes of cells. These chromosomes were known to be made up of two major chemical components: protein and DNA. Moreover, chromosomes were thought to have something to do with the characteristic traits of organisms, but beyond that nothing was known about the physical nature of genetic information.

Oswald Avery was a bacteriologist at the hospital of the Rockefeller Institute in New York City. He was studying pneumonia, a disease that was caused by bacteria and that was a major cause of death in the late nineteenth and early twentieth centuries. Several different strains of pneumococci, the class of bacteria that causes pneumonia, were known to exist; some strains in this class were nonpathogenic (did not cause disease). Avery had demonstrated in 1917 that the blood and urine of patients infected by different pathogenic strains contained distinct soluble substances, specific for each strain. Later, experiments suggested that these specific substances were polysaccharides, starchlike molecules derived from the distinct cell coatings or capsules of these bacteria. Nonvirulent pneumococci were unencapsulated, and the differences in the coats of the encapsulated forms reflected the strain differences among the virulent pneumococci.

In 1928, Frederick Griffith, an English public health officer, reported the results of experiments using different strains of pneumococci to infect mice. Griffith had observed the following: Mice injected with a nonpathogenic (unencapsulated) strain of pneumococci did not contract pneumonia; mice injected with encapsulated pathogenic bacteria that had first been killed by heating also did not contract pneumonia. To bacteriologists, these results were not surprising.

Griffith also inoculated mice with a combination of nonpathogenic bacteria and heat-killed pathogenic pneumococci; by contrast, many of those mice contracted pneumonia and died. Moreover, live bacteria recovered from these animals were encapsulated. The virulence and capsule-forming traits of one strain of bacteria had been transferred to a formerly nonvirulent, unencapsulated strain, thereby transforming the latter into a pathogenic strain. This acquired pathogenicity was maintained in subsequent generations of these bacteria, and the phenomenon was dubbed "genetic transformation." Soon after they were reported, Griffith's experiments were repeated with similar findings in several laboratories.

## Initial Skepticism

Although Avery was at first skeptical of Griffith's work, it was later confirmed by Martin Dawson and James Lionel Alloway, two scientists who happened to be conducting their experiments in Avery's laboratory. Dawson demonstrated that genetic transformation did not require the infection of a host animal—using bacteria cultures in the laboratory instead of live mice, Dawson was able to reproduce Griffith's results. Alloway produced cell-free extracts of broken encapsulated bacteria and showed that such extracts were as effective as heat-killed cells in transforming nonvirulent strains.

The combined weight of evidence for genetic transformation, much of it coming from Avery's own laboratory, was irresistible. Avery set out to identify the agent responsible for it. At the time, many biologists believed this agent to be protein, one of the major chemical components that make up chromosomes.

Together with two new collaborators in his laboratory, Colin Munro MacLeod and Maclyn McCarty, Avery performed the key experiments that first identified deoxyribonucleic acid (DNA) as the active transforming material. The scientists exhaustively fractionated transforming extracts, removing polysaccharides, lipids, and proteins by physical, chemical, and enzymatic treatment without removing the ability to transform. They tested and retested their extracts, using different methods of measurement and different sources of enzymes; their results continued to show that the transforming principle behaved like DNA. Furthermore, their extract was extraordinarily potent: It continued to transform even when diluted to exceedingly low concentrations. In 1944, the three scientists published their evidence that DNA seemed responsible for the transfer of genetic information.

## Consequences

Far from being accepted as an elegant proof of DNA's role, Avery's paper met with resistance and disbelief. One reason was the presumed simplicity (if not monotony) of DNA structure. It was thought to be a polymer of identical repeating units, similar to some starch molecules. Such a structure for DNA seemed incompatible with the variety and specificity of genetic information.

This presumed uniformity was even more striking in comparison with the immense diversity that had been observed among protein molecules, which like DNA were known to be associated with chromosomes. The prevailing view held that proteins, not DNA, were probably the vectors of genetic information. Alternatively, some suggested that transformation of pneumococci was a special case: DNA might be having some other effect on these cells that caused them to begin making capsules and become virulent.

It took several years and two other studies to resolve the doubt about DNA. In 1949, Rollin Hotchkiss, who had begun work in Avery's laboratory in 1935, demonstrated DNA-mediated transfer of an entirely different set of characteristics, related to antibiotic resistance, to a formerly nonresistant strain of pneumococci. This showed conclusively that capsule formation was not a special case.

Then, in 1952, Alfred Hershey and Martha Chase cultured viruses that infected bacteria, reproduced themselves inside, then burst the bacterial cells to release many progeny viruses. Hershey and Chase showed that, in these viral infections, virus proteins remained outside the bacterial cell, while the viral DNA was injected into each bacterium, producing the new viruses. That this DNA alone was responsible for the subsequent production of progeny demonstrated that viral genetic information resided in the same substance that carried the genes of pneumococci. The case for DNA was now irrefutable.

In April, 1953, James D. Watson and Francis Crick published their models of the double, helical structure of DNA, a model that explained how complex genetic information could be carried by a polymer built from simple subunits and how this polymer could be replicated over and over in generation after generation. Watson and others, whose work formed the basis for the new field of molecular biology, traced their interest in nucleic acids to Avery's experiments.

*Jennifer L. Cruise*

# Weizsäcker Completes Theory of Planetary Formation

*Carl Friedrich von Weizsäcker devised a theory of planetary formation based on contemporary theories of high-temperature turbulence and star formation.*

**What:** Astronomy
**When:** 1943-1944
**Where:** Strassburg, Germany
**Who:**
CARL FRIEDRICH VON WEIZSÄCKER
(1912-      ), a German nuclear astrophysicist

## Early Theories

The earliest scientific hypotheses of planetary formation were those of René Descartes (1644), Immanuel Kant (1755), and Pierre-Simon Laplace (1796). All of them proposed nebular (gas cloud) theories which stated that the universe, then not known beyond the sun and five planets, was filled with gas and dustlike particles of matter. Descartes, the French mathematician and philosopher, imagined a large primary gas vortex of circular shape, surrounded by still smaller eddies, from which, respectively, the sun, major planets, and their satellites were to have formed as the result of turbulent collision and condensation.

Likewise, the German philosopher Kant, in his *Allgemeine Naturgeschichte und Theorie des Himmels* (1755; *Universal Natural History and Theories of the Heavens*, 1900), proposed a large rotating gas and dust cloud, which increased speed and flattened, becoming a disk, as it contracted because of gravitational attraction. From this disk, the remaining matter was supposed to have condensed to form the sun and planets.

Laplace, the French astronomer and mathematician, modified Kant's theory by assuming that as the disk-shaped cloud's rotation increased, centrifugal force at its edge also increased until it exceeded gravity forces acting to-

ward the center, thereafter separating into concentric rings, each subsequently condensing to form a planet.

A major problem with these nebular hypotheses became apparent after the 1870's, when scientists began to make further observations of stars and stellar nebulas: If the solar system's nebula increased rotational speed as it contracted, the sun should be rotating much faster than the planets. Its rotational speed should include the bulk of the solar system's angular momentum (speed of rotation); the actual rate, however, was equal to only 2 percent of the solar system's total. James Clerk Maxwell further argued that Laplace's rings would not coalesce directly into planets but would first have to be collected into rings of smaller planetoids, or planetesimals.

In a series of papers published around 1900, American geologist T. C. Chamberlain and astronomer F. R. Moulton argued strenuously against the nebular hypothesis, revived Comte de Buffon's 1745 idea of a catastrophic star-sun encounter, and presented a tidal-collisional planetesimal model. They proposed that the solar system developed from material ejected by huge solar tides raised in a glancing collision of another star or comet. English physicist Sir James Jeans and geophysicist Sir Harold Jeffreys later proposed a similar theory, in which a close encounter withdrew solar gas filaments that coalesced into beadlike strings of protoplanets.

Within two decades, however, several problems arose with collision accounts of planetary origins. For example, the statistical frequency of interstellar encounters was far too low to make this a probable mechanism. Also, no collision hypothesis could explain the current angular momentum distribution. In 1939, the American astrophysicist Lyman Spitzer showed that gases torn from the sun or a passing star/comet would disperse before being able to cool sufficiently for condensation.

## Building a More Perfect Theory

In mid-1943, at the University of Strassburg in Germany, nuclear astrophysicist Carl Friedrich von Weizsäcker was completing a nebular theory paper titled "On the Formation of Planetary Systems." After initially summarizing the history of earlier nebular hypotheses, he addressed the question of how the sun's original mass was distributed within the boundaries of the present solar system. This raised the old question about the sun's low angular momentum.

Weizsäcker assumed that, obeying the laws of momentum and energy conservation, a portion of the original gas nebula would fall into the cloud's center, liberating energy that would carry off most of the sun's angular momentum. Weizsäcker next discussed whether and how it was possible for particles in the rotating disk to form stable and predictable patterns. He concluded that this would be possible if the primary force at work was gravity. The next stage, his theory's core, derived a set of five concentric lenticular (lens-shaped) rings around the sun.

The corresponding diagram of this system was eventually reprinted in many textbooks and publications. This nebula figure, which was ingeniously derived from particle dynamics, revealed a ratio of orbital distances that accorded with the well-known Bode-Titius law of 1772, which predicted how far from the sun each planet should orbit. This provided a major consistency and validity check for the whole theory.

## Consequences

Although most immediate discussions of Weizsäcker's theory were delayed by World War II, almost all initial published reactions were positive. In the spring of 1945, the noted nuclear physicist George Gamow and the cosmologist J. A. Hynek published a short review, "A New The-

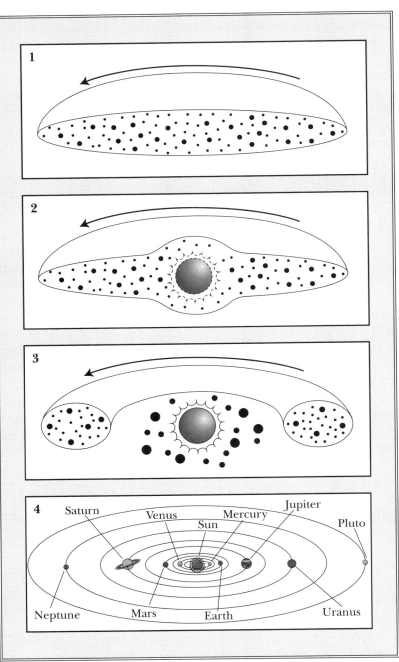

*Current models of planetary formation, such as this one, owe much to Weizsäcker's theory. (1) A large cloud of rotating gas and dust eventually formed (4) the planets and solar system.*

ory by C. F. von Weizsäcker on the Origin of the Planetary System," in the *Astrophysical Journal*. In their opinion, the theory "allowed an interpretation of the Bode-Titius law of planetary distances" and explained "all the principal features of the solar system"—particularly the fact that all the planets orbit along the same plane and in the same direction—and why larger planets have lower densities.

The theory received further attention in 1946, when the noted astrophysicist Subrahmanyan Chandrasekhar published a favorable review in the *Reviews of Modern Physics*. Nevertheless, German astronomer Friedrich Nölke and Dutch astrophysicist D. ter Haar in 1948 independently published criticisms of Weizsäcker's theory, based on rigorous and extensive hydrodynamic considerations of nebular eddies. Nölke showed that serious difficulties remained in the angular momentum problem. According to ter Haar, there was still a thousandfold difference between the actual and the predicted solar mass. Dutch American astronomer Gerhard Peter Kuiper also rejected Weizsäcker's theory of planetary formation, but he redeveloped the nebular theory, proposing a more random formation.

Later theories incorporated the ideas of turbulence, magnetic fields, and planetesimals, maintaining that supersonically turbulent nebular clouds break up into chaotic swarms, or "floccules," that continually disperse and reform according to certain statistical laws. Despite advances in empirical and theoretical astrophysics, Weizsäcker's theory of planetary formation remains, among some scientists, a partial source and model for future theories of solar system formation.

*Gerardo G. Tango*

# Eckert and Mauchly Create ENIAC Computer

*John Presper Eckert and John William Mauchly developed the first general-purpose electronic digital computer, leading directly to modern methods of computation.*

**What:** Computer science
**When:** 1943-1946
**Where:** Philadelphia, Pennsylvania
**Who:**
JOHN PRESPER ECKERT (1919-1995), an electrical engineer
JOHN WILLIAM MAUCHLY (1907-1980), a physicist, engineer, and professor
JOHN VON NEUMANN (1903-1957), a Hungarian American mathematician, physicist, and logician
HERMAN HEINE GOLDSTINE (1913-     ), an army mathematician
ARTHUR WALTER BURKS (1915-     ), a philosopher, engineer, and professor
JOHN VINCENT ATANASOFF (1903-1995), a mathematician and physicist

## A Technological Revolution

The Electronic Numerical Integrator and Computer (ENIAC) was the first general-purpose electronic digital computer. By demonstrating the feasibility and value of electronic digital computation, it initiated the computer revolution. The ENIAC was developed during World War II (1939-1945) at the Moore School of Electrical Engineering by a team headed by John William Mauchly and John Presper Eckert, who were working on behalf of the U.S. Ordnance Ballistic Research Laboratory (BRL) at the Aberdeen Proving Ground in Maryland. Early in the war, the BRL's need to generate ballistic firing tables already far outstripped the combined abilities of the available differential analyzers and teams of human computers.

In 1941, Mauchly had seen the special-purpose electronic computer developed by John Vincent Atanasoff to solve sets of linear equations. Atanasoff's computer was severely limited in scope and was never fully completed. The functioning prototype, however, helped convince Mauchly of the feasibility of electronic digital computation and so led to Mauchly's formal proposal in April, 1943, to develop the general-purpose ENIAC. The BRL, in desperate need of computational help, agreed to fund the project, with Lieutenant Herman Heine Goldstine overseeing it for the U.S. Army.

This first substantial electronic computer was designed, built, and debugged within two and one-half years. Even given the highly talented team, it could be done only by taking as few design risks as possible. The ENIAC ended up as an electronic version of prior computers: Its functional organization was similar to that of the differential analyzer, while it was programmed via a plugboard (which was something like a telephone switchboard), much like the earlier electromechanical calculators made by the International Business Machines (IBM) Corporation. Another consequence was that the internal representation of numbers was decimal rather than the now-standard binary, since the familiar electromechanical computers used decimal digits.

Although the ENIAC was completed only after the end of the war, it was used primarily for military purposes. In fact, the first production run on the system was a two-month calculation needed for the design of the hydrogen bomb. John von Neumann, working as a consultant to both the Los Alamos Scientific Laboratory and the ENIAC project, arranged for the production run immediately prior to ENIAC's formal dedication in 1946.

## A Very Fast Machine

The ENIAC was an impressive machine: It contained 18,000 vacuum tubes, weighed 27 metric tons, and occupied a large room. The final cost to the U.S. Army was about $486,000. For this price, the Army received a machine that computed up to a thousand times faster than its electromechanical precursors; for example, addition and subtraction required only 200 microseconds (200 millionths of a second). At its dedication ceremony, the ENIAC was fast enough to calculate a fired shell's trajectory faster than the shell itself took to reach its target.

The machine also was much more complex than any predecessor and employed a risky new technology in vacuum tubes; this caused much concern about its potential reliability. In response to this concern, Eckert, the lead engineer, imposed strict safety factors on all components, requiring the design to use components at a level well below its specified limits. The result was a machine that ran for as long as three days without a hardware malfunction.

Programming the ENIAC was effected by setting switches and physically connecting accumulators, function tables (a kind of manually set read-only memory), and control units. Connections were made via cables running between plugboards. This laborious and error-prone process often required a one-day set time.

The team recognized this problem, and in early 1945, Eckert, Mauchly, and von Neumann worked on the design of a new machine. Their basic idea was to treat both program and data in the same way, and in particular to store them in the same high-speed memory; in other words, they planned to produce a stored-program computer. Neumann described and explained this design in his "First Draft of a Report on the EDVAC" (EDVAC is an acronym for Electronic Discrete Variable Automatic Computer). In his report, von Neumann contributed new design techniques and provided the first general, comprehensive description of the stored-program architecture.

After the delivery of the ENIAC, von Neumann suggested that it could be wired up so that a set of instructions would be permanently available and could be selected by entries in the function tables. Engineers implemented the idea, providing sixty instructions that could be invoked from the programs stored into the function tables. Despite slowing down the computer's calculations, this technique was so superior to plugboard programming that it was used exclusively thereafter. In this way, the ENIAC was converted into a kind of primitive stored-program computer.

## Consequences

The ENIAC's electronic speed and the stored-program design of the EDVAC posed a serious engineering challenge: to produce a computer memory that would be large, inexpensive, and fast. Without such fast memories, the electronic control logic would spend most of its time idling. Vacuum tubes themselves (used in the control) were not an effective answer because of their large power requirements and heat generation.

The EDVAC design draft proposed using mercury delay lines, which had been used earlier in radars. These delay lines converted an electronic signal into a slower acoustic signal in a mercury solution; for continuous storage, the signal picked up at the other end was regenerated and sent back into the mercury. Maurice Vincent Wilkes at the University of Cambridge was the first to complete such a system, in May, 1949. One month earlier, Frederick Calland Williams and Tom Kilburn at Manchester University had brought their prototype computer into operation, which used cathode-ray tubes (CRTs) for its main storage.

In the meantime, Eckert and Mauchly formed the Electronic Control Company (later the Eckert-Mauchly Computer Corporation). They produced the Binary Automatic Computer (BINAC) in 1949 and the Universal Automatic Computer (UNIVAC) I in 1951; both machines used mercury storage.

The memory problem that the ENIAC introduced was finally resolved with the invention of the magnetic core in the early 1950's. Core memory was installed on the ENIAC and soon on all new machines. The ENIAC continued in operation until October, 1955, when parts of it were retired to the Smithsonian Institution. The ENIAC proved the practicality of digital electronics and led directly to the development of stored-program computers.

*Kevin B. Korb*

**927**

# Frederick Sanger Determines Structure of Insulin

*Frederick Sanger developed a series of methods for determining the unique order of the amino acid building blocks of the protein hormone insulin. He showed that proteins have a definite sequence.*

**What:** Chemistry
**When:** 1943-1955
**Where:** Cambridge, England
**Who:**
FREDERICK SANGER (1918-        ), a British biochemist

## What Are Proteins?

Since the early nineteenth century, the word "protein" has been applied to complex organic materials found in all plant and animal cells. These large molecules contain carbon, hydrogen, oxygen, nitrogen, and sulfur. Much work had been done in developing ways to separate different proteins. There was also progress in determining what proteins do and what their functions are. Proteins transport molecules, provide support and protection, and control chemical reactions.

By the early 1940's, it was well established that proteins were made of some twenty different amino acids. How these amino acids were arranged was a major question. Were they linked in repeating units or was there a definite constant order for each protein? Finding the building plan for these complicated giant molecules was one of the greatest of all research problems.

Frederick Sanger had just finished his doctorate in biochemistry when he began working to answer this question. Sanger had the opportunity to work with Professor Albert C. Chibnall at Cambridge University. Chibnall's group was studying insulin, the hormone needed for the treatment of diabetes. Because of its medical use, the hormone was available in pure form. Chibnall had already determined the relative amounts of the various amino acids in insulin.

## Insulin Structure Established

In 1943, Sanger began a twelve-year study that established the unique amino acid sequence of insulin. He showed that the building blocks of proteins have a definite, specified order. Sanger developed strategies and techniques that have been applied to many other proteins.

Sanger's work can be divided into several steps. First, he determined how many chains there were on the insulin molecule. The first amino acid in any chain has a free amino end. Sanger attached a colored dye, fluorodinitrobenzene (FDNB), to the insulin molecule. The

*Frederick Sanger.*

dyed insulin was then broken into individual amino acids. Sanger found that the dye had reacted with two different amino acids. Therefore, he concluded that insulin had two chains. Sanger separated the two chains by oxidizing (combining with oxygen) the sulfurs that joined them.

When a protein is treated with strong acid, it comes apart into its individual amino acids. Sanger, however, gently treated each insulin chain with a weaker acid solution. The chain came apart but broke into bigger pieces of three, four, or five amino acids. Sanger separated these fragments and used his dye to find the identity of the first amino acid in each piece. He also determined the amino acid composition of each small piece. Gradually, he began to put the pieces together.

Besides using the acid solution that broke the chain in various places, Sanger also used digestive enzymes. These enzymes, chymotrypsin and trypsin, cut the amino acid chains in very specific places after certain amino acids. The order of the chain was determined by overlapping the various fragments.

Once Sanger had successfully found the order of the two chains of the insulin molecule, he had to figure out how they were joined. The smaller chain, the amino acid with a sulfur group, had four cysteines (crystalline amino acids derived from cystine). The larger chain had two cysteines. Therefore, there were several different ways in which the chains could be linked. The problem was complicated because the sulfur-to-sulfur bonds tended to rearrange in acidic solution. After many efforts, the rearrangement was finally avoided.

By 1955, Sanger had determined the amino acid sequence of the 21-unit A chain and the 30-unit B chain of bovine insulin. He also found that insulin from other animals differed very slightly in its composition. He showed that these two chains were linked by sulfur bridges joining the cysteines at position 7 in both chains. A second disulfide bridge linked position 20 of the first chain with 19 of the second. This landmark work provided conclusive evidence that a protein had a very specific sequence of amino acids.

Sanger was awarded the Nobel Prize in Physiology or Medicine in 1958. When he accepted the prize, he expressed the hope that his work eventually would be helpful in fighting diseases involving proteins.

## Consequences

Sanger's work in determining the sequence of the amino acid units in insulin established conclusively that proteins have a definite order in the arrangement of their amino acids. This finding suggested to other scientists that there must be a genetic code that provides the information for the amino acid sequence. Sanger's communication with other scientists through his papers and conferences helped others to find the code.

Sanger provided other protein chemists with the strategies and tools needed for analyzing other proteins and amino acid chains. The amino acid sequence of a protein is very important information in understanding the molecular basis of that protein's activity in the body. The sequences of hundreds of different proteins have been studied since Sanger's initial work. The order of the various amino acids dictates the three-dimensional shape of a protein. Substitutions of a single amino acid sometimes will dramatically change the shape or function of a protein and cause it to be harmful. Amino acid sequencing was essential to the understanding of many diseases, such as sickle cell anemia and phenylketonuria.

By studying the differences in the insulin of different animal species, Sanger provided a way to study the evolutionary relationships among species. The closer the amino acid composition of their various proteins, the closer the organisms are to one another. Much work on enzyme function, the use of monoclonal antibodies for cancer treatment, and genetically engineered proteins can be traced to the sequencing studies done by Sanger.

*Helen M. Burke*

# Allied Leaders Meet at Casablanca

*At Casablanca, the political and military leaders of the United States and Great Britain met to plan their strategy for winning World War II.*

**What:** International relations
**When:** January 14-24, 1943
**Where:** Casablanca, Morocco
**Who:**
FRANKLIN DELANO ROOSEVELT (1882-1945), president of the United States from 1933 to 1945
GEORGE CATLETT MARSHALL (1880-1959), chief of staff of the U.S. Army from 1939 to 1945
ERNEST J. KING (1878-1956), chief of U.S. Naval Operations from 1942 to 1945
SIR WINSTON CHURCHILL (1874-1965), prime minister of Great Britain from 1940 to 1945
HENRI-HONORÉ GIRAUD (1879-1949), French commander in North Africa
CHARLES DE GAULLE (1890-1970), leader of the French Committee of National Liberation

## The Meeting Convenes

Late in 1942, after Axis forces were defeated at the Battle of Alamein and the Allies invaded French North Africa, United States president Franklin D. Roosevelt suggested a meeting of the Allied leaders to plan further strategy for World War II. British prime minister Winston Churchill accepted the invitation, but Soviet dictator Joseph Stalin did not, for the Battle of Stalingrad had just begun and he could not leave his country. Casablanca, Morocco, which had recently come under Allied control, was selected for the conference.

On January 12, 1943, Roosevelt was present along with his chief civilian adviser, Harry Hopkins, and his two military advisers: General George C. Marshall, chief of staff of the U.S. Army, and Admiral Ernest J. King, chief of U.S. Naval Operations. Churchill arrived with his mil-

itary and naval advisers, including General Sir Alan Brooke, chief of the imperial general staff.

It was not hard for the Allied leaders to agree on a number of military matters. The bombing raids on Germany would continue, and navy ships would continue to help transport war supplies across the Atlantic Ocean.

They also agreed about how to deal with France. Once they had taken French North Africa, the Allies had had to find a leader for the French. France itself was under German domination, and the Vichy government of southern France was cooperating with the Nazis. It would have been logical to choose General Charles de Gaulle, who had been leading the Free French Forces in resisting Germany; yet Churchill and Roosevelt did not trust him.

Instead, they had chosen General Henri-Honoré Giraud as Supreme French Commander for North Africa. Angered by this decision, de Gaulle refused to have anything to do with Giraud or his government. At Casablanca, Roosevelt and Churchill invited de Gaulle to make peace with Giraud. The two French leaders reluctantly made a temporary reconciliation, and de Gaulle was named commander alongside Giraud.

## Strategy Arguments

The real disagreements at Casablanca had to do with future strategy. The British wanted to continue fighting Germany and Italy in the Mediterranean area until a large number of troops could be brought together to cross from Great Britain to France and retake Europe. Churchill hoped that Italy could be invaded and forced to surrender. This would bring Turkey into the war on the Allied side, and perhaps the Allies could invade the Balkan states and force the Germans out. Then Germany could be attacked through the "soft underbelly of Europe."

British planners believed that this strategy

would bring the fewest risks for the Allies. They were sure that the struggle to liberate France from German control would be difficult, and they wanted to avoid that struggle until plenty of troops were trained and ready.

The Americans, especially Marshall and King, did not like this plan. To them, fighting in the Mediterranean was not particularly important. The only way to defeat Germany, they argued, was by sending a large invasion force across the English Channel into France.

If the British plan was chosen, King said, he would make sure that most new U.S. landing boats were sent to the Pacific Ocean to help in the war against Japan, rather than to Europe. Roosevelt overruled him, however, because in 1940 the U.S. Joint Chiefs of Staff had agreed that if the United States had to enter World War II and fight both in the Far East and in Europe, the European conflict would take priority.

Roosevelt agreed to the British plan of continuing the Mediterranean war, at least for a time, and General Dwight D. Eisenhower was appointed Supreme Commander of the Allied Forces in the Mediterranean. British general Sir Henry Alexander was named Eisenhower's deputy. They were told to plan an invasion of Sicily, off the coast of Italy, after the Axis Powers were driven out of North Africa. Meanwhile, Allied leaders would continue to make plans for a cross-Channel invasion of France.

At a press conference on the last day of the Casablanca meetings, Roosevelt announced that the "elimination of German, Japanese, and Italian war power means the unconditional surrender by Germany, Italy, and Japan." He went on to say that he did not mean the destruction of the people of those countries, but rather "the destruction of the philosophies which are based on conquest and the subjugation of other people."

## Consequences

Whether the Allies chose the best course of action at Casablanca was debated long after

World War II was over. Critics of the British strategy have argued that delaying the cross-Channel invasion meant delaying the end of the war for a year. Those who defend the strategy say that the Allies needed that year to build up their forces.

The most controversial decision made at Casablanca, however, was Roosevelt's press conference statement. One of Roosevelt's main goals in making the statement was to reassure Stalin that Great Britain and the United States would not make peace with Germany as long as it was still fighting the Soviet Union. With the cross-Channel invasion postponed, Stalin feared that he would have to fight Germany alone; Roosevelt was attempting to calm this fear.

Yet some historians believe that Roosevelt's harsh words made it hard later for the Allies to bring an end to the war. The message to the people of the Axis nations was that the Allies were intent on total defeat and required unconditional surrender. Without Roosevelt's statement, resistance groups within the Axis nations might have tried to overthrow their governments in the hope of bargaining with the Allies. Other historians argue, however, that Roosevelt's statement did not have much effect on the war.

*José M. Sánchez*

# Cousteau and Gagnan Develop Aqualung

*Jacques-Yves Cousteau and Émile Gagnan developed the Aqualung, a device that allowed divers to descend hundreds of meters below the surface of the ocean.*

**What:** Earth science; Sports
**When:** Spring, 1943
**Where:** Paris, France
**Who:**
JACQUES-YVES COUSTEAU (1910-1997), a French navy officer, undersea explorer, inventor, and author
ÉMILE GAGNAN, a French engineer who invented an automatic air-regulating device

## The Limitations of Early Diving

Undersea dives have been made since ancient times for the purposes of spying, recovering lost treasures from wrecks, and obtaining natural treasures (such as pearls). Many attempts have been made since then to prolong the amount of time divers could remain underwater. The first device, described by the Greek philosopher Aristotle in 335 B.C.E., was probably the ancestor of the modern snorkel. It was a bent reed placed in the mouth, with one end above the water.

In addition to depth limitations set by the length of the reed, pressure considerations also presented a problem. The pressure on a diver's body increases by about one-half pound per square centimeter for every meter ventured below the surface. After descending about 0.9 meter, inhaling surface air through a snorkel becomes difficult because the human chest muscles are no longer strong enough to inflate the chest. In order to breathe at or below this depth, a diver must breathe air that has been pressurized; moreover, that pressure must be able to vary as the diver descends or ascends.

Few changes were possible in the technology of diving until air compressors were invented during the early nineteenth century. Fresh, pressurized air could then be supplied to divers. At first, the divers who used this method had to wear diving suits, complete with fishbowl-like helmets. This "tethered" diving made divers relatively immobile but allowed them to search for sunken treasure or do other complex jobs at great depths.

## The Development of Scuba Diving

The invention of scuba gear gave divers more freedom to move about and made them less dependent on heavy equipment. ("Scuba" stands for *s*elf-*c*ontained *u*nderwater *b*reathing *a*pparatus). Its development occurred in several stages. In 1880, Henry Fleuss of England developed an outfit that used a belt containing pure oxygen. Belt and diver were connected, and the diver breathed the oxygen over and over. A version of this system was used by the U.S. Navy in World War II spying efforts. Nevertheless, it had serious drawbacks: Pure oxygen was toxic to divers at depths greater than 9 meters, and divers could carry only enough oxygen for relatively short dives. It did have an advantage for spies, namely, that the oxygen—breathed over and over in a closed system—did not reach the surface in the form of telltale bubbles.

The next stage of scuba development occurred with the design of metal tanks that were able to hold highly compressed air. This enabled divers to use air rather than the potentially toxic pure oxygen. More important, being hooked up to a greater supply of air meant that divers could stay under water longer. Initially, the main problem with the system was that the air flowed continuously through a mask that covered the diver's entire face. This process wasted air, and the scuba divers expelled a continual stream of air bubbles that made spying difficult. The solution, according to Axel Madsen's *Cousteau* (1986), was "a valve that would allow inhaling and exhaling through the same mouthpiece."

Jacques-Yves Cousteau's father was an executive for Air Liquide—France's main producer of industrial gases. He was able to direct Cousteau to Émile Gagnan, an engineer at the company's Paris laboratory who had been developing an automatic gas shutoff valve for Air Liquide. This valve became the Cousteau-Gagnan regulator, a breathing device that fed air to the diver at just the right pressure whenever he or she inhaled.

With this valve—and with funding from Air Liquide—Cousteau and Gagnan set out to design what would become the Aqualung. The first Aqualungs could be used at depths of up to 68.5 meters. During testing, however, the dangers of Aqualung diving became apparent. For example, unless divers ascended and descended in slow stages, it was likely that they would get "the bends" (decompression sickness), the feared disease of earlier, tethered deep-sea divers. Another problem was that, below 42.6 meters, divers encountered nitrogen narcosis. (This can lead to impaired judgment that may cause fatal actions, including removing a mouthpiece or developing an overpowering desire to continue diving downward, to dangerous depths.)

Cousteau believed that the Aqualung had tremendous military potential. During World War II, he traveled to London soon after the Normandy invasion, hoping to persuade the Allied Powers of its usefulness. He was not successful. So Cousteau returned to Paris and convinced France's new government to use Aqualungs to locate and neutralize underwater mines laid along the French coast by the German navy. Cousteau was commissioned to combine minesweeping with the study of the physiology of scuba diving. Further research revealed that the use of helium-oxygen mixtures increased to 76 meters—the depth to which a scuba diver could go without suffering nitrogen narcosis.

*Jacques-Yves Cousteau wearing an Aqualung.*

Library of Congress

## Consequences

One way to describe the effects of the development of the Aqualung is to summarize Cousteau's lifetime achievements. In 1946, he and Philippe Tailliez established the Undersea Research Group of Toulon to study diving techniques and various aspects of life in the oceans. They studied marine life in the Red Sea from 1951 to 1952. From 1952 to 1956, they engaged in an expedition supported by the National Geographic Society. By that time, the Research Group had developed many techniques that enabled them to identify life-forms and conditions at great depths.

Throughout their undersea studies, Cousteau and his coworkers continued to develop better

**933**

techniques for scuba diving, for recording observations by means of still and television photography, and for collecting plant and animal specimens. In addition, Cousteau participated (with Swiss physicist Auguste Piccard) in the construction of the deep-submergence research vehicle, or bathyscaphe. In the 1960's, he directed a program called Conshelf, which tested a human's ability to live in a specially built underwater habitat. He also wrote and produced films on underwater exploration that attracted, entertained, and educated millions of people.

Cousteau won numerous medals and scientific distinctions. These include the Gold Medal of the National Geographic Society (1963), the United Nations International Environment Prize (1977), membership in the American and Indian academies of science (1968 and 1978, respectively), and honorary doctor of science degrees from the University of California, Berkeley (1970), Harvard University (1979), and Rensselaer Polytechnical Institute (1979).

*Sanford S. Singer*

# Waksman Discovers Streptomycin Antibiotic

*Selman Abraham Waksman searched for an antibacterial substance in soil microorganisms, discovering eighteen antibiotics, including streptomycin, the first effective drug against tuberculois.*

**What:** Medicine
**When:** September, 1943-March, 1944
**Where:** New Brunswick, New Jersey
**Who:**
SELMAN ABRAHAM WAKSMAN (1888-1973), a Soviet-born American soil microbiologist and winner of the 1952 Nobel Prize in Physiology or Medicine
RENÉ DUBOS (1901-1982), a French-born American microbiologist
WILLIAM HUGH FELDMAN (1892-1974), an American pathologist

## Curing Cows

Some microbiologists in the late nineteenth century believed that a struggle for survival occurred in the microbial world. They thought that microbes might contain substances that inhibited the growth of other microbes. There were attempts to isolate chemotherapeutic agents from such microbial substances as molds and bacteria, but the field was abandoned in the early twentieth century until the reawakening of interest in such agents by René Dubos in the 1930's.

Dubos was a student of Selman Abraham Waksman, whose area of expertise was the population of microorganisms that inhabit the soil. Waksman specialized in one type of soil microbe, the "actinomycetes," organisms intermediate between bacteria and fungi. His research included a study on how the tubercle bacillus fared when introduced into soil. From 1932 to 1935, Waksman established that the germ could not survive because of the antagonism of soil microbes. The finding substantiated the fact that pathogenic germs do not survive when introduced into soil. At the time, his finding did not seem to lead to anything new; it was only another example of microbes inhibiting other microbes.

Dubos wondered what would happen if soil were enriched with pathogenic germs. He pondered if perhaps their presence would encourage soil microbes antagonistic to them to flourish. In February, 1939, Dubos announced that he had tracked down such an antagonistic microorganism, *Bacillus brevis*, and from it had isolated two antibacterial substances, tyrocidine and gramicidin. The latter proved to be the first true antibiotic drug, attacking pneumococcus, staphylococcus, and streptococcus germs. Too toxic for human therapy, it became useful in treating animals. It aroused public interest when, at the 1939 New York World's Fair, sixteen of the Borden cow herd developed a streptococcal udder infection and gramicidin cured twelve of the cows of the bacteria.

## Curing Tuberculosis

Dubos's discovery alerted scientists to the possibility of finding other powerful drugs in microorganisms, and the central figure in exploiting this field was Waksman. He seized on Dubos's work and converted his research on soil actinomycetes into a search for the antibacterial substances in them. The actinomycetes proved to be the most fertile source for antibiotics. Waksman coined and defined the term "antibiotic" in 1941 to describe the novel drugs found in microbes. He developed soil enrichment methods and discovered eighteen antibiotics between 1940 and 1958. He cultured thousands of soil microbes in artificial media and screened them for activity. The promising ones were then chemically processed to isolate the antibiotics.

Streptomycin was the most important of Waksman's discoveries. In September, 1943, with his students Elizabeth Bugie and Albert Schatz, he isolated a soil actinomycete, *Streptomyces griseus*, which contained an antibiotic he named "streptomycin." It was antagonistic to certain types of bacteria. His report appeared in January,

The Nobel Foundation

*Selman Abraham Waksman.*

1944, and two months later, another article claimed that streptomycin was active against the deadly tubercle germ, *Mycobacterium tuberculosis.*

In the 1940's, tuberculosis was not fully under control. There was no cure, only prolonged bed rest and a regimen of nutritious food. The tubercle germ could invade any organ of the body, and in its various forms, the disease took a horrifying toll. A diagnosis of tuberculosis entailed lifelong invalidism, and patients died because the available treatment was so limited.

As the search for a cure progressed, the medical world took notice of the clinical tests conducted by William Hugh Feldman and H. Corwin Hinshaw at the Mayo Clinic. They had been investigating the chemotherapy of tuberculosis since the 1930's. Many scientists believed that such therapy was unattainable, but Feldman and Hinshaw refused to accept this verdict. They worked with sulfa drugs and sulfones and found some effect in suppressing the growth of tubercle bacilli, but not their eradication. Feldman had visited Waksman before the discovery of streptomycin and indicated a desire to try any promising antibiotics.

When Waksman found antitubercular effects in 1944, he wrote at once to Feldman to offer streptomycin for his studies. Feldman and Hinshaw had developed a practical system to determine the ability of a drug to slow the course of tuberculosis in guinea pigs. They used streptomycin on guinea pigs inoculated with the tubercle germ. In December, 1944, they issued their first report. The tests revealed streptomycin's ability to reverse the lethal course of the inoculations, and they concluded that it was highly effective in inhibiting the germ, exerting a striking suppressive effect, and was well tolerated by the animals.

Feldman and Hinshaw were now ready to test human patients. Hinshaw enlisted two physicians from a nearby sanatorium. On November 20, 1944, and for the next six months, a twenty-one-year-old woman with advanced pulmonary tuberculosis received streptomycin. In June of 1945, she was discharged, her tuberculosis arrested; she married eventually and reared three children.

This happy ending was followed by many more. Feldman and Hinshaw deserve the credit for proving that streptomycin could be used against tuberculosis. They demonstrated its value in carefully conducted trials. Some observers believe that they should have shared the 1952 Nobel Prize with Waksman.

## Consequences

Waksman did more than discover a major antibiotic—his work encouraged others to attempt to isolate new antibiotics by means of screening programs similar to those he devised. The 1950's witnessed a large increase in the number of antibiotics, and antibiotics became a large industry with total production of more than nine million pounds in 1955.

Streptomycin was not perfect. As early as 1946, reports appeared on the resistance of bacilli to the drug. Such resistant strains could be responsible for the failure of therapy. New drugs came to the rescue; Swedish investigators found that a drug consisting of para-aminosalicylic acid would inhibit the tubercle bacillus, although not as effectively as streptomycin. In 1949, the Veterans Administration combined the two drugs. After that, "combination" therapy proved to be the key to the future of chemotherapy, as the combination delayed the appearance of resistant strains. By 1970, using available drugs, and by the judicious use of combinations, physicians could achieve recovery in nearly all cases of pulmonary tuberculosis.

*Albert B. Costa*

# Western Allies Invade Italy

*After Italy's Fascist dictator, Benito Mussolini, was overthrown, Italy surrendered to the Allies, but the invading Allied forces faced a long, hard fight to oust German troops from the Italian peninsula.*

**What:** War
**When:** September 3-9, 1943
**Where:** Reggio, Taranto, and Salerno, Italy
**Who:**

FRANKLIN DELANO ROOSEVELT (1882-1945), president of the United States from 1933 to 1945

GEORGE CATLETT MARSHALL (1880-1959), chief of staff of the U.S. Army from 1939 to 1945

DWIGHT D. EISENHOWER (1890-1969), supreme allied commander in the western Mediterranean in 1943

MARK WAYNE CLARK (1896-1984), U.S. Army commander in North Africa and Italy

SIR WINSTON CHURCHILL (1874-1965), prime minister of Great Britain from 1940 to 1945

BERNARD MONTGOMERY (1887-1976), commander of the British Eighth Army from 1942 to 1944

BENITO MUSSOLINI (1883-1945), dictator of Italy from 1925 to 1943

VICTOR EMMANUEL III (1869-1947), king of Italy from 1900 to 1946

PIETRO BADOGLIO (1871-1956), premier of Italy from 1943 to 1944

## The Allies Make Plans

At the Casablanca Conference of January, 1943, American and British leaders agreed to send their forces to invade Sicily once the Axis Powers were defeated in North Africa. In May, 1943, British prime minister Winston Churchill traveled to Washington, D.C., to talk strategy again with U.S. president Franklin D. Roosevelt.

Pointing out how quickly the Axis forces had been driven out of North Africa, Churchill proposed that the Allies invade Italy as soon as possible. This move would enable them to take advantage of Italy's war-weariness, and perhaps the Italians would surrender.

Sending large numbers of troops to invade Italy was not the preferred plan of Roosevelt and General George C. Marshall, U.S. Army chief of staff. To them, Italy was not the major enemy in World War II; instead, they argued (as Marshall had at Casablanca) that troops should be sent to Great Britain to prepare for crossing the English Channel and facing German armies in France.

Roosevelt and Marshall soon gave in to Churchill, however, and agreed to a limited invasion of Italy. They did insist that troops from the Mediterranean be transported to Great Britain on schedule, no matter what happened in Italy. General Dwight D. Eisenhower, Supreme Allied Commander in the Mediterranean, was ordered to command the operation.

Sicily was invaded in July, 1943, and within a month the Italian and German forces on the island were defeated. This success encouraged the Allies to hope that their invasion of the Italian mainland would be quick and relatively painless.

The Allies' conquest of Sicily helped bring about a shake-up in the Italian government. On July 24-25, Benito Mussolini, Italy's Fascist dictator, met with the Fascist Grand Council to discuss the problem of defeatism. The Italian people were clearly tired of the war, and a recent bombing of Rome had raised calls for a peace settlement.

In response, the council recommended that Mussolini be dismissed. When the dictator asked for the help of King Victor Emmanuel III, the monarch dismissed him, and he was arrested. Marshal Pietro Badoglio was appointed premier of Italy, and the Fascist dictatorship was overthrown.

The king and Badoglio wanted to surrender Italy to the Allies, but they needed to take account of German plans. German chancellor Adolf Hitler had no intention of letting the Allies capture Italy by the king's surrender. He told the German commander in Italy, Field Marshal Albert Kesselring, to look into the situation. Kesselring recommended that if Italy surrendered or the Allies invaded, large numbers of German troops should be brought into Italy to fight. Hitler agreed, and troops were sent off to reinforce Kesselring's army.

## Surrender and Invasion

Aware that German troops were concentrating in Italy, Badoglio still wished to surrender. He was not willing to sign a surrender, however, until the Allies landed an invasion force—if Allied troops did not enter Italy, it would be entirely occupied by the Germans. So he contacted the Allies and expressed his desire to surrender if the Allies would move into Italy immediately.

Since Roosevelt had stated publicly at Casablanca that the Allies would end the war only with the Axis Powers' complete and unconditional surrender, the Allies were not ready to accept Badoglio's request right away. There were secret

## WORLD WAR II: THE EUROPEAN THEATER, 1943-1944

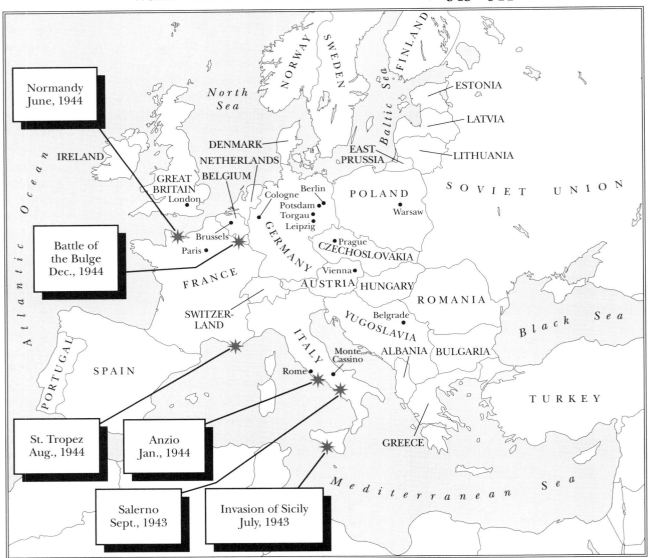

negotiations for more than a month; meanwhile, German troops continued to pour into Italy and prepare to do battle.

Italy's document of surrender was signed on September 3, 1943. On the same day, parts of the British Eighth Army, commanded by General Bernard L. Montgomery, landed at Reggio, a southern Italian port. The largest invasion took place on September 9, when the American Fifth Army, commanded by General Mark Wayne Clark, landed at Salerno, south of Naples.

The fighting bogged down immediately. Between late July and early September, the Germans had increased their troops in Italy from three divisions to nineteen—enough to prevent the Allies from moving forward quickly. On the hilly, rugged land of southern Italy, the Allied forces advanced quite slowly.

## Consequences

Unwilling to abandon his ally after the Fascist dictatorship was overthrown, Hitler sent a parachute raid to rescue Mussolini from arrest and set him up in a puppet government in northern Italy. In 1945, however, Mussolini was recaptured and shot.

The month of surrender negotiations had been a serious setback for the British and American forces. The Allies were not able to capture Rome until June, 1944. Yet that was not the end: When Germany surrendered in May, 1945, the Allied armies were still fighting in northern Italy. Taking the Italian peninsula had proved to be one of the most difficult campaigns of the entire war.

*José M. Sánchez*

# First Nuclear Reactor Is Activated

*The reactor at Oak Ridge National Laboratory produced the first substantial quantities of plutonium, making it practical to produce usable amounts of energy from a chain reaction.*

---

**What:** Energy
**When:** November 4, 1943
**Where:** Oak Ridge, Tennessee
**Who:**
ENRICO FERMI (1901-1954), an American physicist
MARTIN D. WHITAKER (1902-1960), the first director of Oak Ridge National Laboratory
EUGENE PAUL WIGNER (1902-1995), the director of research and development at Oak Ridge

---

## The Technology to End a War

The construction of the nuclear reactor at Oak Ridge National Laboratory in 1943 was a vital part of the Manhattan Project, the effort by the United States during World War II (1939-1945) to develop an atomic bomb. The successful operation of that reactor was a major achievement not only for the project itself but also for the general development and application of nuclear technology. The first director of the Oak Ridge National Laboratory was Martin D. Whitaker; the director of research and development was Eugene Paul Wigner.

The nucleus of an atom is made up of protons and neutrons. "Fission" is the process by which the nucleus of certain elements is split in two by a neutron from some material that emits an occasional neutron naturally. When an atom splits, two things happen: A tremendous amount of thermal energy is released, and two or three neutrons, on the average, escape from the nucleus. If all the atoms in a kilogram of "uranium 235" were to fission, they would produce as much heat energy as the burning of 3 million kilograms of coal. The neutrons that are released are important, because if at least one of them hits another

atom and causes it to fission (and thus to release more energy and more neutrons), the process will continue. It will become a self-sustaining chain reaction that will produce a continuing supply of heat.

Inside a reactor, a nuclear chain reaction is controlled so that it proceeds relatively slowly. The most familiar use for the heat thus released is to boil water and make steam to turn the turbine generators that produce electricity to serve industrial, commercial, and residential needs. The fissioning process in a weapon, however, proceeds very rapidly, so that all the energy in the atoms is produced and released virtually at once. The first application of nuclear technology, which used a rapid chain reaction, was to produce the two atomic bombs that ended World War II.

## Breeding Bomb Fuel

The work that began at Oak Ridge in 1943 was made possible by a major event that took place in 1942. At the University of Chicago, Enrico Fermi had demonstrated for the first time that it was possible to achieve a self-sustaining atomic chain reaction. More important, the reaction could be controlled: It could be started up, it could generate heat and sufficient neutrons to keep itself going, and it could be turned off. That first chain reaction was very slow, and it generated very little heat; but it demonstrated that controlled fission was possible.

Any heat-producing nuclear reaction is an energy conversion process that requires fuel. There is only one readily fissionable element that occurs naturally and can be used as fuel. It is a form of uranium called uranium 235. It makes up less than 1 percent of all naturally occurring uranium. The remainder is uranium 238, which does not fission readily. Even uranium 235, however, must be enriched before it can be used as fuel.

**940**

The process of enrichment increases the concentration of uranium 235 sufficiently for a chain reaction to occur. Enriched uranium is used to fuel the reactors used by electric utilities. Also, the much more plentiful uranium 238 can be converted into plutonium 239, a form of the human-made element plutonium, which does fission readily. That conversion process is the way fuel is produced for a nuclear weapon. Therefore, the major objective of the Oak Ridge effort was to develop a pilot operation for separating plutonium from the uranium in which it was produced. Large-scale plutonium production, which had never been attempted before, eventually would be done at the Hanford Engineer Works in Washington. First, however, plutonium had to be produced successfully on a small scale at Oak Ridge.

The reactor was started up on November 4, 1943. By March 1, 1944, the Oak Ridge laboratory had produced several grams of plutonium. The material was sent to the Los Alamos laboratory in New Mexico for testing. By July, 1944, the reactor operated at four times its original power level. By the end of that year, however, plutonium production at Oak Ridge had ceased, and the reactor thereafter was used principally to produce radioisotopes for physical and biological research and for medical treatment. Ultimately, the Hanford Engineer Works' reactors produced the plutonium for the bomb that was dropped on Nagasaki, Japan, on August 9, 1945.

The original objectives for which Oak Ridge had been built had been achieved, and subsequent activity at the facility was directed toward peacetime missions that included basic studies of the structure of matter.

## Consequences

The most immediate impact of the work done at Oak Ridge was its contribution to ending World War II. When the atomic bombs were dropped, the war ended, and the United States emerged intact. The immediate and long-range devastation to the people of Japan, however, opened the public's eyes to the almost unimaginable death and destruction that could be caused by a nuclear war. Fears of such a war remain to this day, especially as more and more nations develop the technology to build nuclear weapons.

On the other hand, great contributions to human civilization have resulted from the development of nuclear energy. Electric power generation, nuclear medicine, spacecraft power, and ship propulsion have all profited from the pioneering efforts at the Oak Ridge National Laboratory. Currently, the primary use of nuclear energy is to produce electric power. Handled properly, nuclear energy may help to solve the pollution problems caused by the burning of fossil fuels.

*John M. Shaw*

*The first nuclear reactor (white building, right center) at the Oak Ridge National Laboratory around 1944.*

**941**

# British Scientists Create Colossus Computer

*A secret team of specialists developed Colossus, the first all-electronic calculating device, in order to decipher German military codes during World War II.*

**What:** Computer science
**When:** December, 1943
**Where:** Bletchley Park, England
**Who:**
THOMAS H. FLOWERS, an electronics expert
MAX H. A. NEWMAN (1897-1984), a mathematician
ALAN MATHISON TURING (1912-1954), a mathematician
C. E. WYNN-WILLIAMS, a member of the Telecommunications Research Establishment

## An Undercover Operation

In 1939, during World War II (1939-1945), a team of scientists, mathematicians, and engineers met at Bletchley Park, outside London, to discuss the development of machines that would break the secret code used in Nazi military communications. The Germans were using a machine called "Enigma" to communicate in code between headquarters and field units. Polish scientists, however, had been able to examine a German Enigma and between 1928 and 1938 were able to break the codes by using electromechanical codebreaking machines called "bombas." In 1938, the Germans made the Enigma more complicated, and the Polish were no longer able to break the codes. In 1939, the Polish machines and codebreaking knowledge passed to the British.

Alan Mathison Turing was one of the mathematicians gathered at Bletchley Park to work on codebreaking machines. Turing was one of the first people to conceive of the universality of digital computers. He first mentioned the "Turing machine" in 1936 in an article published in the *Proceedings of the London Mathematical Society.* The Turing machine, a hypothetical device that can solve any problem that involves mathematical computation, is not restricted to only one task—hence the universality feature.

Turing suggested an improvement to the Bletchley codebreaking machine, the "Bombe," which had been modeled on the Polish bomba. This improvement increased the computing power of the machine. The new codebreaking machine replaced the tedious method of decoding by hand, which in addition to being slow, was ineffective in dealing with complicated encryptions that were changed daily.

## Building a Better Mousetrap

The Bombe was very useful. In 1942, when the Germans started using a more sophisticated cipher machine known as the "Fish," Max H. A. Newman, who was in charge of one subunit at Bletchley Park, believed that an automated device could be designed to break the codes produced by the Fish. Thomas H. Flowers, who was in charge of a switching group at the Post Office Research Station at Dollis Hill, had been approached to build a special-purpose electromechanical device for Bletchley Park in 1941. The device was not useful, and Flowers was assigned to other problems.

Flowers began to work closely with Turing, Newman, and C. E. Wynn-Williams of the Telecommunications Research Establishment (TRE) to develop a machine that could break the Fish codes. The Dollis Hill team worked on the tape driving and reading problems, and Wynn-Williams's team at TRE worked on electronic counters and the necessary circuitry. Their efforts produced the "Heath Robinson," which could read two thousand characters per second. The Heath Robinson used vacuum tubes, an uncommon component in the early 1940's. The vacuum tubes performed more reliably and rapidly than the relays that had been used for counters. Heath

Robinson and the companion machines proved that high-speed electronic devices could successfully do cryptoanalytic work (solve decoding problems).

Entirely automatic in operation once started, the Heath Robinson was put together at Bletchley Park in the spring of 1943. The Heath Robinson became obsolete for codebreaking shortly after it was put into use, so work began on a bigger, faster, and more powerful machine: the Colossus.

Flowers led the team that designed and built the Colossus in eleven months at Dollis Hill. The first Colossus (Mark I) was a bigger, faster version of the Heath Robinson and read about five thousand characters per second. Colossus had approximately fifteen hundred vacuum tubes, which was the largest number that had ever been used at that time. Although Turing and Wynn-Williams were not directly involved with the design of the Colossus, their previous work on the Heath Robinson was crucial to the project, since the first Colossus was based on the Heath Robinson.

Colossus became operational at Bletchley Park in December, 1943, and Flowers made arrangements for the manufacture of its components in case other machines were required. The request for additional machines came in March, 1944. The second Colossus, the Mark II, was extensively redesigned and was able to read twenty-five thousand characters per second because it was capable of performing parallel operations (carrying out several different operations at once, instead of one at a time); it also had a short-term memory. The Mark II went into operation on June 1, 1944. More machines were made, each with further modifications, until there were

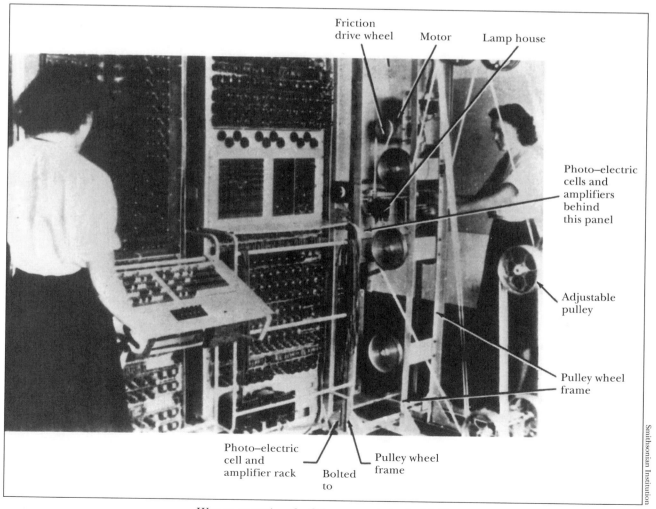

*Women operating the Colossus computer in 1943.*

Smithsonian Institution

ten. The Colossus machines were special-purpose, program-controlled electronic digital computers, the only known electronic programmable computers in existence in 1944. The use of electronics allowed for a tremendous increase in the internal speed of the machine.

**Consequences**

The Colossus machines gave Britain the best codebreaking machines of World War II and provided information that was crucial for the Allied victory. The information decoded by Colossus, the actual messages, and their influence on military decisions would remain classified for decades after the war.

The later work of several of the people involved with the Bletchley Park projects was important in British computer development after the war. Newman's and Turing's postwar careers were closely tied to emerging computer advances. Newman, who was interested in the impact of computers on mathematics, received a grant from the Royal Society in 1946 to establish a calculating machine laboratory at Manchester University. He was also involved with postwar computer growth in Britain.

Several other members of the Bletchley Park team, including Turing, joined Newman at Manchester in 1948. Before going to Manchester University, however, Turing joined Britain's National Physical Laboratory (NPL). At NPL, Turing worked on an advanced computer known as the Pilot Automatic Computing Engine (Pilot ACE). While at NPL, Turing proposed the concept of a stored program, which was a controversial but extremely important idea in computing. A "stored" program is one that remains in residence inside the computer, making it possible for a particular program and data to be fed through an input device simultaneously. (The Heath Robinson and Colossus machines were limited by utilizing separate input tapes, one for the program and one for the data to be analyzed.) Turing was among the first to explain the stored-program concept in print. He was also among the first to imagine how subroutines could be included in a program. (A subroutine allows separate tasks within a large program to be done in distinct modules; in effect, it is a detour within a program. After the completion of the subroutine, the main program takes control again.)

# German V-2 Rockets Enter Production

*The world's first operational flying bomb, the V-1, was used against the Allies in World War II, while the V-2, the world's first long-range, liquid-fueled rocket, went into production.*

**What:** Military technology; War; Space and aviation
**When:** 1944
**Where:** Germany
**Who:**
WERNHER VON BRAUN (1912-1977), chief engineer and prime motivator of rocket research in Germany
WALTER ROBERT DORNBERGER (1895-1980), commander of the Peenemünde Rocket Research Institute
ING FRITZ GOSSLAU, head of the V-1 development team
PAUL SCHMIDT, designer of the impulse jet motor

## The "Buzz Bomb"

On May 26, 1943, in the middle of World War II, key German military officials were briefed by two teams of scientists, one representing the air force and the other representing the Army. Each team had launched its own experimental aerial war craft. The military chiefs were to decide which project merited further funding and development. Each experimental craft had both advantages and disadvantages, and each counterbalanced the other. Therefore, it was decided that both craft were to be developed. They were to become the V-1 and the V-2 aircraft.

The impulse jet motor used in the V-1 craft was designed by Munich engineer Paul Schmidt. On April 30, 1941, the motor had been used to assist power on a biplane trainer. The development team for the V-1 was headed by Ing Fritz Gosslau; the aircraft was designed by Robert Lusser.

The V-1, or "buzz bomb," was capable of delivering a one-ton warhead payload. While still in a late developmental stage, it was launched, under Adolf Hitler's orders, to terrorize inhabited areas of London in retaliation for the damage that had been wreaked on Germany during the war. More than one hundred V-1's were launched daily between June 13 and early September, 1944. Because the V-1 flew in a straight line and at a constant speed, Allied aircraft were able to intercept it more easily than they could the V-2.

Two innovative systems made the V-1 unique: the drive operation and the guidance system. In the motor, oxygen entered the grid valves through many small flaps. Fuel oil was introduced and the mixture of fuel and oxygen was ignited. After ignition, the expanded gases produced the reaction propulsion. When the expanded gases had vacated, the reduced internal pressure allowed the valve flaps to reopen, admitting more air for the next cycle.

The guidance system included a small propeller connected to a revolution counter that was preset based on the distance to the target. The number of propeller revolutions that it would take to reach the target was calculated before launch and punched into the counter. During flight, after the counter had measured off the selected number of revolutions, the aircraft's elevator flaps became activated, causing the craft to dive at the target. Understandably, the accuracy was not what the engineers had hoped.

## Vengeance Weapon 2

According to the Treaty of Versailles (1919), world military forces were restricted to 100,000 men and a certain level of weaponry. The German military powers realized very early, however, that the treaty had neglected to restrict rocket-powered weaponry, which did not exist at the end of World War I (1914-1918). Wernher von Braun was hired as chief engineer for developing the V-2 rocket.

The V-2 had a lift-off thrust of 11,550.5 new-

Library of Congress

*A captured V-2 rocket at White Sands, New Mexico.*

tons and was propelled by the combustion of liquid oxygen and alcohol. The propellants were pumped into the combustion chamber by a steam-powered turboprop. The steam was generated by the decomposition of hydrogen peroxide, using sodium permanganate as a catalyst. One innovative feature of the V-2 that is still used was regenerative cooling, which used alcohol to cool the double-walled combustion chamber.

The guidance system included two phases: powered and ballistic. Four seconds after launch, a preprogrammed tilt to 17 degrees was begun, then acceleration was continued to achieve the desired trajectory. At the desired velocity, the engine power was cut off via one of two systems. In the automatic system, a device shut off the engine at the velocity desired; this method, however, was inaccurate. The second system sent a radio signal to the rocket's receiver, which cut off the power. This was a far more accurate method, but the extra equipment required at the launch site was an attractive target for Allied bombers. This system was more often employed toward the end of the war.

Even the 907-kilogram warhead of the V-2 was a carefully tested device. The detonators had to be able to withstand 6 g's of force during lift-off and reentry, as well as the vibrations inherent in a rocket flight. Yet they also had to be sensitive enough to ignite the bomb upon impact and before the explosive became buried in the target and lost power through diffusion of force.

The V-2's first successful test was in October of 1942, but it continued to be developed until August of 1944. During the next eight months, more than three thousand V-2's were launched against England and the Continent, causing immense devastation and living up to its name: *Vergeltungswaffe zwei* (vengeance weapon 2). Unfortunately for Hitler's regime, the weapon that took fourteen years of research and testing to perfect entered the war too late to make an impact upon the outcome.

## Consequences

The V-1 and V-2 had a tremendous impact on the history and development of space technology. Even during the war, captured V-2's were studied by Allied scientists. American rocket scientists were especially interested in the technology, since they too were working to develop liquid-fueled rockets.

After the war, German military personnel were sent to the United States, where they signed contracts to work with the U.S. Army in a program known as "Operation Paperclip." Testing of the captured V-2's was undertaken at White Sands Missile Range near Alamogordo, New Mexico. The JB-2 Loon Navy jet-propelled bomb was developed following the study of the captured German craft.

The Soviet Union also benefited from captured V-2's and from the German V-2 factories that were dismantled following the war. With these resources, the Soviet Union developed its own rocket technology, which culminated in the launch of Sputnik 1, the world's first artificial satellite, on October 4, 1957. The United States was not far behind. It launched its first satellite, Explorer 1, on January 31, 1958. On April 12, 1961, the world's first human space traveler, Soviet cosmonaut Yuri A. Gagarin, was launched into Earth orbit.

*Ellen F. Mitchum*

**946**

# IBM Completes Development of Mark I Calculator

> *Howard H. Aiken's distaste for solving differential equations inspired him to construct an automatic digital calculator that was a forerunner of modern computers.*

**What:** Computer science
**When:** 1944
**Where:** Endicott, New York
**Who:**

HOWARD H. AIKEN (1900-1973), Harvard University professor and architect of the Mark I

CLAIR D. LAKE (1888-1958), a senior engineer at IBM

FRANCIS E. HAMILTON (1898-1972), an IBM engineer

BENJAMIN M. DURFEE (1897-1980), an IBM engineer

## The Human Computer

The physical world can be described by means of mathematics. In principle, one can accurately describe nature down to the smallest detail. In practice, however, this is impossible except for the simplest of atoms. Over the years, physicists have had great success in creating simplified models of real physical processes whose behavior can be described by the branch of mathematics called "calculus."

Calculus relates quantities that change over a period of time. The equations that relate such quantities are called "differential equations," and they can be solved precisely in order to yield information about those quantities. Most natural phenomena, however, can be described only by differential equations that can be solved only approximately. These equations are solved by numerical means that involve performing a tremendous number of simple arithmetic operations (repeated additions and multiplications). It has been the dream of many scientists since the late 1700's to find a way to automate the process of solving these equations.

In the early 1900's, people who spent day after day performing the tedious operations that were required to solve differential equations were known as "computers." During the two world wars, these human computers created ballistics tables by solving the differential equations that described the hurling of projectiles and the dropping of bombs from aircraft. The war effort was largely responsible for accelerating the push to automate the solution to these problems.

## A Computational Behemoth

The ten-year period from 1935 to 1945 can be considered the prehistory of the development of the digital computer. (In a digital computer, digits represent magnitudes of physical quantities. These digits can have only certain values.) Before this time, all machines for automatic calculation were either analog in nature (in which case, physical quantities such as current or voltage represent the numerical values of the equation and can vary in a continuous fashion) or were simplistic mechanical or electromechanical adding machines.

This was the situation that faced Howard Aiken. At the time, he was a graduate student working on his doctorate in physics. His dislike for the tremendous effort required to solve the differential equations used in his thesis drove him to propose, in the fall of 1937, constructing a machine that would automate the process. He proposed taking existing business machines that were commonly used in accounting firms and combining them into one machine that would be controlled by a series of instructions. One goal was to eliminate all manual intervention in the

**947**

process in order to maximize the speed of the calculation.

Aiken's proposal came to the attention of Thomas Watson, who was then the president of International Business Machines Corporation (IBM). At that time, IBM was a major supplier of business machines and did not see much of a future in such "specialized" machines. It was the pressure provided by the computational needs of the military in World War II that led IBM to invest in building automated calculators. In 1939, a contract was signed in which IBM agreed to use its resources (personnel, equipment, and finances) to build a machine for Howard Aiken and Harvard University.

IBM brought together a team of seasoned engineers to fashion a working device from Aiken's sketchy ideas. Clair D. Lake, who was selected to manage the project, called on two talented engineers—Francis E. Hamilton and Benjamin M. Durfee—to assist him.

After four years of effort, which was interrupted at times by the demands of the war, a machine was constructed that worked remarkably well. Completed in January, 1943, at Endicott, New York, it was then disassembled and moved to Harvard University in Cambridge, Massachusetts, where it was reassembled. Known as the IBM automatic sequence controlled calculator (ASCC), it began operation in the spring of 1944 and was formally dedicated and revealed to the public on August 7, 1944. Its name indicates the machine's distinguishing feature: the ability to load automatically the instructions that control the sequence of the calculation. This capability was provided by punching holes, representing the instructions, in a long, ribbonlike paper tape that could be read by the machine.

Computers of that era were big, and the ASCC I was particularly impressive. It was 51 feet long by 8 feet tall, and it weighed 5 tons. It contained more than 750,000 parts, and when it was running, it sounded like a room filled with sewing machines. The ASCC later became known as the Harvard Mark I.

**Consequences**

Although this machine represented a significant technological achievement at the time and contributed ideas that would be used in subsequent machines, it was almost obsolete from the start. It was electromechanical, since it relied on relays, but it was built at the dawn of the electronic age. Fully electronic computers offered better reliability and faster speeds. Howard Aiken continued, without the help of IBM, to develop successors to the Mark I. Because he resisted using electronics, however, his machines did not significantly affect the direction of computer development.

For all its complexity, the Mark I operated reasonably well, first solving problems related to the war effort and then turning its attention to the more mundane tasks of producing specialized mathematical tables. It remained in operation at the Harvard Computational Laboratory until 1959, when it was retired and disassembled. Parts of this landmark computational tool are now kept at the Smithsonian Institution.

*Paul G. Nyce*

# Hodgkin Uses Computer to Find Structure of Penicillin

*Dorothy Crowfoot Hodgkin used an electronic computer to work out the X-ray data of penicillin, becoming the first to use a computer in direct application to a biochemical problem.*

**What:** Chemistry; Computer science
**When:** 1944-1949
**Where:** Oxford, England
**Who:**

DOROTHY CROWFOOT HODGKIN (1910-1994), an English crystallographer who won the 1964 Nobel Prize in Chemistry

BARBARA WHARTON LOW (1920- ), an English chemist who assisted in the penicillin project

C. W. BUNN and

A. TURNER-JONES, the English crystallographers who collaborated on the penicillin project with Hodgkin

L. D. COMRIE and

G. B. HEY, the English engineers who developed the computing methods used in the penicillin determination

## Solving an Important Problem

The method Dorothy Crowfoot Hodgkin used to solve the problem of the molecular structure of the antibiotic penicillin involved three scientific disciplines that had, until that time, little in common: synthetic organic chemistry, X-ray crystallography, and the new field of computer technology.

Penicillin was discovered in 1928 by the Scottish bacteriologist Sir Alexander Fleming at St. Mary's Hospital in London. *Time* magazine brought worldwide attention to the new wonder drug in 1941 but noted that because penicillin was difficult to extract, it had limited usefulness until it could be prepared less expensively or synthesized.

Because of the potential antibiotic activity of penicillin and its low toxicity, it was invaluable on the battlefields of World War II (1939-1945). An enormous effort that involved scientists in academia, industry, and government was undertaken by the British and American governments in 1942 to increase the supply of penicillin. The research effort moved in two directions: improved means of producing natural penicillins and the synthesis of an artificial penicillin.

The development of a synthetic route to penicillin required a knowledge not only of the chemical formula of penicillin but also of its molecular structure. The usual method of structure determination by organic chemists at that time was to subject the compound to harsh treatment and then study the degradation products. The analysis of these fragments would make it possible to reconstruct the original molecule.

After several failed attempts, however, it seemed that the traditional organic method of structure determination was not working. To settle the issue, a sample of penicillin was sent to Hodgkin at the end of 1944 for a single-crystal X ray study.

## A Significant New Tool

The discipline of X-ray crystallography provided a new tool for examining the internal arrangements of atoms in crystals and provided solutions, almost immediately, for many inorganic structural questions. The object of a crystal-structure determination is to ascertain the positions of all atoms in the basic unit of the crystal. The process involves collection of data, solution of the phase relations among the scattered X rays (determination of a trial structure), and refinement of the structure.

Hodgkin applied her particular qualities of precision, astute mathematical analysis, and special imagination to the penicillin problem as soon as crystals were prepared for photographing. Three crystals were prepared for study: sodium benzylpenicillin and the isomorphous potassium and rubidium salts of benzylpenicillin. Potassium and rubidium benzylpenicillin were chosen for the X-ray analysis in the hope that their crystal structures would be amenable to a special type of mathematical analysis called Fourier analysis. If this were true, the analysis could be completed without any detailed chemical information. After some initial work at the University of Oxford, the data on the sodium salt were sent to C. W. Bunn and A. Turner-Jones at Imperial Chemical Industries, where the analysis was completed.

The early electron density projections indicated the presence of a heavy scattering center, identified as sulfur, in the crystal structure. Armed with these few clues and the tentative structures, Hodgkin and Barbara Wharton Low proceeded with a trial-and-error analysis. They constructed scale models using heavy wire, and by studying the shadows made by illuminating the model with parallel light beams, they were able to record new atomic coordinates. Hodgkin and Low frequently compared their results with those of Bunn and Turner-Jones. Eventually, the two groups agreed on the beta-lactam structure as the correct one for the penicillin molecule.

Hodgkin then proceeded to attempt refinement of the crystal structure in three dimensions following the same scheme as that of two-dimensional projections: calculating structure factors, then using three-dimensional Fourier series. Although the calculations involve no theoretical difficulty, there is considerable practical difficulty because of the cumbersome nature of the formulas involved. L. D. Comrie and G. B. Hey of the Scientific Computing Service devised methods of executing the necessary calculations. Their use of large-scale computing methods allowed the analysis to be carried much further than was usual in such a structure analysis.

## Consequences

X-ray crystallographic analysis of even the simplest molecules requires many mathematical computations. When the work on penicillin began, the computing equipment available was grossly inadequate. By the mid-1940's, Hodgkin was able to borrow an old International Business Machines (IBM) Corporation card-punch machine; without it, the refinement of the penicillin structure would have been almost impossible.

Hodgkin continued to make use of computers of varying degrees of complexity to analyze even larger molecules. It was becoming clear that understanding the chemistry of life processes required a detailed smaller knowledge of the compounds involved. By 1956, Hodgkin had shown by the penicillin work and by solving the structure of vitamin $B_{12}$ that it was possible to use X-ray crystallography alone to analyze even very complex structures. As the size and complexity of the molecules to be analyzed increased, the need for more advanced computers increased also. Fortunately, computer technology kept pace with the demands of other fields of science.

The use of the computer in modeling chemical compounds has gone far beyond the computational stage. The development of graphics capability allows models of molecules to be viewed on screen and manipulated in many ways. New drugs are frequently computer-designed to mimic natural molecules and to act as positive or negative inhibitors of many body processes. The introduction of the computer truly has revolutionized chemistry.

*Grace A. Banks*

*Dorothy Crowfoot Hodgkin.*

# Ryle Locates First Known Radio Galaxy

*Martin Ryle's interferometric radio telescope detected and provided details on the structure of the first identifiable radio galaxy.*

**What:** Astronomy
**When:** 1944-1952
**Where:** Cambridge, England
**Who:**
SIR MARTIN RYLE (1918-1984), an English radio physicist and astronomer
FRANCIS GRAHAM SMITH (1923-        ), an English radio physicist and astronomer
WALTER BAADE (1893-1960), a German American astronomer
RUDOLF MINKOWSKI (1895-1976), a German American physicist

## Improving Radio Telescopes

The initial measurements of cosmic radio emission by the American radio engineers Karl Jansky and Grote Reber between 1932 and 1940 showed reasonable similarity between the universe as it was revealed by radio waves and as it was seen by optical telescopes. This led many astronomers to conclude that most, if not all, celestial radio emissions came from interstellar gas evenly distributed throughout the universe. Until the post-World War II period, the greatest drawback of the new discipline of radio astronomy was its limited accuracy in determining the celestial position and structural detail of an object detected from radio signals. A higher degree of accuracy was necessary in order to give optical astronomers a small enough "window" in which to look for an object discovered by radio telescopes.

Immediately following World War II, J. S. Hey used receivers from the Army Operational Radar Unit to study some of the extraterrestrial radio emissions reported earlier by Jansky and Reber. In 1946, Hey and his colleagues reported an observational discovery of particular import: that a radio source in the constellation Cygnus varied significantly in strength over very short time periods. In contrast to Reber, who had concluded that interstellar hydrogen between the stars was the source of all celestial radio signals, Hey argued that the fluctuations were too localized for interstellar gas and suggested instead the existence of a localized, starlike object.

In Australia, a similar group was formed under the direction of J. L. Pawsey. In 1946, the Australian group verified Hey's observations of a localized radio source in Cygnus, using one of the earliest radio interferometers. An interferometer is a series of radio telescopes connected over a wide area. The resolution, or receiving power, of this array is equal to that of a single radio telescope with a diameter equal to the distance between the farthest single radio telescopes. Further "radio stars" were discovered by Pawsey's group in June, 1947, using an improved Lloyd's interferometer developed by L. McCready, Pawsey, and R. Payne-Scott. Incorporating an antenna mounted on top of a high cliff, this interferometer was able to use the ocean to reflect radio waves. Almost simultaneously, using a different type of interferometer, Sir Martin Ryle at Cambridge found another intense localized radio source in the constellation Cassiopeia.

Ryle and others had extensive wartime experience in developing airborne radar detectors, radar countermeasures, underwater sonar arrays, and signal detection and localization equipment. Ryle was joined in 1946 by Francis Graham Smith, who helped him rearrange his interferometer to cover a wider receiving area. Ryle's cosmic radio "pyrometer" was used successfully in June, 1946, to measure a large sunspot.

**951**

## Puzzling Results

Hey and his colleagues remained unable to determine the accurate position of their radio source to better than 2 degrees. The successful resolution of solar sunspots of small diameter, however, suggested the possibility of additional radio telescope improvements to Ryle. In 1948, Ryle, Smith, and others made the first detailed radio observations of Cygnus A, using an improved version of their pyrometric radio telescope. Ryle and Smith subsequently published an improved position for Cygnus A. Ryle and Smith's measurements included the discovery of short-period radio bursts, which they (incorrectly) used to argue that the ultimate radio source must be a radiating star of some unrecognized type.

In 1949, astronomers decided to establish the locations of Cygnus A and Cassiopeia by constructing a very large interferometric radio telescope, with a maximum separation of 160 kilometers. In 1951, Ryle developed a new "phase-switching" receiver based on 1944 sonar detection efforts. Phase switching permits the radio receiver to reject sources with large size and thus improves the receiver's ability to emphasize and locate weaker sources. After completing the prototypes, in 1951 researchers again made measurements of Cygnus A and Cassiopeia. They discovered that the radio sources were clearly not stars and that the Cygnus A source was actually two distinct sources. Ryle's phase-switched records were such improvements that his colleagues compiled the first radio object catalog, which listed more than fifty cosmic radio sources.

By late 1951, Smith had further localized the coordinates of Ryle's two radio stars, reducing the original error windows of Hey and others by a factor of sixty. Smith then approached the director of the Cambridge Observatory to seek visual identification of the two radio sources. While part of the Cassiopeia source, a supernova remnant, was found in 1951, the poor atmospheric observing conditions in England prevented the Cambridge observers from making a complete identification. Shortly thereafter, Smith sent his data to Walter Baade and Rudolf Minkowski of the Palomar Observatory in Southern California.

The objects of Baade and Minkowski's visual search were discovered only after many difficulties; they were hidden among many other stars and faint galaxies. In 1952, Baade wrote to Smith that the result of his visual search was puzzling. He had found a cluster of galaxies, and the position of the radio source coincided closely with the position of one of the brightest members of the cluster. Moreover, the source seemed to be receding at a high velocity. These findings suggested that the source of the radio transmissions was outside of the galaxy.

## Consequences

At first, there was notable skepticism over the notion of extragalactic radio sources. Because of this climate of disbe-

*Sir Martin Ryle.*

lief, Baade and Minkowski did not publish their results until 1954. Subsequent research, however, confirmed the extragalactic location of these objects, forcing cosmologists to find a place for these sources in their theories of the universe. Perhaps the most decisive radio data came from Ryle and Scheuer in 1955. They found that the most "normal" galaxies emit radio signals comparable in intensity to those emitted by the Milky Way galaxy. Nevertheless, there were many other galaxies—many not different in their optical appearances from normal galaxies—which are much more powerful sources; these became known as "radio galaxies." By the time of the 1958 Solvay Conference on the Structure and Evolution of the Universe, the existence of at least eighteen extragalactic radio objects had been confirmed, opening a new era in cosmology.

*Gerardo G. Tango*

# Soviets Invade Eastern Europe

*During 1944 and 1945 the Soviets drove Nazi forces out of the Soviet Union and then took over Eastern Europe from Poland to the Balkans.*

**What:** War
**When:** April, 1944 to mid-1945
**Where:** Eastern Europe
**Who:**
JOSEPH STALIN (1879-1953), dictator of the Soviet Union from 1924 to 1953
GEORGI KONSTANTINOVICH ZHUKOV (1895-1974), commander of all Soviet field armies in the Ukrainian regions
ADOLF HITLER (1889-1945), chancellor of Germany from 1933 to 1945
HEINZ GUDERIAN (1886-1954), chief of the German general staff from 1944 to 1945
ERICH VON MANSTEIN (1887-1973), German army group commander in the Ukrainian regions

## German Hopes Fade

In late 1942, the brutal Battle of Stalingrad had turned the tide of war in the Soviet Union against the Nazi invaders. The Battle of Kursk in July, 1943, fought with tanks, was another disaster for the Germans. Large numbers of armored units and experienced combat troops had been wasted, and the Soviets continued to press eastward.

For German chancellor Adolf Hitler and his general staff, the question now became how to continue to fight a defensive war. There were still many German troops in the Soviet Union, and they still had control of large stretches of territory. Yet victory was no longer a realistic hope.

Hitler and his generals had already disagreed many times over how to conduct the war. The generals wanted to defend only the most important areas and to let the others go. Both General Heinz Guderian, who was appointed chief of the general staff in 1944, and Field Marshal Erich von Manstein, who commanded German troops in the southern Soviet Union, wanted to concentrate their efforts on building up defense forces that could move quickly to respond to Soviet attacks.

Hitler remained stubbornly opposed to such a plan. Although even he had to admit that the Germans could no longer hope to mount a real offensive in the Soviet Union, he was determined to cling to whatever territory his armies still held.

Hitler was sure that his "no-withdrawal" policy was right, and he could point to the past as evidence. When the German armies lay outside Moscow in late 1941, Hitler had insisted that they dig in and keep their position until spring, though his field commanders had objected. Obeying Hitler's orders, the Germans did manage to hold their front—and Hitler concluded that he was wiser than his trained officers. He intended to continue his policy at all costs.

By April, 1944, the Soviets were at the foothills of the Carpathian Mountains on the Hungarian frontier, and they had gained control of the important railroad line that ran between Odessa, Ukraine, and Warsaw, Poland. Other Soviet forces were coming near to Brest-Litovsk, a Belorussian city on the Bug River, where Hitler had launched his invasion in 1941.

The oil fields of Romania were now within the Soviets' reach, and Hitler expected that his forces in that area would suffer more attacks. Instead, around the time of the D-Day Allied landings in Normandy, the Soviets attacked the German army group center. Some 300,000 Germans were wounded or killed, and a 250-mile break was made in the German front.

East Prussia and the Baltic area now lay open to the Soviets, who had already seized both Belorussia and northeastern Poland.

*This sign, posted July, 1944, in Romania, tells Romanians that the Germans, not the Russians, are their enemies.*

## The Soviets Push Forward

By July 26, 1944, the Soviets reached the Vistula River in Poland, and five days later they reached the outskirts of Warsaw. During the month of July the Germans had lost two hundred miles of territory.

The mastermind behind the Soviet successes was Marshal Georgi Zhukov. More than any other Soviet commander, Zhukov had been responsible for bringing the first major Soviet victory in the war, with his surprising counteroffensive at Stalingrad. Now his forces were ready to move into the Eastern European countries.

As early as 1942, Soviet dictator Joseph Stalin had made it clear to British prime minister Winston Churchill that the Soviet Union would insist upon major changes in Eastern Europe once the war was over. By the fall of 1944, Stalin was in a position to begin fulfilling this ambition.

He was motivated partly by the desire to conquer new lands for Communism, but he was also concerned to protect his country. In earlier centuries, the Swedes and the French had moved through Eastern Europe to invade Russia; Germany had used the same route in its invasions during World Wars I and II. In bringing Eastern Europe under Soviet control, Stalin hoped to prevent such attacks in the future.

In August, 1944, Soviet forces began driving the Germans from the Balkans. Galati, Romania, fell on August 27, the Ploesti oil fields of Romania were seized on August 30, and Bucharest was taken on August 31. During a twelve-day period in August, the Germans had retreated about 250 miles, and in the next six days they had pulled back another 200 miles as the Soviets moved to the Yugoslavian border at Turnu Severin.

With the German army group center de-

**955**

stroyed, the Nazis could no longer mount a real resistance to the Soviets. Belgrade, Yugoslavia, fell to Soviet forces in October. In early 1945, Soviet armies had advanced to within about 300 miles of Berlin. A large number of armored units kept Budapest, Hungary, under German control for a time, but that city finally fell in February, 1945.

The Soviet successes resulted partly from the building up of equipment and forces. By 1945, Soviet forces outnumbered the German defenders eleven to one in infantry, seven to one in tanks, and twenty to one in heavy guns and artillery. Hitler refused to believe such statistics, however, and called them "rubbish."

During April, 1945, Soviet forces occupied the rest of Hungary, Czechoslovakia, and about half of Austria, including its capital, Vienna. On May 2, Berlin fell to the Red Army, and the war in Europe was formally ended several days later.

## Consequences

The war on the Eastern Front had been a bloody one. Between 1941 and 1945, more than 1 million Germans died in the East, 4 million were wounded, and 1.3 million were missing. Estimates of Soviet losses include 10 million deaths and 4 million other casualties.

With the Soviet occupation of Eastern Europe, Nazi domination was exchanged for Communist control. As part of the peace settlement, Finland and Romania both gave up territories to the Soviet Union. More important, Communist governments soon came into power in Czechoslovakia, Yugoslavia, Bulgaria, Poland, Hungary, and Romania. In 1949, Germany was split into a democratic capitalist nation in the west and a Communist nation in the east. The economies and governments of the new Communist states came under heavy Soviet influence, creating what came to be called the Eastern Bloc.

*Howard M. Hensel*

# Allies Invade France

*Allied forces landed on the beaches of Normandy, France, and broke down German defenses so that they could sweep on to liberate Paris from Nazi control.*

**What:** War
**When:** June 6, 1944
**Where:** Normandy, France
**Who:**

SIR WINSTON CHURCHILL (1882-1945), prime minister of Great Britain from 1940 to 1945

BERNARD MONTGOMERY (1887-1976), commander of the Allied armies in northern France

FRANKLIN DELANO ROOSEVELT (1882-1945), president of the United States from 1933 to 1945

GEORGE CATLETT MARSHALL (1880-1959), chief of staff of the U.S. Army from 1939 to 1945

DWIGHT D. EISENHOWER (1890-1969), supreme commander of the Allied Expeditionary Force

JOSEPH STALIN (1879-1953), dictator of the Soviet Union from 1924 to 1953

ADOLF HITLER (1889-1945), chancellor of Germany from 1933 to 1945

KARL RUDOLF GERD VON RUNDSTEDT (1875-1953), German commander in chief on the Western Front

ERWIN ROMMEL (1891-1944), commander of German army Group B and inspector of Coastal Defenses

## Operation Overlord

After the German victory over France in 1940, one of the Allies' main goals was to invade France and drive the German invaders out. When the United States entered the European conflict in 1941, American generals proposed that the British and U.S. forces begin gathering as soon as possible to cross the English Channel and take France.

British plans, however, were different. British prime minster Winston Churchill feared that an invasion then would fail because the Allies lacked the boats they needed to land large numbers of soldiers on the French beaches. He proposed that Allied troops first attack German and Italian forces in the Mediterranean area.

American president Franklin D. Roosevelt and U.S. Army chief of staff George C. Marshall continued to argue for an invasion of France. So did Joseph Stalin, the Soviet dictator, who wanted the Allies to draw German troops and resources away from the battlefields of the Soviet Union—the Eastern Front.

The Americans had enough troops, but because new landing craft had to go to the Pacific area, where the Americans were fighting Japan, as well as to Europe, it was not until 1944 that there were enough boats to make an invasion of France possible.

Serious planning for the invasion, dubbed Operation Overlord, began in 1943. British and American troops would be shipped from the Mediterranean, where an invasion of Italy had already taken place. They would congregate in southern England to await the invasion, which would take place in May or June of 1944.

General Dwight D. Eisenhower, who commanded American forces in Europe, was selected as Supreme Commander of the Allied Expeditionary Force, and Great Britain's General Bernard L. Montgomery, the hero of the Battle of Alamein, was appointed to command the Twenty-first Army Group, the main Allied land force.

The main invasion would take place on the beaches of Normandy. Eisenhower knew that the

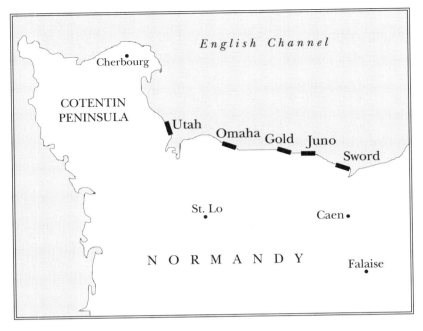

France. Hitler also made Rommel Inspector of Coastal Defenses.

Rommel concluded that the Allies would be able to land anywhere along the coast, so that his best defense would be to hold tanks in reserve and send them to the battlefield once the Allies landed. The Allies guessed that he would choose this strategy, however, and in the spring of 1944 they bombed roads and railways along the entire coast to prevent the armored units from moving quickly. Guessing that the Allies would land somewhere in the vicinity of the Pas de Calais, Rommel built up his strongest defenses in that area.

invasion would have to be successful on the first day; the Allies would have to gain firm control of an area of land, or beachhead, that would allow them to continue landing troops. It would take six weeks to land enough men to outnumber the German defenders. This first crucial day was called D-Day.

Though the first plans called for three divisions to be landed, Eisenhower and Montgomery insisted that the force must be increased to five divisions. The day of the attack would have to be preceded by a night with a full moon, so that parachute forces could be dropped. The attack would have to begin at dawn, so that the troops would have a full day to establish their beachhead, and the Allies would need a low tide at dawn and a calm sea. Tide and moon requirements could be met on June 4, 5, and 6, 1944; Eisenhower picked June 4 as D-Day.

### The Invasion

German chancellor Adolf Hitler was fully aware that the Allies were planning an invasion. The Germans had built coastal fortifications in France in 1940. In 1944 Hitler appointed Field Marshal Gerd von Rundstedt as German commander in the West, and Field Marshal Erwin Rommel became Commander of Army Group B, the main German army in northern

The Allies succeeded in keeping their plans secret. High winds on June 4 and 5 made the invasion impossible then, but the weather improved on the morning of June 6, and Allied troops were ready. More than five thousand Allied ships, carrying nearly a hundred thousand troops and escorted by more than one thousand heavy bombers and more than two thousand fighter planes, landed on the beaches of Normandy. It was the largest combined air, sea, and land operation in history.

All the troops went ashore successfully, and after a day of bitter fighting the Allies established their beachhead. Believing that the Normandy invasion was meant to distract the Germans from an upcoming larger invasion at Calais, Rommel and Hitler did not send large numbers of troops to defend Normandy. As the Allied leaders had hoped, Rommel was also unable to move his tanks quickly enough to push the Allied forces back.

Nevertheless, the battles of the following weeks were difficult. The port of Cherbourg was taken by the Allies on June 27, but the U.S. First Army struggled hard to move past that point. The city of Caen, Normandy, was not captured until mid-July.

It was Allied warplanes that finally broke the German line of defense. An intense bombardment on July 25 allowed the U.S. First Army to break out. By August 3, General George Patton's U.S. Third Army was sweeping forward through France. On August 25, troops of that army liberated Paris.

## Consequences

By September 14, Allied forces had moved to the French-German border and had advanced into Belgium past Antwerp, Brussels, and Liege. Shortages of equipment and fuel then brought the offensive to a halt. In December, Hitler moved some of his troops from the Eastern Front to the West and began trying to retake Belgium. Though at first the Germans were able to push the Allies back, the Battle of the Bulge used up their strength.

In late April, 1945, when Hitler finally gave up and committed suicide, Allied forces were moving into Germany from the west, while Soviet forces were entering Berlin from the east. The Normandy invasion had succeeded.

*José M. Sánchez*

# U.S. B-29's Bomb Japan

> *The first use of B-29 bombers, at Yawata, Japan, was not a military success, but it was a valuable test of a weapon that would later prove its destructive capacity.*

**What:** War; Weapons technology
**When:** June 15, 1944
**Where:** Chengtu, China; and Yawata, Japan
**Who:**
FRANKLIN DELANO ROOSEVELT (1882-1945), president of the United States from 1933 to 1945
HENRY HARLEY "HAP" ARNOLD (1886-1950), commander of U.S. Army Air Forces from 1942 to 1946
KENNETH B. WOLFE (1896-1971), commander of the Twentieth Bomber Command
LA VERNE "BLONDIE" SAUNDERS (1903-1988), commander of the Fifty-eighth Bombardment Wing

## Planning for the B-29

After the Doolittle air raid on Tokyo in April, 1942, American bombs were not dropped on the Japanese home islands for two years. Japanese forces advanced quickly in the Pacific area, and because they kept a tight hold on the Asian mainland, the Americans were not able to set up bases close enough to Japan to allow air raids. The available heavy bombers, the B-17 Flying Fortress and the B-24 Liberator, did not have a long enough range.

Yet United States military planners believed that the B-29 Superfortress, a new bomber, was promising. The Boeing Company had begun to develop this long-range, high-altitude bomber in 1938. Although it was not test-flown until September, 1942, the Air Corps had already authorized Boeing to begin production as soon as possible.

Much larger than the B-17, the Superfortress measured 99 feet in length, with a wing span of 141 feet. Fully loaded, it weighed more than sixty tons, and at top speed it flew 363 miles per hour. Powered by four 2,200-horsepower Wright "Double Cyclone" engines, it could fly combat missions of up to 1,600 miles. Three separate pressurized compartments meant that its crew of eleven could cruise at 33,000 feet without needing oxygen masks.

The bomber's fuselage was armed with ten .50-caliber machine guns in five turrets, and there were two more .50's and a 20-millimeter cannon in the tail. The Superfortress could carry a bomb load of ten tons.

Air Force leaders had considered using the B-29 in Europe, but by the time large numbers of these planes could be ready, other British and American bombers flying from England were already raiding Germany. By the end of 1943, Air Force chief general Hap Arnold had decided that the B-29 should be used against Japan.

The major problem was air bases. American bases in the Aleutian Islands were too far from Japan. Saipan, Tinian, and Guam, in the Mariana Island group, were not expected to be in American hands until the winter of 1944. So Air Force planners looked to China.

President Franklin D. Roosevelt had already been wishing to "do something" for China. Japan had invaded that country in the 1930's, before World War II had even broken out, but China had still not received much in the way of supplies and support from the Allies. Roosevelt feared that unless the Chinese leader, Chiang Kai-shek, received some real help, he would quit the war.

When the Air Force leaders suggested basing the new B-29's in India but "staging" them—sending them on missions—from fields in China, Roosevelt approved. This plan, called Operation Matterhorn, became official in November, 1943.

American fliers had long dreamed of an independent, powerful bombing force that could beat any enemy into surrender without a land invasion. Perhaps the Superfortress was the needed weapon.

## Mounting the Raid

General Arnold, representing the U.S. Joint Chiefs of Staff, kept command of the newly created Twentieth Bomber Command. The local British commander, Lord Louis Mountbatten, and U.S. Army commander Lieutenant General Joseph W. Stilwell had no say over the use of the B-29's in their region. Operation Matterhorn's B-29 bases were built at Chengdu (Chengtu), China.

The Twentieth Bomber Command, under Brigadier General Kenneth B. Wolfe, was originally composed of the Fifty-eighth Bombardment Wing and the Seventy-third Bombardment Wing. In April, 1944, however, the Seventy-third Wing was separated in order to be sent to the Mariana Islands (which were now due to be taken by Allied forces in June).

A wing contained 112 bombers, along with replacement ships, slightly over three thousand officers, and eight thousand enlisted men. Engineers and service personnel brought the total strength of the Twentieth Bomber Command to about twenty thousand persons.

Since all supplies for the Chinese bases had to be flown in, stockpiling in preparation for the raids was difficult. B-29's from India had to fly seven round trips to bring enough gasoline and other supplies to make possible one mission over Japan. With the loss of the Seventy-third Wing, the Fifty-eighth Wing could not bring in enough supplies for raids of one hundred planes or more—the hoped-for number—more than a few times each month.

A large number of engine failures and many losses of planes—to be expected when an aircraft is new and the crews are not experienced in using it—further delayed the first raid on Japan. It was finally set for June 15, 1944.

The Air Force's Committee of Operations Analysts had suggested that a good target for the

*B-29's drop their bombs over Yokohama, Japan.*

National Archives

B-29's would be the coke ovens that supplied Japan's steel mills. The first strike would be at the coke ovens of the Imperial Iron and Steel Works at Yawata, on the island of Kyushu.

Beginning on June 13, ninety-two planes left the Bengal fields in India. Seventy-nine of them reached the Chengtu bases. Each came loaded with two tons of five-hundred-pound bombs. The Joint Chiefs in Washington, D.C., ordered a night mission, with bombs to be dropped from between eight thousand and eighteen thousand feet.

On June 15, sixty-eight planes, led by Wing Commander Brigadier General La Verne "Blondie" Saunders, left the Chengdu fields. Four were forced back by engine trouble, and one crashed immediately after takeoff.

Forty-seven Superfortresses bombed Yawata that night—most using radar because haze and smoke had combined to black out the city. Antiaircraft fire was light, and there was little opposition from enemy fighters.

Thirteen B-29's had not made it to Yawata for various reasons, mostly mechanical. Five planes were lost, one of them being hit by enemy fighters on the return trip.

**Consequences**

Photo reconnaissance after the raid showed that the B-29's had caused very little damage. There had been one hit on a shop at the steel plant, and none on the ovens.

These high-altitude, precision bombings were different from the firebomb raids that would begin in May, 1945, from the Mariana Islands. Meanwhile, the Fifty-eighth Wing averaged two raids a month until March, 1945, when it was moved to Saipan.

Though bringing supplies into China had been difficult, Operation Matterhorn had given important encouragement to the Chinese and had been a learning experience for the B-29 crews. The first raid on Japan was not a success, nor was Operation Matterhorn as a whole. Yet the Superfortress had proved itself to be a superb plane, and it would later be used very effectively to drop firebombs.

*Charles W. Johnson*

# Congress Passes G.I. Bill

*The G.I. Bill, which gave federal aid to returning World War II veterans so that they could pursue their education, marked the beginning of the United States government's involvement in higher education.*

**What:** Education; Social reform
**When:** June 22, 1944
**Where:** United States
**Who:**
FRANKLIN DELANO ROOSEVELT (1882-1945), president of the United States from 1933 to 1945
GEORGE F. ZOOK (1885-1951), president of the American Council on Education

## Evaluating the Problem

Soon after the Japanese bombed Pearl Harbor, bringing the United States into World War II, President Franklin D. Roosevelt and other American leaders began thinking about the fact that many young men had to interrupt their education to serve as soldiers and sailors. In 1942, a report called *Statement of Principles Relating to the Educational Problems of Returning Soldiers, Sailors, and Displaced War Industry Workers* was published by the Institute of Adult Education of Teachers College at Columbia University, New York. The report said that the federal government should pay for some kind of job training and educational program for veterans once the war was over.

In July, 1942, Roosevelt discussed the matter with President George F. Zook and other officials of the American Council on Education. Convinced that the federal government did have a responsibility to help, Roosevelt appointed the Conference on Postwar Readjustment of Civilian and Military Personnel. This committee wrote a set of recommendations called "Demobilization and Reconversion," saying that action was needed in several areas: setting up a job-training program for returning veterans; establishing special courses at regular colleges and universities for veterans who returned to finish their degrees; getting funds for these programs from the federal government; and making these programs a cooperative effort among local and state agencies and the federal government.

Next, Roosevelt enlisted the help of the War and Navy departments to make more specific guidelines and proposals to meet these goals. The report that the Committee of the War and Navy Departments issued on October 27, 1943, became the basis for the G.I. Bill.

As this proposal came before Congress, it had several factors in its favor. As often happens during wartime, patriotic feelings were especially intense. Many people were reading the newspaper reports of Ernie Pyle, a journalist who traveled with American troops and emphasized the courage of ordinary soldiers. Pyle insisted that the United States owed something to those who were risking their lives on the battlefields of Europe.

At the same time, statistics from Selective Service records were painting a dim picture of American education. To the dismay of teachers and other school officials, the government announced that more than 676,000 men were disqualified from serving in the military because they did not have even four years of schooling. Furthermore, it was becoming clear that American education had failed to produce enough qualified mathematicians, scientists, and foreign-language experts to meet the country's wartime needs.

Not enough schools, not enough skilled teachers, not enough courses—for many educators and politicians, it all added up to one thing. Only large amounts of federal money could allow American schools and colleges to live up to the ideals of this land where opportunity was supposed to have no limits.

## The New Law

Congress had already begun to respond to the need. On March 24, 1943, President Roosevelt signed into law the Vocational Rehabilitation Bill, which provided federal funds to retrain and rehabilitate veterans and workers who had been disabled in the civilian defense industry.

On June 22, 1944, the Serviceman's Readjustment Act, better known as the G.I. Bill of Rights, became law. It defined veterans as individuals who had served actively for more than ninety days after September 16, 1940, and had received an honorable discharge from the military. These persons would receive help to continue their education at any "approved educational or training institution" that they chose. They could study full- or part-time, and the bill covered every level of study from elementary to graduate school.

The federal government would pay all tuition and fees required by the university. In addition, the veteran was entitled to receive a subsistence check each month. The amount of this check varied according to the number of dependents the veteran had, and it would be adjusted to keep up with inflation. By 1947, a veteran with one or more dependents was entitled to receive ninety dollars a month, while a single veteran received sixty-five dollars.

The G.I. Bill also provided money for new veterans hospitals and guaranteed that returning veterans who were having trouble finding jobs would receive unemployment compensation of twenty dollars per week for up to one year. Also, veterans who wanted to buy a home or farm or to set up a new business could receive low-interest loans.

## Consequences

In passing the G.I. Bill, Congress was guaranteeing educational opportunity to more than 15 million citizens. The responsibility for that education would be shared by the individual, the school, and the national government. Not surprisingly, the new law had far-reaching effects.

During the war, many colleges and universities struggled to survive, for large numbers of students and faculty left to participate in the war effort. The war years were especially hard for small private colleges.

When the veterans began returning, however, a flood of students quickly swamped public and private institutions throughout the United States. Schools that had been facing bankruptcy soon found themselves erecting new dormitories and other buildings. Special dormitories for married students sprang into existence.

At the crest of this flood in 1947, there were 1 million veterans out of a total student population of 2.5 million. In 1952, a new G.I. Bill was passed for veterans of the Korean War, and later there was a similar law to help veterans of the Vietnam War.

*George Q. Flynn*

# Arab League Is Formed

*A number of Middle Eastern states banded together to promote their Arabic cultural unity and discuss matters of concern in their region of the world.*

**What:** International relations
**When:** September 25, 1944-March 22, 1945
**Where:** Alexandria and Cairo, Egypt
**Who:**
ROBERT ANTHONY EDEN (1897-1977), British secretary of state for foreign affairs from 1940 to 1945
MUSTAFA EL-NAHAS PASHA (1876-1965), prime minister of Egypt from 1942 to 1944
NURI AS-SAID (1888-1958), prime minister of Iraq
ABDULLAH IBN HUSEIN (1882-1951), emir of Transjordan from 1921 to 1946
SHUKRI AL-KUWATLI (1891-1967), president of Syria from 1943 to 1949

## Proposals for Arab Unity

After World War I, the British and French occupied the Levant and the Fertile Crescent in the Middle East. Arab leaders in that area had already been working for unity among the various Arab states, but that goal was frustrated by the presence of Europeans.

The situation began to change, however, in 1940, when France surrendered to Germany and had to let go of some of its power in the Middle East. The British had already decided to encourage Arab unity by supporting regional organizations—though British officials still wanted to influence those organizations.

Realizing that Middle Easterners were chafing under British control, Axis leaders encouraged Arab nationalists to seek independence. In Iraq there was a brief fascist uprising; though it failed, it made British leaders realize that they needed to rethink their policy in the Middle East.

On May 29, 1941, Robert Anthony Eden,

Great Britain's secretary of state for Foreign Affairs, spoke of his country's "long tradition of friendship with the Arabs" and said that Great Britain would support independence for the French mandates of Syria and Lebanon. "It seems to me both natural and right," he continued, "that the cultural and economic ties between the Arab countries, and the political ties too, should be strengthened."

In November, 1942, Mustafa el-Nahas Pasha, prime minister of Egypt, called on the Arabs to form a "powerful and cohesive bloc." The next month, Iraqi prime minister Nuri as-Said visited Cairo and announced that he was ready to work for Arab unity.

In February, 1943, Eden again expressed Great Britain's willingness to support an organization for Arab unity, though he mentioned that the British had not received any specific proposals. In response, Emir Abdullah of Transjordan (which later became Jordan) proposed that a "United Syrian State" be formed by the combination of Syria, Lebanon, Palestine, and his own country. Another plan was to include Iraq in an "Arab Federation" along with the other countries.

Other Arab leaders were not especially interested in Abdullah's plan. President Shukri al-Kuwatli of Syria declared that he would reject any unification plan that did not call for all the Arab states to be united: "Syria will refuse to have raised in her sky any flag higher than her own, save that of an Arab Union."

Egypt also hoped to take the lead in the movement for Arab unity; in the spring of 1943, Arab leaders interested in unification were invited to come to Cairo. In July and August, Nuri of Iraq visited the Egyptian capital after brief talks with Arab leaders in Syria, Lebanon, Transjordan, and Palestine. With Nahas, he called for a conference on Arab unity to be held in Egypt.

In fall, 1943, Nahas was also visited by representatives of Transjordan and Saudi Arabia. Lebanon expressed its interest in January, 1944, although (partly because of its large Christian population) Lebanese officials stressed that each country must keep its independence.

### The Conferences

The next step was the General Arab Conference in Alexandria, which began on September 25, 1944, and was attended by the prime ministers of Egypt, Transjordan, Syria, Iraq and Lebanon. Delegates from Saudi Arabia and Palestine and an observer from Yemen also arrived.

The participants expressed many different opinions, so that the conference could not produce a final answer to the question of Arab unity. The delegates considered proposals for a unitary state with central political authority and for a federated state with more independence for member nations, but several did not like the idea of giving up their national sovereignty and merging into a larger state.

Though Syria supported the ideal of a union of all Arab countries, it was suspicious of plans such as that of Abdullah to bring only certain Arab states together. Lebanon preferred a loose confederation that would involve no political or military obligations, and Yemen agreed.

Egypt attempted to make a compromise: Each state would keep its sovereignty within a league of Arab nations. Resolutions for cooperation among the Arab states would have to be approved unanimously—except that disputes among individual Arab states would be resolved by a majority vote of the other league members. With minor changes, this plan was approved by Iraq.

*President Franklin D. Roosevelt (seated, upper right) meets with Arab leaders on a U.S. warship in February, 1945, before the creation of the League of Arab States.*

At the conclusion of the Alexandria conference on October 7, 1944, delegates of the participating states signed a protocol, or document, expressing their intention to form a League of Arab States. The league would be led by a council, on which each state would be equally represented. The council would mediate disputes between member states, and the league would work for cooperation on economic and cultural matters.

After the Alexandria conference, however, the prime ministers of Egypt, Syria, and Transjordan were all removed from their positions,

**966**

and some Egyptian leaders were quite critical of the proposed league. Leaders of the Lebanese Christian community claimed that the league would weaken Lebanon's sovereignty.

Because of these criticisms, Arab leaders who met in Cairo in the early spring of 1945 considered only the formation of a loose confederation. The Arab League Pact, signed on March 22, 1945, confirmed most of the Alexandria protocol and established a permanent secretary general for the organization. The new league would meet twice each year. Unlike the Alexandria protocol, however, this agreement did not mention the goal of bringing the Arab nations into a closer unity in the future. Instead, it emphasized that each nation would remain sovereign and independent.

## Consequences

Since it was established, the Arab League has been a symbol of the common national ties of the Arab peoples. Middle Eastern and international affairs have been discussed at Arab League meetings over the decades. One important result in the early 1970's was the oil embargo imposed by a coalition of Arab Middle Eastern states. This unified action allowed Arab nations to gain increased power in the world economy.

# Allies Gain Control of Pacific in Battle for Leyte Gulf

*The Battle for Leyte Gulf, in which the Japanese first used kamikaze warplanes, left the Japanese navy unable to continue its involvement in World War II battles.*

**What:** War
**When:** October 19-25, 1944
**Where:** Leyte Gulf, Surigao Strait, and the San Bernardino Strait, off the Philippine Islands
**Who:**
WILLIAM F. HALSEY (1882-1959), commander of the U.S. Seventh Fleet
TAKEO KURITA, commander of the Japanese Central Force
TAKIJIRO OHNISHI (1891-1945), commander of the First Japanese Air Fleet

## The Battle Begins

The invasion of Leyte Island in October, 1944, was the first step in an Allied campaign to liberate the Philippine Islands from the Japanese. Lying in the middle of the Philippine archipelago, Leyte was a particularly important island.

The Japanese planned to defend the island through air attacks against the invading fleet, forces and equipment to fight off the Americans on land, and strikes by the Imperial Navy against the Allied beachhead and its forces. The Home Fleet was Japan's most powerful weapon, and it would try to trap the Allied invasion army on the beaches of Leyte.

The Allied armada that moved toward Leyte in mid-October was made up of more than 700 ships, including 157 warships. The U.S. Third Fleet was also available to support the invasion. Under the command of Admiral William F. Halsey, the Third Fleet was given two tasks—to provide cover for the Leyte landings and, if the opportunity came, to destroy the Japanese fleet.

Opposing the Allied forces were three Japanese naval units. The Northern Force, consisting of five battleships, twelve cruisers, and numerous destroyers, was assigned to decoy Halsey's Third Fleet toward the north, away from the beaches. The Central Force would slip through the San Bernardino Strait and combine with the third group, the Southern Force, sailing through the Surigao Strait to attack the Allied fire-support ships off Leyte and the beachhead itself.

The main strength of the Japanese fleet was its battleships, since its carriers and land-based aircraft had mostly been destroyed in earlier battles and raids. The Japanese had decided that their only chance for success against the Leyte invasion was to maneuver their huge battleships into positions where they could shell the Leyte beaches.

The first contact between the enemy forces was made on October 23. American submarines sighted the Central Force and sank two cruisers, one of which was the flagship *Atago*. When Halsey received reports of another Japanese armada, the Southern Force, he and his fellow officers reacted quickly. The fire-support portion of the Seventh Fleet, commanded by Vice Admiral Thomas C. Kinkaid, moved into a blocking position at the lower end of Leyte Gulf to keep the Japanese from pushing through Surigao Strait. Seventh Fleet escort carriers guarded the entrance to Leyte Gulf, and Halsey's carriers prepared to launch air strikes against the enemy units that were streaming through the San Bernardino Strait.

## The Japanese Gamble Fails

As the Americans made their preparations, three waves of land-based Japanese aircraft struck the U.S. carriers. This began the first phase of the Battle for Leyte Gulf: the Battle of the Sibuyan Sea. The Japanese damaged the carrier *Princeton* so badly that it had to be scuttled. American carrier planes, however, found the Central Force, sank the huge battleship *Mushashi*, and damaged several other ships.

Nevertheless, the Central Force kept moving ahead. On the afternoon of October 24, Halsey stopped the attack on the Central Force in order to chase what seemed to be a better target. American scout planes had sighted the Northern Force, and Halsey believing it to be the most powerful Japanese threat, turned his carrier task forces northward—just as the Japanese had intended. It was a serious mistake for the Americans, for it allowed the Japanese Central Force to sail undisturbed through the San Bernardino Strait toward the landing area.

Late in the evening of October 24, battleships and cruisers of the Seventh Fleet began attacking the Southern Force. The Japanese were defeated badly. Only two ships of their leading group sur-

vived, and the rest fled under heavy sea and air attacks that continued the next morning.

In this phase, the Battle of Surigao Strait, the Allies won a great victory at little cost to themselves. When this part of the battle was over, the only survivors of the Southern Force were five destroyers and a heavy cruiser.

Early on October 25, however, the Central Force emerged from the San Bernardino Strait, heading for Leyte Gulf, and surprised the Americans. The slow American escort carriers, skillfully commanded by Rear Admiral Thomas L. Sprague, fought a brave but apparently hopeless battle. One escort carrier was sunk by enemy shellfire, and Japanese guns scored hit after hit.

Suddenly, however, Admiral Takeo Kurita, who commanded the Central Force, broke off all contact with the battered carriers and retreated north toward the San Bernardino Strait. Because of this puzzling action, the American transports and troops at the beachhead escaped destruc-

*General Douglas MacArthur (center) leads the troops ashore at Leyte.*

tion. Admiral Kurita later explained that he decided to turn away from his goal because he believed that the Northern Force's decoy maneuver had failed, and that a large task force of Allied carriers was preparing to trap his force in the gulf and destroy it.

The same day, Vice Admiral Takijiro Ohnishi, who commanded Japanese air forces based on Luzon, another Philippine island, decided to salvage something from the Japanese defeat. He ordered into battle the first kamikaze squadrons—special suicide units whose pilots allowed their planes to dive and crash into American ships. This type of attack was so unexpected that one escort carrier was sunk and others were seriously damaged.

Soon Halsey's Third Fleet caught the Japanese Northern Force off Cape Engaño. Now greatly outnumbering the Japanese in ships and planes, the American force sank four Japanese aircraft carriers and a destroyer; many other ships were badly damaged as well. Undoubtedly, the whole Northern Force would have been destroyed if Halsey had not received urgent appeals to turn back and intercept the Central Force. The Third Fleet failed to catch up with Kurita, however, and the Northern Force was also able to get away.

**Consequences**

Though the Central and Northern forces had escaped, these units of the Japanese Home Fleet were never again able to strike. The Battle for Leyte Gulf weakened the Japanese navy so much, through its losses of ships and morale, that it could no longer mount a real challenge to Allied control of the Pacific.

The Japanese had been encouraged, however, by the first successes of their kamikaze attacks, and they used this form of warfare again during later World War II battles.

*Theodore A. Wilson*

# Blalock Performs First "Blue Baby" Operation on Heart

*Alfred Blalock developed the first surgical method of correcting cyanosis, which is caused by congenital defects in the heart.*

**What:** Medicine
**When:** November 29, 1944
**Where:** Baltimore, Maryland
**Who:**
ALFRED BLALOCK (1899-1964), an American surgeon and physiologist
HELEN BROOKE TAUSSIG (1898-1986), an American physician
SANFORD EDGAR LEVY (LEEDS) (1909-    ), an American physician
VIVIEN THOMAS (1910-1985), a surgical laboratory technician

## Heart Experiments

Cyanosis, often called the "blue baby" syndrome, is usually caused by congenital defects in the heart or blood vessels. (A congenital defect is a physical problem that exists at birth but is not inherited from the parents.) In a "blue baby," blood does not circulate properly, and the blood does not carry enough oxygen. Symptoms include blue lips and fingertips, shortness of breath, fainting, poor physical growth, skin problems, and deformities in the fingers and toes. Before the 1940's, children born with these problems often died at an early age; if they did survive, they suffered much pain. About seven out of every thousand babies are born with some kind of congenital heart disease.

Alfred Blalock was a skillful surgeon who studied the circulatory system. At Vanderbilt University, Blalock conducted experiments on dogs. Wanting to understand the effects of high blood pressure, his team of researchers linked the left subclavian artery (a major branch of the aorta) to the left pulmonary artery (which is connected to the lungs). They found that this did not in-

crease blood pressure in the lungs very much. Five years later, Blalock would use a similar technique to correct blue baby disease.

Blalock did another surgical experiment on dogs, connecting the aorta to the left subclavian artery to correct blockage in the aorta. This operation was also successful, but Blalock did not feel ready to try it on human beings. He was afraid that clamping major blood vessels during surgery would cut off circulation to the brain and other organs for too long. In 1942, however, a Swedish surgeon, Clarence Crafoord, reported that clamping the aorta for twenty-eight minutes during surgery did not cause damage. This was reassuring to Blalock.

## The "Blue Baby" Operation

During a conference on children's medicine at The Johns Hopkins University, the cardiologist Helen Brooke Taussig asked whether Blalock's procedure could be used to improve lung circulation in children with congenital heart defects. Taussig believed that children suffering from a condition known as "tetralogy of Fallot" had a narrowed pulmonary valve and artery and that this caused poor pulmonary circulation, so that not enough oxygen passed through the body.

Blalock and Taussig began working together to design a way to join the left subclavian artery to the pulmonary artery in children so that more blood would flow to the lungs. With Thomas's help, they performed more experiments on dogs to test the procedure.

On November 29, 1944, Blalock performed the first blue baby operation on a fifteen-month-old girl who suffered from tetralogy of Fallot. Taussig and Vivien Thomas assisted him, along with resident surgeon William P. Longmire and anesthesiologist Merel Harmel. The surgery

**971**

lasted three hours. Blalock clamped the left subclavian artery, cut through it several centimeters away from where it branched off the aorta, and tied off the useless upper end. He pulled the now-open end down toward the left pulmonary artery, which had also been clamped, and made an opening in the wall of the left pulmonary artery so the two could be stitched together. When the clamps were released, blood flowed out of the aorta through the left subclavian artery and into the left pulmonary artery. The lungs quickly began receiving a greater flow of blood.

The child soon was much healthier than she had been before. Sadly, she died nine months later. Yet Blalock and Taussig were encouraged to find that their surgical treatment could work, and within two months after the first blue baby surgery, they performed two more. By December, 1945, Blalock had performed sixty-five of these operations, with a success rate of 80 percent. Doctors came from all around the world to learn how to perform the new surgery. Blalock

was praised as a hero, and newspapers published many reports of how he had saved many children who otherwise might have died.

## Consequences

The Blalock-Taussig Shunt, as the operation came to be called, has saved thousands of lives and allowed many children to lead normal lives. The procedure also led to other experimental treatments for heart problems. A synthetic shunt to connect blood vessels in this type of surgery was first used by Frank Redo and Roger Ecker in 1963; this procedure is called the modified Blalock-Taussig Shunt.

Open-heart surgery, first developed in the 1950's, is now used to correct tetralogy of Fallot and other heart deformities. Because infants are usually too small for open-heart surgery, however, the Blalock-Taussig Shunt is still used as the first step in a series of operations to treat heart problems in very young children.

*Rodney C. Mowbray*

# Allies Defeat Germany at Battle of the Bulge

*After Germany's defeat at the Battle of the Bulge in Belgium, the Nazis' final surrender was only a matter of time.*

**What:** War
**When:** December 16, 1944-January 28, 1945
**Where:** Ardennes, Belgium
**Who:**
ADOLF HITLER (1889-1945), chancellor of Germany from 1933 to 1945
KARL RUDOLF GERD VON RUNDSTEDT (1875-1953), German commander on the Western Front
DWIGHT D. EISENHOWER (1890-1969), supreme commander of the Allied Expeditionary Force
OMAR BRADLEY (1893-1981), commander of the Allied Twelfth Army Group
BERNARD MONTGOMERY (1887-1976), commander of the Allied Twenty-first Army Group

## Hitler Makes Plans

In September, 1944, German chancellor Adolf Hitler realized that the German armies were being pushed back on all fronts. In the East, the Red Army stood just outside Warsaw; it had gained back all the ground the Germans had taken in 1941 and was now threatening the Balkans. In northern Italy, it was becoming more and more difficult for the Germans to hold their defensive position, the Gothic Line.

Allied bombers were raiding German cities daily, and the Luftwaffe (German air force) had few working defensive planes left. The German navy could not move out of its ports, and only a few submarines could be fueled to go on voyages of destruction. In the West lay the greatest threat of all: the Allied army that had swept from Normandy into Belgium and stood ready to invade Germany.

At a conference with his generals in September, Hitler revealed his plans to launch a huge attack on the British and American army. On the principle that the best defense is a good offense, he proposed to throw all of his reserves into battle against the Allies.

Some questioned this strategy, warning that the Soviet Union was a greater and more pressing threat to Germany. Hitler replied that he did not have the force to move immediately against the Red Army. The Allied forces were weaker, he pointed out, with fewer men than the Red Army, and their supply lines were long. Since June they had been bringing supplies for their army from the Normandy ports. Though they had captured Antwerp, Belgium, in September, that did not solve their supply problem, for Antwerp's port was blocked by debris.

Therefore, Hitler planned a strike aimed at splitting the Allied army. The attack would begin in the Ardennes Forest in Belgium, which was not well defended, and would push forward until

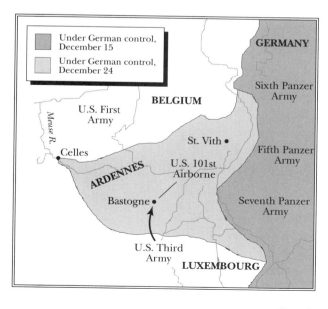

National Archives

*U.S. soldiers from the 289th Infantry march along a snowy road in Belgium in January, 1945.*

the Germans captured Antwerp. With this success, the Allies would be so devastated that their leaders would quickly agree to an armistice. Then the German armies could be turned eastward to defeat the Soviets.

Along with his army thrust, Hitler intended to use V-1 and V-2 rockets to bombard both London and Antwerp. The offensive in Belgium would have air cover from Hitler's new jet airplanes, which were much faster than any Allied warplane.

### The Battle

The German forces were gathered with great secrecy. On December 16, 1944, the weather forecast showed that the Allied planes would be hampered because of poor visibility. Quickly, the

German forces struck. The Allies were caught completely off guard, and the Germans made an important breakthrough.

Field Marshal Gerd von Rundstedt commanded the entire offensive, and two German armies—the Sixth Panzer Army, commanded by General Sepp Dietrich, and the Fifth Panzer Army, under General Hasso von Manteuffel—spread to form a pair of pincers aimed at Antwerp. The American First Army, under General Courtney Hicks Hodges, was defending the Ardennes and was the hardest hit.

General Dwight D. Eisenhower, Supreme Commander of the Allied Expeditionary Force, decided on a defensive strategy. He divided command of the Allied forces around the "bulge" created by the German thrust: The northern armies were to be commanded by General Bernard L. Montgomery, British Commander of the Twenty-first Army group, and the southern armies by General Omar N. Bradley, American Commander of the Twelfth Army Group. They were told to allow their defensive line to give way, but not to break.

In the meantime, defensive reserves would be brought in to help drive back the Germans. General George S. Patton's Third Army would swing in from the south to relieve the defenders around the bulge.

The main fighting soon centered on the town of Bastogne. American defenders of the town were besieged by Manteuffel's army, but a fortunate clearing of the skies allowed Allied planes to drop supplies. By late December, the Germans had to give up their siege of the town, and they began to withdraw.

Throughout January, 1945, Dietrich's and Manteuffel's armies were pushed back, although Hitler continued to send his last remaining reserves to the bulge. By January, 1928, the German armies were in full retreat, and the Allies had recovered all the areas they had previously lost.

## Consequences

This last major German offensive of World War II confirmed that Germany could not continue to fight. If Hitler had saved his troops to defend the Siegfried Line—the German frontier fortifications—the Germans could have put up a strong defense, and the Allies would have suffered many more casualties.

As it was, the Germans had more than 100,000 casualties in the Battle of the Bulge, compared with the Allies' 75,000. When the battle was over, German forces on the Western Front had no fuel reserves and were almost out of ammunition. Meanwhile, on the Eastern Front, German armies had not received vital supplies since September, when the Battle of the Bulge was first planned.

Hitler had no reserves left to send against either the Allied forces or the Soviets; still, he resolved to defend Germany to the very end. It was not until late April, 1945, that he gave up and committed suicide. By that time Allied forces had advanced as far as the Elbe River in Germany, and the Red Army was entering Berlin. The Germans surrendered on May 7, 1945.

*José M. Sánchez*

# Duggar Discovers First Tetracycline Antibiotic, Aureomycin

*Benjamin Minge Duggar directed the research that led to the discovery, production, and application of Aureomycin, the first all-purpose antibiotic that was both safe and effective.*

**What:** Medicine
**When:** 1945
**Where:** Pearl River, New York
**Who:**
BENJAMIN MINGE DUGGAR (1872-1956), an American botanist, professor, and consultant in mycology (the study of fungi) at Lederle Laboratories
YELLAPRAGADA SUBBAROW (1896-1948), an Indian American biochemist
ALBERT CARL DORNBUSH (1914-     ), an American microbiologist

## Back to the Soil

On July 21, 1948, at a conference arranged by the prestigious New York Academy of Sciences in the Museum of Natural History, the new antibiotic Aureomycin was introduced to the public. It was uncommon for a medical breakthrough to be trumpeted to the world as this drug was. About twenty clinicians were present to broadcast the spectacular results they had achieved with it. Also on hand were several scientists from Lederle Laboratories, where Aureomycin had first been isolated by Benjamin Minge Duggar.

Before the auspicious debut of Aureomycin, several antibiotics were already available to health professionals. By far the most effective of the new "miracle drugs" was penicillin, successful against 40 percent of bacteria-caused diseases, and streptomycin, effective against 30 percent. These therapeutic agents had given physicians the weapons they needed to conquer many terminal and disabling infections. Nevertheless, many dangerous infections refused to surrender to either agent,

and these now stood out. Several were caused by viruses and by small, viruslike bacteria.

The discovery of penicillin, a drug derived from molds formed in the soil, had sent scientists back to the soil in a quest for natural substances with antibiotic properties. Lederle Laboratories was a pharmaceutical company that specialized in drugs for treating infectious diseases. Its success owed much to the biochemist Yellapragada SubbaRow, who had joined Lederle in 1940. As director of research, he supervised Lederle's chemical, medicinal, and pharmaceutical research. He was responsible for bringing Duggar to Lederle. Mindful of Duggar's extensive knowledge of molds, SubbaRow invited the former professor of several of Lederle's scientists to take a position with the company as a consultant in mycological research and production.

## Out of Retirement

Although Duggar's name did not become familiar outside his field until 1948, his work had been known to botanists for a long time. For several decades, he had been recognized internationally as an authority on molds and fungi. Nevertheless, upon reaching the age of seventy, he was forced to retire from teaching. Still quite active physically and mentally, Duggar was not content to live in retirement, especially in the midst of a devastating world war. Hence, in 1944, he accepted SubbaRow's offer to join Lederle.

At first, Duggar's work at Lederle was concerned with plant sources of antimalarial drugs. He was impressed deeply by the success of penicillin, which had just come into widespread use. He perceived that in the field of antibiotics the surface had only been scratched. He soon initiated an immense soil-screening program. His ob-

**976**

jective was to discover a superior antibiotic—one that would combat diseases that completely resisted penicillin and other available antibiotics.

Lederle scientists began their project by gathering from all over the country more than six hundred soil samples, which they screened for strains that might have microbe-killing potential. The molds were tested by putting them in petri dishes, along with specific microorganisms, and then observing their ability to inhibit the growth of neighboring organisms.

The screening was a long, tedious task. Duggar and his team suffered disappointment after disappointment. More than thirty-five hundred strains were scrutinized and rejected.

Eventually, the investigators came to a petri dish labeled "A 377," which contained a golden-colored mold obtained from a soil sample taken from the campus of the University of Missouri. It was one of the several samples sent to Duggar in the summer of 1945 by his former colleague, William Albert Albrecht, chairperson of Missouri's Department of Soils. Duggar was pleased to observe that A-377 exhibited antibiotic potential.

Test-tube experiments were conducted in September, 1945, to gauge the mold's effectiveness against some fifty pathogenic organisms. Albert Carl Dornbush, from the University of Wisconsin, was in charge of the important in vitro work. The results were astounding. The mold arrested the growth of staphylococci, streptococci, and bacilli. The fact that the mold, later named *Streptomyces aureofaciens* by Duggar, resisted bacilli signified that it was producing an antibiotic that might be able to treat more diseases than either penicillin or streptomycin. The antibiotic substance extracted from *Streptomyces aureofaciens* was christened "Aureomycin," a name derived from the Latin word *aureus* (gold) and the Greek *mykes* (fungus). Both the mold and the antibiotic had a golden hue. Aureomycin was later given the generic name of "chlortetracycline." The tetra-

cyclines are a family of antibiotics that, next to penicillin, many consider to be the safest and best of the "wonder drugs."

## Consequences

Antibiotics revolutionized the therapy of infectious diseases in the 1940's, and Aureomycin brought extraordinary assets to the new age of medicine. One was its versatility. Possessing a much broader application than penicillin or streptomycin, it was effective against 90 percent of bacteria-caused infections. Also, unlike the two premier antibiotics, which were usually administered by needle, Aureomycin was effective when taken by mouth. Consequently, it could be administered quickly and painlessly either in the physician's office or in the patient's home.

Most important, Aureomycin proved to be remarkably effective against certain infections that had failed to respond to either penicillin or streptomycin. One of these illnesses was Rocky Mountain spotted fever, recognized as one of the most severe of all infectious diseases. The tick-transmitted malady killed one out of every five victims. The number of fatalities declined sharply, however, after Aureomycin became available. In its first clinical trials, the drug dramatically restored to health a boy who was in a coma from tick fever. Numerous other diseases caused by atypical bacteria yielded to the new microbe-fighter.

Since 1948, Aureomycin and other tetracyclines have been widely—and often inappropriately—used to treat a wide range of diseases. This unrestricted application caused many pathogenic microorganisms, notably streptococcus strains, to become resistant. As a result, tetracyclines are no longer the antibiotics of choice for treating most common respiratory or urinary tract infections.

*Ronald W. Long*

# Americans Return to Peace After World War II

President Harry S. Truman led the United States in trying to avoid unemployment and inflation as the nation struggled to return to a peacetime economy.

**What:** Economics; Social change
**When:** 1945-1946
**Where:** United States
**Who:**

HARRY S. TRUMAN (1884-1972), president of the United States from 1945 to 1953

JOHN W. SNYDER (1895-1985), director of the Office of War Mobilization and Reconversion

## The Challenge of Peace

World War II came to an end in August, 1945. Yet the victory celebrations in the United States were hardly over before some Americans began to worry about the effects of peace. More than twelve million Americans in the armed forces were eager to come home. For five years the nation had been geared for war; the American gross national product had more than doubled since 1940, and a majority of that product had been war equipment and supplies. Because factories had converted to wartime production, consumer goods such as clothing, shoes, and household items had been scarce or rationed during the war. Americans had saved record amounts to buy those goods once they were again available.

Though peace was welcomed by Americans, it brought with it ominous economic questions. Would there be enough jobs for the returning members of the armed forces? Could factories convert from wartime to peacetime production fast enough to supply the demand for consumer goods? Could Americans hold back on spending long enough to prevent a dangerous rise in prices and wages? If not, what would prevent a return to the breadlines of the 1930's Great Depression?

In the late summer of 1945, Harry S. Truman was an untried president. He had served as a senator during Franklin D. Roosevelt's New Deal and had spent a few months as president as World War II came to an end. Yet he served in the shadow of his predecessor, Roosevelt.

How would Truman respond to the challenge of peace? His actions would test his abilities as a president, as well as the Democratic Party's commitment to continuing Roosevelt's New Deal.

In August, 1945, John W. Snyder, director of the Office of War Mobilization and Reconversion, prepared a report on the steps needed to win a "peacetime victory." According to Snyder, adjusting to peace would indeed be difficult. His report predicted "dislocation" in the economy and high unemployment.

## The Fair Deal

Following Snyder's advice, the president called Congress into special session and began hammering out the first parts of what would later be known as the Fair Deal. On September 6, 1945, Truman asked Congress to help "hold the line" in the American economy through continued government control of prices and wages. He also requested "full employment" laws, a new housing program, an increase in the minimum wage, and more.

When Truman finished his sixteen-thousand-word message to Congress, no one could doubt where he stood. As one member of Congress said, "It's just a plain case of out-dealing the New Deal."

Congress responded by passing the Full Employment Act; though it was weakened by a number of amendments, it did place the responsibility for employment in the hands of the federal government and the president's Council of Economic Advisers. Yet Truman had to work hard to get other reforms. He lacked Roosevelt's prestige, and the changes he was proposing did not seem as urgent as the New Deal measures; he was trying to prevent another depression rather than trying to stop one that was already happening.

During World War II, the Office of Price Administration (OPA) had had major responsibility for regulating the economy. To prevent soaring inflation of prices and wages, Truman asked that OPA's powers be extended for eighteen months. Late in 1945, Congress granted a six-month extension, and through the spring of 1946 OPA's fate was debated.

OPA director Chester Bowles displayed charts, graphs, and statistics to show that his agency was still needed, while Republicans led by Robert A. Taft and lobbyists led by the National Association of Manufacturers argued that OPA should be closed down. Finally, an OPA bill went to the president in late June, carrying so many amendments that it was practically worthless. Truman vetoed it, accepted the fact that he had failed to persuade Congress to continue government regulation, and watched prices climb.

As the fight for OPA raged in Congress, the Truman administration had faced a challenge from organized labor over "holding the line" on wages. As returning servicepeople took advantage of the so-called G.I. Bill of Rights, passed in 1944, and reclaimed their former jobs, there were suddenly many more workers. Faced with less overtime pay and higher prices, workers began demanding higher wages.

President Truman had always been known as a "friend of labor," but now he desperately needed to keep wages down so that prices would not rise even more. When government mediators became involved in contract disputes be-tween management and labor, Truman urged that wages be raised only slightly.

On May 18, 1946, Alexander F. Whitney, president of the Brotherhood of Railroad Trainmen, and Alvanley Johnston, president of the Brotherhood of Locomotive Engineers, called on their union members to strike; this action was sure to bring the nation's railroads to a standstill. Johnston and Whitney agreed to delay their strike, but they did not back down on their demands.

Faced with economic paralysis, President Truman was furious. He went before Congress to ask for the right to run the railroads by military command. While he was still speaking before Congress on May 25, 1946, however, word came that Johnston and Whitney had signed a contract; the strike had been avoided.

*A returning sailor kisses a woman in Times Square in New York City on August 14, 1945.*

Digital Stock

During 1946, the Truman administration involved itself with strikes in the oil, automobile, steel, and meat-packing industries. Truman's greatest trouble with labor, however, came with the coal miners led by John L. Lewis, president of the United Mine Workers' Union. The government seized the mines for a time to bring about one settlement; when the miners remained on strike against the government, the administration tried to get a court order to force them back to work. After a long court battle, Truman finally reopened the mines.

**Consequences**

In the end, Truman was helped by time and the flexibility of the American economy. The administration had failed to "hold the line" on prices and had not seen great success in keeping wages low. Although in the spring of 1946 about eight million Americans were unemployed, unemployment never became as severe as it had been in the Great Depression. Inflation was not too extreme, and soon a return to peacetime production made consumer goods more available.

In the congressional elections of 1946, the Democrats suffered some losses because of their policies of "reconversion." On the other hand, Truman had proved himself as a president, the New Deal still lived, and the nation had geared itself for peace.

# Marshall Mission Attempts to Unify China

Sent to China to prevent a civil war, George C. Marshall worked hard to bring the Nationalists and the Communists together but did not succeed.

**What:** International relations
**When:** 1945-1947
**Where:** China
**Who:**
GEORGE CATLETT MARSHALL (1880-1959), president Harry S. Truman's special representative in China
CHIANG KAI-SHEK (1887-1975), head of the Nationalist Party (Kuomintang)
MAO ZEDONG (1893-1976), chairman of the Chinese Communist Party

## Marshall's Mission

After leaving Washington, D.C., for his first vacation after six hard years as United States Army chief of staff during World War II, General George C. Marshall had been home in Leesburg, Virginia, only a few minutes before he received a telephone call from President Harry S. Truman. "General," said the president, "I want you to go to China for me."

Truman wanted to send Marshall as his special representative to help bring about the "unification of China by peaceful, democratic methods." Few people would have accepted this almost impossible mission, but Marshall packed his bags and left for China on December 15, 1945.

Marshall realized that he would face a very difficult situation in China. Since the United States had entered World War II, military and diplomatic officials had kept him informed about China's internal problems. The main antagonists threatening to plunge China into a civil war were the Chinese Communist Party, led by Mao Zedong, and the Nationalist Guomindang (Kuomintang) Party, under the leadership of Chiang Kai-shek. These two political parties had been fighting each other since the 1920's.

Even after the Japanese invaded China, Chiang had considered the Communists a greater threat; he was determined to eliminate them before he began to fight the Japanese. Eventually, however, he had been kidnapped by some of his own troops and forced to agree to uniting with the Communists to fight Japan. This brought about an uneasy truce between the two parties.

Through the years of World War II, however, the Communists and Nationalists continued to build up their own strength; each intended to seize power after the war. After the Japanese surrender, both sides scrambled to take control of as much territory as possible and to accept the surrender of Japanese weapons and supplies for their own use. The Nationalists managed to gain much of central, southern, and eastern China, while the Communists were strong in northern China and parts of Manchuria. Civil war was in the making, and Marshall's job was to mediate a peace settlement.

## A Frustrating Task

Since both sides at first claimed that they were willing to negotiate, Marshall suggested that they try to reach agreement on three goals: a cease-fire, a conference to decide how to form a coalition government, and a merger of the Nationalist and Communist forces into a national army.

Both sides agreed to an immediate cease-fire, and the two parties began once again to communicate with each other. They also agreed to hold the Political Consultative Conference. In the meantime, an Executive Headquarters composed of one Nationalist, one Communist, and one American was organized to deal with the many problems involved in taking these actions. Teams were also set up to settle any cease-fire disputes that might arise.

As Marshall pleaded, threatened, and persuaded, the two sides reached two major agree-

Library of Congress

*Chiang Kai-shek.*

Relieved by this success, Marshall decided to return to the United States on March 11, 1946, to arrange a loan for China from the Export-Import Bank. All throughout the negotiations, however, the Nationalists and the Communists had been preparing their troops to strike in case the other side should betray the peace process.

Taking advantage of Marshall's absence, each side ordered its troops to begin fighting to gain new territory. Marshall returned to China but could not get either side to come back into serious negotiations.

Finally, in January, 1947, Marshall decided that the situation was hopeless and asked President Truman to recall him to the United States.

**Consequences**

Once the negotiations had failed, a storm of blaming broke loose. Marshall blamed the Communists and Nationalists equally for having failed to come to a real agreement. The Nationalists accused the United States of trying to force them to give in to the Communists politically when they could have beaten Mao's forces easily in combat. Meanwhile, the Communists said that the United States had used the negotiations as a smoke screen while actually helping Chiang Kai-shek to continue the war against them.

In the civil war, the Communist forces soon gained the upper hand. After the Nationalists were finally driven out of mainland China in 1949, Marshall was blamed once again. This time the accusations came from U.S. senator Joseph R. McCarthy and his supporters, who loudly accused Marshall and the Truman administration of selling out the Nationalists and "giving" China to the Communists.

*Beatrice Spade*

ments. At the Political Consultative Conference, which included not only Nationalist and Communist representatives but also independent delegates, the members agreed that the country should be ruled by a State Council that would include members from all major political parties; it would have both legislative and executive powers. Half of the State Council's members would come from the Nationalist Party, and the rest from other parties. The Communists and Nationalists also reached an agreement about how to reduce their armies and combine the forces that remained into one army.

# Municipal Water Supplies Are Fluoridated

*The artificial fluoridation of municipal water supplies to prevent dental decay began in 1945.*

**What:** Medicine
**When:** January, 1945
**Where:** United States
**Who:**

FREDERICK S. MCKAY (1874-1959), an American dentist

HERMAN R. CHURCHILL, an American chemist

H. TRENDLEY DEAN (1893-1962), a U.S. Public Health Service employee who promoted research into dental fluorosis and tooth-decay reduction

## Mottled Teeth

Since 1945, the fluoridation of water supplies has shown how politically and socially controversial "science" can be. Proponents of fluoridation have insisted that it inhibits tooth decay with little or no health risk. Their opponents have accused them of being Communist agents, of attempting to deny people the freedom of choice, and of destroying the environment and endangering people's health.

Doctors had long been aware of mottled, or spotted, teeth (a condition known as "dental fluorosis") in the mouths of patients, but it was not known what caused the condition. In 1901, Frederick S. McKay, a graduate of the University of Pennsylvania Dental School, began a lifelong search for the cause of dental fluorosis. He began his research in Colorado Springs, Colorado. By 1916, McKay had come to believe that something in the water was the cause. Over time, he also noticed that many patients with mottled teeth nevertheless had exceptionally healthy teeth. By 1928, he had become convinced that mottling and the absence of tooth decay were somehow related.

McKay and other investigators finally recognized fluorine as the cause of mottling when studying the teeth of Bauxite, Arkansas, residents who grew up after the town's water-supply source was changed in 1909. Officials of the nearby Aluminum Company of America (ALCOA) became concerned that the company might somehow be responsible and set about to study the problem. In 1931, ALCOA's chief chemist, Herman R. Churchill, wrote to McKay that the high levels of fluorine in the area's drinking water could be the cause. McKay verified this in samples sent from Colorado Springs and sites in South Dakota and Idaho.

H. Trendley Dean of the U.S. Public Health Service eventually demonstrated a clear connection between naturally fluoridated water and the absence of tooth decay. He called for a full study of the effects of adding fluorides to water supplies. In the early 1940's, Dean led a research team that determined the fluorine level that was high enough to inhibit tooth decay but not high enough to cause mottling: one part fluorine per million parts water.

After toxicity studies indicated that levels of fluorine at 1 to 1.5 parts per million did not harm human health, field tests were instituted in Grand Rapids and Muskegon, Michigan. Sodium fluoride was added to the Grand Rapids water supply; Muskegon served as the experimental "control," meaning that its water supply was left unaltered in order to provide a comparison with Grand Rapids. The study was scheduled to last ten years.

## Plowing Ahead

As early as 1941, however, a determined group of Wisconsin dentists decided that the fluoridation of water supplies was too beneficial to public health to wait for ten-year studies to be completed. They formed the Fluorine Study Committee and enlisted several prominent backers. On March 19, 1945, two months after the Michigan study began, the Wisconsin dental society

**983**

agreed that fluorine should be added to all Wisconsin water supplies. By April, 1945, the State Board of Health decided to allow fluoridation experiments in Wisconsin.

Until 1950, the U.S. Public Health Service chose to await the study results, recommending the direct application of fluorine to teeth rather than the fluoridation of water supplies. In June, 1950, however, bowing to public pressure, the Public Health Service reversed its position and recommended that communities fluoridate. The American Dental Association followed in November, 1950, and in December, 1951, the American Medical Association endorsed fluoridation. Once the leading health organizations accepted the value of fluoridation, the process became a standard health procedure. Eight states made fluoridation mandatory; in the other states, approximately 40 percent of the water districts were fluoridated by the 1990's. Worldwide, about 250 million people drink fluoridated water.

## Consequences

Almost everything connected with fluoridation has remained controversial. The average number of decayed, missing, and filled teeth in American school-age children has declined from seven in 1945 to three. Foes of fluoridation, however, claim that this decline has resulted from such factors as improvements in nutrition, oral hygiene, and dental care rather than from the fluoridation of water supplies. Proponents claim that fluoridation reduces tooth decay by as much as 25 percent, but some researchers insist that no significant difference exists in the tooth-decay rates of those who drink fluoridated water and those who do not.

The possible economic benefits of fluoridation have also caused debate. The National Institute of Dental Research has stated that each dollar spent on fluoridation (which typically costs twenty cents to fifty cents per person per year) reduces dental costs by fifty dollars. Opponents, however, have argued that when actual costs of dental care are compared, no benefits are apparent for drinkers of fluoridated water.

In 1977, the National Academy of Sciences issued a report that recommended additional research regarding the possible connection of fluoride with dental fluorosis and with other health problems, including cancer, birth defects, kidney disease, and genetic mutations. Proponents argue that claims that fluoridation causes serious illness have not been proven. Opponents counter that other health benefits claimed for fluoridation, such as a reduction in the incidence and severity of osteoporosis, have not been proven either.

By the latter decades of the twentieth century, some environmental organizations had begun to include fluoride in lists of pollutants. In 1986, the National Resources Defense Council sued the Environmental Protection Agency in order to block the agency's institution of a maximum legal level of four parts per million for fluoride in drinking water. Although the agency's action was upheld in court, environmentalists continued to call for further studies.

*William F. Steirer, Jr.*

# Battle of Germany Ends World War II in Europe

> *Pushing into Germany from the west as the Red Army entered from the east, Allied forces defeated the Nazis and brought World War II to an end in Europe.*

**What:** War
**When:** January-May, 1945
**Where:** Germany
**Who:**

ADOLF HITLER (1889-1945), chancellor of Germany from 1933 to 1945

KARL RUDOLF GERD VON RUNDSTEDT (1875-1953), German commander in chief on the Western Front until March 10, 1945

ALBERT KESSELRING (1887-1960), German commander in chief on the Western Front, appointed to succeed Rundstedt on March 10, 1945

DWIGHT D. EISENHOWER (1890-1969), supreme commander of the Allied Expeditionary Force

BERNARD MONTGOMERY (1887-1976), commander of the Allied Twenty-first Army Group

OMAR BRADLEY (1893-1979), commander of the Allied Twelfth Army Group

GEORGI KONSTANTINOVICH ZHUKOV (1895-1974), commander of the First White Russian Front

KONSTANTIN K. ROKOSSOVSKI (1896-1968), commander of the Second White Russian Front

IVAN STEPANOVICH KONEV (1897-1973), commander of the First Ukrainian Front

## The Soviets Advance

At the beginning of 1945, Adolf Hitler's Third Reich tottered on the brink of total defeat. In Western Europe, British, American, and Canadian forces, under the supreme command of General Dwight D. Eisenhower, were slowly pushing the Germans back from the "bulge" they had created in the front in mid-December, 1944, when they launched their surprise attack in the Ardennes region of Belgium. On January 8, realizing that his troops and armor were being overcome, Hitler authorized Field Marshal Gerd von Rundstedt, commander in chief on the Western Front, to pull back toward the Siegfried Line (fortifications the Germans had constructed along their western frontier before the war).

The next day, Hitler's intelligence officers brought in a new set of ominous reports about the serious threat in the East: the heavy buildup of Soviet forces along the Vistula River in Poland. According to the reports, the Red Army would soon launch a major attack on the Reich from their positions on the Vistula, which they had held since August, 1944. Toward the south, most of the Balkans had also fallen to Soviet armies, which now stood only 120 miles from Vienna. Yet Hitler refused to take these warnings seriously.

Only three days after Hitler was warned, the Soviet armies, organized into "fronts" (army groups), surged across the Vistula, quickly overrunning the German line of defense. Led by the First White Russian Front, commanded by Marshal Georgi Zhukov, and the First Ukrainian Front, commanded by Marshal Ivan Konev, the Soviets made rapid gains all along the battlefront that stretched from East Prussia to the Carpathian Mountains.

Zhukov's forces took Warsaw on January 17, and those of Konev crossed the German frontier in Silesia only three days later. To the north, the Second White Russian Front, under the command of Marshal Konstantin Rokossovski, swept to the Gulf of Danzig and, by January 26, had encircled all German forces in East Prussia.

Hulton Archive

*Russian soldiers raise the red flag of the Soviet Union over the ruins of the Reichstag in Berlin in April, 1945.*

Meanwhile, Zhukov crossed the German frontier into Pomerania and Brandenburg, and on January 31 his lead troops reached the Oder River—only forty miles from Berlin. Early in February, Konev moved his forces quickly across Silesia to catch up with Zhukov.

By the end of February, the Soviet push had lost its momentum. This gave the Germans time to bring reinforcements to defend the Eastern Front along the Oder and Neisse rivers. Many of these troops, however, were taken from German units holding the Western Front, where the Battle of the Rhineland had already begun.

### The Allies Advance

Eisenhower had decided to clear the entire western bank of the Rhine River before attempting to cross that last great barrier to the German heartland. This "broad front" strategy was opposed by Field Marshal Bernard L. Montgomery, the British Commander of the Allied Twenty-first Army Group. Marshal wanted to send a single powerful striking force across the Rhine, past the Ruhr Valley industrial area, and on to Berlin.

Eisenhower stuck to his plan, however, and the Battle of the Rhineland began on February 8. The major obstacle for the Allies was the heavily fortified Siegfried Line. Throughout February and March, the Allied forces, made up of three army groups, pressed steadily forward. By mid-March, Montgomery's Twenty-first Army Group had cleared the southern Rhineland, and General Omar Bradley's Twelfth Army Group had pushed through the middle.

On March 7, an American unit fighting under Bradley's command seized a bridge at Remagen which the Germans had failed to destroy and thus became the first Allied combat force to cross the Rhine. They quickly established a bridgehead on the east bank. Three days later, Hitler replaced Rundstedt with Field Marshal Albert Kesselring as commander in chief in the West.

In the south, German forces were putting up stronger resistance, but part of Bradley's forces joined with the Sixth Army Group and were able to wipe out German opposition in the Saar Basin. So it was that by March 25, the Allies had met Eisenhower's goal of clearing the region west of the Rhine.

Montgomery's forces crossed the lower Rhine on March 23-24, while other crossings were made to the south. On March 26, American units broke out of the Remagen bridgehead, and the great dash of American armies into the German heartland was under way.

Following Eisenhower's plan, Montgomery's forces on the north and Bradley's on the south encircled the Ruhr Valley in the pincer movement. The two forces met east of the Ruhr on April 1. There was now a two-hundred-mile hole in the Western Front which Kesselring could not hope to close. About 325,000 German troops were taken prisoner on April 18, when the conquest of the Ruhr was complete.

Meanwhile, American forces reached the Elbe River on April 11, and Eisenhower ordered them to halt. British prime minister Winston Churchill had urged Eisenhower to take Berlin in order to prevent the Soviets from doing so. Yet Eisenhower refused to do so, for he knew that moving into Berlin would cost the Allies many casualties—and anyway, the Allies had already agreed that after the war half of Berlin would be turned over to the Soviets.

Following his "broad front" strategy, Eisenhower decided to concentrate on destroying Nazi forces in the northern and southern regions of Germany. There was a rumor that the Nazis were going to fortify a "National Redoubt" in the Bavarian and Austrian Alps to the south and use it to make a last stand. Eisenhower took this threat seriously and sent some of his forces toward Bavaria; others moved into northern Germany, to prevent German forces from withdrawing into Denmark.

On April 16, the Soviets launched their offensive against Berlin; on April 25, the capital was surrounded. Hitler committed suicide in the ruins of the Reich Chancellery on April 30, just two days before the city fell to the Red Army. On May 7, representatives of the German high command signed a surrender.

## Consequences

World War II in Europe had finally come to an end; it had been a long and terribly bloody conflict. In the Pacific, the war with Japan did not end for another four months. Soon afterward, the Nuremberg Trials began to shock the world with revelations of the Nazis' hideous treatment of the Jews. The Holocaust became a deeply disturbing sign of human beings' capacity for evil. This war had left wounds that would not heal quickly.

*Edward P. Keleher*

# Churchill, Roosevelt, and Stalin Meet at Yalta

*In a spirit of cordiality, leaders of the United States, Great Britain, and the Soviet Union made plans together for the United Nations and the approaching end of World War II.*

**What:** International relations
**When:** February 4-11, 1945
**Where:** Yalta, Crimea, Soviet Union
**Who:**
FRANKLIN DELANO ROOSEVELT (1882-1945), president of the United States from 1933 to 1945
SIR WINSTON CHURCHILL (1874-1965), prime minister of Great Britain from 1940 to 1945
JOSEPH STALIN (1879-1953), dictator of the Soviet Union from 1924 to 1953

## The U.N. and Eastern Europe

Early in 1945, as Soviet armies advanced on Germany through Eastern Europe and American and British armies moved in western Germany, the leaders of the Allied nations met at Yalta in the Crimea, a peninsula in southern Russia. They had several goals: to think through the problems that would arise once Germany was defeated, to plan an occupation policy for the conquered nations, and to discuss the problems of the United Nations, Eastern Europe, and the Far East.

At the conference, the United States was represented by President Franklin D. Roosevelt; his closest adviser was Harry Hopkins, and Secretary of State Edward Stettinius came along with them. The British were represented by Prime Minister Winston Churchill and Foreign Secretary Anthony Eden. The Soviet Union, the host country, was represented by Joseph Stalin and Commissar of Foreign Affairs Vyacheslav Molotov.

The "Big Three"—Churchill, Roosevelt, and Stalin—had met together once before, at Tehran, Iran; however, they had postponed many decisions, agreeing to discuss them later at Yalta. For the most part, the atmosphere at Yalta was pleasant and friendly. Underneath the cordiality, however, Churchill and Stalin were suspicious of each other's motives, and Roosevelt, who died two months later, was a sick man.

Many of the decisions reached at Yalta simply ratified agreements worked out in advance by the foreign ministers. Some were reached only after much bargaining.

One issue of concern was the United Nations (U.N.). This organization had been proposed as far back as 1941, and in 1944 the foreign ministers had established its organizational structure. Yet there were conflicts. The Soviets insisted that the U.N. Security Council have a veto on all matters—even those having to do with procedures. In order to give themselves equality with the Pan-American and British Commonwealth blocs, they argued that each Soviet republic was autonomous and demanded that the Soviet Union

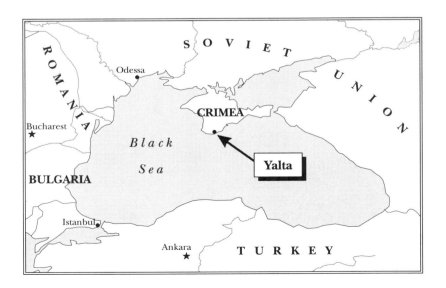

*Sir Winston Churchill, Franklin Delano Roosevelt, and Joseph Stalin (seated, from left to right) at the Livadia Palace, Yalta.*

be given sixteen seats in the General Assembly.

After much discussion, a compromise was reached: The Soviets would have three General Assembly seats. Molotov would be sent to the U.N.'s founding meeting, which would be held in San Francisco in April, 1945.

Poland was a special concern for two reasons: its frontiers and its government. When Poland was conquered by the Germans and the Soviets in 1939, a Polish government-in-exile had been established in London. Matters were further muddied after the Soviet Union joined the Allies in 1941 and began liberating Poland from Nazi rule in 1944, for Stalin then formed the Polish Committee of National Liberation—a Communist government-in-exile.

Stalin wanted the Communists to be recognized as Poland's legitimate government. He

also wanted Poland's eastern frontier to be drawn along the Curzon Line, so that the Soviet Union would regain territory that it had lost after World War I. After much discussion and bargaining between Churchill and Stalin, the Soviet leader's proposals were accepted. To compensate for the land lost to the Soviet Union, Poland's western frontier would be moved farther west, into Germany. The Communist government and the London Polish government would be brought together, and open elections would be held "as soon as possible."

The Allies agreed to help all temporary governments in other recently liberated Eastern European countries until elections could be held. At a meeting in October, 1944, Stalin and Churchill had already settled matters having to do with the Balkans. The Americans and British

**989**

would have a "sphere of influence" in Greece, and the Soviets in Romania and Bulgaria; each side would have an equal share in Hungary and Yugoslavia. The future of Czechoslovakia, Finland, and the Baltic States was not discussed.

## Germany and the Far East

The future of Germany was the knottiest problem faced at Yalta. The foreign ministers, who had been discussing the matter since 1943, made a number of recommendations. Eventually the Big Three agreed that Germany would be divided into four zones of military occupation for Great Britain, the United States, the Soviet Union, and France. An Allied "control commission" in Berlin would decide exactly how the occupation would be carried out, and Berlin itself would be divided and occupied by all four powers.

As for German reparations (payment of war damages), Stalin wanted to impose a very severe punishment for the twenty million Soviets killed in the war and the thousand Soviet towns and cities that the Germans had destroyed. He approved of a plan proposed earlier by U.S. treasury secretary Henry Morgenthau, Jr. Under this plan, all Germany's industry would be dismantled and given to the Allies as reparations; Germany would have to return to an economy based on agriculture.

At Yalta, neither Churchill nor Roosevelt favored this plan; Churchill argued that the health of Europe depended upon a prosperous Germany. Furthermore, they could not come to an agreement with Stalin on the exact amount of reparations that should be demanded from Germany. They decided only that a reparations commission would be appointed to study the problem.

Finally, the question of the Far East was discussed. Roosevelt wanted the Soviets to join the Allies in the war against Japan once the fighting in Europe had ended. He and Stalin made official the agreement they had already reached at Tehran: the Soviet Union would declare war on Japan within three months after Germany surrendered. In return, the Soviet Union would be given control over certain areas in the Far East: the Kurile and southern Sakhalin. Though Chinese leaders had not been consulted, Roosevelt also agreed that the Soviets would have privileges in Manchuria; Dairen, a port city in northern China, would be opened to international traffic, and the nearby Port Arthur would be leased to the Soviet Union.

## Consequences

The agreements made at Yalta were considered to be temporary; after World War II ended and the Cold War began, however, they became permanent. The compromises with Stalin, which had been mostly Roosevelt's responsibility, began to be criticized by Westerners who were concerned about the Soviet Union's expansion into Eastern Europe. Others have claimed, however, that at Yalta Roosevelt actually helped limit Soviet imperialism. Clearly, by insisting upon restoring Germany rather than destroying it, Roosevelt and Churchill influenced the new East-West balance of power that developed in the years after World War II.

*Burton Kaufman*

# United Nations Is Created

*As World War II drew to a close, representatives from many countries gathered to create an organization for world peace and security.*

**What:** International relations
**When:** April 26-June 26, 1945
**Where:** San Francisco, California
**Who:**
FRANKLIN DELANO ROOSEVELT (1882-1945), president of the United States from 1933 to 1945
HARRY S. TRUMAN (1884-1972), president of the United States from 1945 to 1953
CORDELL HULL (1871-1955), U.S. secretary of state from 1933 to 1944
JOSEPH STALIN (1879-1953), dictator of the Soviet Union from 1922 to 1953
SIR WINSTON CHURCHILL (1874-1965), prime minister of Great Britain from 1940 to 1945

## Planning for Peace

The League of Nations, which had been founded at the end of World War I to promote world peace, had fallen apart in the 1930's, as Nazis and Fascists gained strength in Europe. Even as World War II began, however, some world leaders were dreaming of a new organization in which nations would work together for peace and well-being. One of these leaders was Cordell Hull, American secretary of state since 1933.

Partly as a result of Hull's hard work, an important step was taken in August, 1941, when the major Allied leaders—British prime minister Sir Winston Churchill, American president Franklin D. Roosevelt, and Soviet leader Joseph Stalin—met on ships in the Atlantic Ocean to talk about how nations should behave once the war was over. Churchill wanted Roosevelt to promise that the United States would join an international political organization after the war. But Roosevelt remembered that Congress had voted against bringing the United States into the League of Nations, and he refused to promise. So the Atlantic Charter of August 14, 1941, did not include plans for an organization, but it did set guidelines for international relations in peacetime.

On January 1, 1942, Roosevelt, Churchill, the Soviet ambassador to the United States, and representatives of twenty-three other nations signed a declaration that they would continue the fight against the Axis Powers—Germany, Italy, and Japan—and, once the war was over, would abide by the Atlantic Charter. Churchill wished to call this gathering of nations "the Grand Alliance," but Roosevelt persuaded him that a better name would be "the United Nations."

During the next year and a half, Hull worked to prepare Congress and the American people to accept a new world peace organization. Then, in October, 1943, his dream came closer to becoming reality when he met with British foreign minister Anthony Eden and Soviet foreign minister Vyacheslav M. Molotov in Moscow. Together, these men issued the Moscow Declaration on General Security, which included the following statement: "[These nations] recognize the necessity of establishing . . . a general international organization, based on the principle of the sovereign equality of all peace-loving states and open to membership by all such states, large and small, for the maintenance of international peace and security."

## Building the U.N.

In August, 1944, representatives of the "Big Four"—the United States, Great Britain, the Soviet Union, and China—met in Dumbarton Oaks, near Washington, D.C., to make more specific plans for the new organization. By October, they had agreed that the United Nations would include a General Assembly led by a Security

**991**

Council, whose permanent members would be Great Britain, the United States, the Soviet Union, China, and France; they also made plans for a new World Court and an Economic and Social Council.

There were some disagreements at the Dumbarton Oaks meetings. The Soviets insisted that the permanent members of the Security Council should be able to veto any decisions that involved themselves, while the Americans and British realized that this would give too much power to a few nations. Also, the Soviet Union demanded that each of its member republics be given a vote in the General Assembly, even though the rest of the world did not see the republics as separate nations.

Roosevelt was able to win some compromises through careful negotiations, especially when he met with Stalin and Churchill at the Yalta Conference of February, 1945. Stalin accepted an American suggestion that the veto could be used only in certain cases. For his part, Roosevelt agreed to support a total of three votes for the Soviet Union in the General Assembly.

Unfortunately, by the time representatives of fifty-one nations gathered in San Francisco for the United Nations Charter Convention, two important American leaders were no longer involved. Poor health had forced Hull to resign as secretary of state in 1944, and President Roosevelt died of a stroke on April 12, 1945. But the new president, Harry S. Truman, was eager to support the United Nations, and the new secretary of state, Edward Stettinius, Jr., led the American delegation to the Charter Convention.

Molotov, who led the Soviet delegation in San Francisco, tried to change the decision made at Yalta; he insisted that each "Big Power" should be able to use the veto to keep an issue from being discussed in the Security Council. To resolve this conflict, Truman contacted Stalin directly and was finally able to convince him that this would be an improper use of the veto.

The new U.N. Charter did set up the Security Council as the main focus of power; along with the permanent members, it was to include six other nations elected for two-year terms by the General Assembly. The Security Council could enforce its decisions on the General Assembly members—but there was a safeguard. Any important decisions had to be approved by a seven-vote majority of the Council, including all five permanent members.

## Consequences

The Charter of the United Nations was signed by fifty-one nations in June, 1945. Within the next few decades, its membership grew to more than one hundred.

As World War II ended, wartime cooperation between the Soviet Union and the Western Allies soon came to an end as well. Over the following years, the Soviet Union often used its veto to block the Security Council from taking action. This was frustrating for American leaders. In the 1970's and early 1980's, there was even some talk of pulling the United States out of the United Nations.

Through the 1980's, however, respect for the United Nations grew around the world. U.N. peacekeeping forces were effective in decreasing violence in some areas of the Middle East and Africa, and in some important cases the United Nations was able to take a strong united stand—for example, against apartheid in South Africa. Meanwhile, U.N. projects spread across Africa, Latin America, and Asia—projects to vaccinate children, dig wells, and help poor people develop their own businesses.

*George Q. Flynn*

# France Quells Algerian Nationalist Revolt

> *Mass violence broke out in eastern Algeria, as Algerians gathered to celebrate the end of World War II in Europe and to demand independence from French rule.*

**What:** Civil strife; Political independence
**When:** May 8, 1945
**Where:** Sétif, Algeria
**Who:**

FERHAT ABBAS (1899-1985), president of the provisional Algerian government-in-exile in 1958

MESSALI HADJ (1898-1974), a radical Algerian Nationalist

## Growing Tension

Since before World War II, there had been tension between French colonists and the Muslims of Algeria. There were three groups of Algerian nationalists. The first, led by Messali Hadj, demanded total independence from French rule; this group was represented by the Algerian People's Party (PPA). The second group, which was led by Ferhat Abbas and the Friends of the Manifesto and Liberty (AML, a group founded in 1944), wanted full French citizenship and self-government for Algerians. The third group, the Islamic religious leaders, had the greatest influence over the Muslim masses of Algeria.

There was another group in Algeria: the *colons*, European (mostly French) settlers. They did not want to give in either to the moderates (those who demanded French citizenship) or to the radicals (who insisted upon independence for Algeria). The *colons* were in alliance with powerful leaders of the French government and military, and they controlled the government of Algeria through influence, money, and the press.

In 1936, the French government attempted to respond to the educated Algerians' request for citizenship with the Blum-Violette proposals, which would have given citizenship at first to a few thousand Muslims. Yet the *colons* opposed and prevented even this small step.

The economy of Algeria mainly served the *colons*, who controlled railroads, shipping, and other major industries. Farmlands in the northern plains were also controlled by *colons*; these lands produced almost ten times as many crops per acre as the less fertile lands to the south, where the Muslims had been driven in the early 1900's. There was high population growth among Muslim farmers, and since their southern lands could not support them, more and more of them were forced to become sharecroppers on *colon* farmlands in the north.

British and American forces landed in Algeria in 1942, and the Algerian people began to hope that this signaled a major change in their land. The Atlantic Charter, which had been signed in 1941, seemed to stand against colonialism, and this also gave hope to Algerian nationalists. In late 1944 the Arab League was formed, and Muslims in Algeria were roused to greater patriotic and nationalistic feelings.

More and more followers of Messali had been joining the AML, which had previously been a moderate organization. At the AML congress in March, 1945, delegates passed a resolution demanding complete independence for Algeria.

## The Sétif Incident

The Communist Party in Algeria was calling for confrontation with the French government, but the AML leaders feared that this would lead to a serious backlash against Algerians, and they publicly attacked the Communist policy in May, 1945. On May Day, there had been clashes between police and demonstrators in Algiers and other cities, and some people had been hurt. The AML did not want more violence.

A celebration of V-E Day, which had brought World War II to an end in Europe, was planned for May 8 in Sétif, in the Department of Constantine. AML leaders promised that they would not use political slogans during this celebration; they planned to lay a wreath at a war memorial and then send the crowd home.

When the parade reached the center of town, however, some people in the crowd brought out small British, American, and French flags and began shouting, "Long live Messali!" "Long live free and independent Algeria!" and "Down with colonialism!" At least one man waved the green-and-white Algerian national flag.

The commissioner of police had only a few officers to call on, but he reluctantly decided to intervene. There was a scuffle, and one spectator was shot, possibly by a panic-stricken police officer.

By noon, rumors of war—some called it a "holy war"—were spreading east as well as north to the Babor Mountains. During the night, armed Muslims raided villages and farms, cutting power lines, breaking into railroad cars, and setting fire to public buildings.

Guelma, a large town about 165 miles east of Sétif, was encircled by a mob on May 9. The Muslim mob went out of control and began attacking Europeans of all ages; sometimes they even mutilated the victims' bodies.

The prefect of Constantine asked Governor General Yves Chataigneau to send in the army. Violence continued for five days throughout the area. More than one hundred European men, women, and children were killed; about one-third of them were government officials.

Martial law was declared in Constantine, and an army of about ten thousand Senegalese and French Legionnaires was given orders to "clean up" the area between Sétif and Guelma, stretching north to the coast. The force rounded up hundreds of Muslims, and anyone who was suspected of crime or who lacked an identification armband was quickly executed. There were brutal public interrogations.

There were also bombardments from the air and sea; many innocent people were killed by bombs. Many *machtas* (Muslim settlements) between Guelma and the coast were totally destroyed.

The *colons* were given weapons, and they quickly formed vigilante groups and roamed the countryside, killing children, women, and elderly men. More than two hundred Muslims were shot in Chevreul. Some of the victims' bodies were mutilated.

No one was sure how many people were killed. Some estimated that only a few hundred people died during these days of violence, while other estimates reached fifty thousand.

Ferhat Abbas, who had visited the governor general to congratulate him on the Allied victory over Germany, was arrested inside the governor general's mansion. Soon afterward, the AML was banned. In all, about forty-five hundred people were arrested after the Sétif incident; about a hundred of them were condemned to death, and sixty-four were given life sentences.

France, along with the rest of Europe, was busy with victory celebrations, and major French newspapers did not make much of the violence in Constantine. French president Charles de Gaulle said that what was happening in Algeria was not very significant.

## Consequences

Muslim nationalists in Algeria were deeply distressed by the violence that followed the Sétif incident. Few of them believed any longer that they could gain independence through peaceful reform. Many young people became determined to fight colonialism in every way possible.

The French tried to satisfy the Algerians by giving them some new political rights. French citizenship was extended to more people, voting rights were given to women, Arabic was recognized as an official language along with French, and the tribes in the southern plains were brought under civilian rule for the first time. Yet the *colons* managed to keep control of the Algerian national assembly, and in the elections of 1948 they used fraud to make sure that their candidates won.

Dissatisfied with the French reforms, the Algerian nationalists continued their struggle. Algeria finally gained independence from France in 1962.

*Asit Kumar Sen*

# Germany Surrenders on V-E Day

*Two weeks after Allied armies coming from the west and Soviet armies advancing from the east met near Torgau, Germany, German officials signed an official surrender.*

**What:** War
**When:** May 8, 1945
**Where:** Germany and Rheims, France
**Who:**
DWIGHT D. EISENHOWER (1890-1969), supreme commander of the Allied Expeditionary Force
BERNARD MONTGOMERY (1887-1976), British commander of the Twenty-first Army Group
OMAR BRADLEY (1893-1981), American commander of the Twelfth Army Group
SIR WINSTON CHURCHILL (1874-1965), prime minister of Great Britain from 1940 to 1945
FRANKLIN DELANO ROOSEVELT (1882-1945), president of the United States from 1933 to 1945
HARRY S. TRUMAN (1884-1972), president of the United States from 1945 to 1953
KARL DÖNITZ (1891-1980), chancellor of Germany after Adolf Hitler's suicide in April, 1945

## The Allied Advance

After pushing back the German counterattack in the Ardennes Forest during the Battle of the Bulge in December, 1944, General Dwight D. Eisenhower, who commanded Allied forces in Western Europe, prepared for a final advance into the heart of Germany. He planned three crossings of the Rhine River: one in the north, by Field Marshal Bernard L. Montgomery's Twenty-first Army Group, which consisted mostly of British and Canadian troops; one in the south, by General Omar N. Bradley's Twelfth Army Group; and a third crossing in the south by the U.S. Third and Seventh armies.

Adolf Hitler, chancellor of Germany, had ordered his commanders to defend every inch of ground. As a result, Eisenhower's forces were able to destroy much of the German army in battles west of the Rhine in February, 1945. Allied forces were also lucky enough to capture the Ludendorff railroad bridge over the Rhine at Remagen on March 7.

By March 28, Bradley's forces had passed through Remagen and reached Marburg, where they were ready to swing northward to join Montgomery's forces. Montgomery's group had also crossed the Rhine and had cut off German army Group B, which had been assigned to defend Germany's main industrial area, the Ruhr Valley.

Eisenhower told Montgomery that once the German units in his area had been completely encircled, the U.S. Ninth Army (which had been fighting with the Twenty-first Army Group) would be sent to join Bradley's Twelfth Army Group and push into Germany toward Dresden. This was a major change in Eisenhower's plans. Before the bridge at Remagen was captured, he had intended that Montgomery should lead the main military effort in central Germany, with Berlin as his target.

On March 28, Eisenhower informed Soviet dictator Joseph Stalin of his plans. Now the capture of Berlin would be left up to the Soviet Red Army, which was moving into Germany from the east.

## Controversy and Surrender

British prime minister Winston Churchill was furious. He believed that Berlin should remain the main goal for both the British and the Americans. Eisenhower argued, however, that Berlin was no longer important, because no German armies or important government agencies were left in the capital. The point, he said, was to end the war as soon as possible, and to do that he had to destroy the remaining armed forces of Germany, which were concentrated in the south. There were rumors that these forces were gathering in the Bavarian Alps for a final show-down.

Yet Churchill believed that the capture of Berlin was important for political reasons. If the Soviets were allowed to take Germany's capital, they would give themselves the credit for winning the war. Churchill also implied that if the British and American forces took Berlin, they could hold the city as a sort of bargaining chip for making deals with the Soviets after the war.

Actually, leaders of the United States, the Soviet Union, and Great Britain had already decided that Germany would be divided into zones of occupation after the war, and Berlin itself would be divided among the Allies. Since part of the city would have to be handed over to the Sovi-

ets in any case, Eisenhower insisted that it would be foolish to waste British and American lives to conquer the capital.

Churchill could not give orders to Eisenhower; that right was reserved for the combined chiefs of staff of the United States and Great Britain, and for the president of the United States. General George C. Marshall, U.S. Army chief of staff, made sure that Eisenhower was given freedom to plan field operations as he saw fit.

Though Churchill appealed to President Franklin D. Roosevelt, the American president was trying to keep good relations with the Soviet Union. He refused to order Eisenhower to race the Red Army to Berlin. After Roosevelt died on April 12, 1945, the new president, Harry S. Truman, adopted the same policy.

So it was that Eisenhower sent his armies into central and southern Germany, avoiding Berlin.

*Field Marshal Wilhelm Keitel signs the surrender terms for the German army.*

The Soviets captured the German capital in late April, while Eisenhower's forces reached the Elbe River in central Germany between April 19 and May 2. On April 25, American and Soviet patrols met near Torgau. Germany had been cut in half.

On April 30, Hitler committed suicide; his successor, Admiral Karl Dönitz, began negotiating to surrender on May 4. Dönitz wanted to hand over German forces to the Western Allies, for he feared that the Soviets would retaliate for the brutality of the Germans during the German invasion of the Soviet Union. Eisenhower, however, refused to strike a deal.

With his country in ruins, Dönitz agreed to the immediate, unconditional surrender of all Germany's armed forces. On May 7, 1945, German and Allied representatives met at Eisenhower's headquarters in Rheims, France, and signed the documents that made the surrender official as of the following day. The war in Europe was over.

## Consequences

Some historians have argued that Eisenhower's decision to leave Berlin to the Red Army had important consequences after World War II, when the Cold War developed between the Soviet Union and the West and a fortified wall was built to split Berlin. According to this interpretation, if Eisenhower had succeeded in taking Berlin, the Western Allies could have prevented the Soviets from gaining control over the eastern half of Germany.

Others point out, however, that both the Soviets and the Western Allies honored their agreement to divide Germany—including Berlin—into zones of occupation after the war. If Eisenhower had reached Berlin first, he would have allowed the Soviets to occupy the part of the city that had been assigned to them. From this point of view, the question of who arrived first is not especially important. Furthermore, the Soviets paid a heavy price—100,000 casualties—for their success in conquering Berlin.

*Stephen E. Ambrose*

# First Atom Bomb Is Detonated

*J. Robert Oppenheimer directed the development and design of the atomic bomb during World War II, culminating in the first successful nuclear explosion.*

**What:** Weapons technology
**When:** July 16, 1945
**Where:** Alamogordo, New Mexico
**Who:**
J. ROBERT OPPENHEIMER (1904-1967), an American physicist
LESLIE RICHARD GROVES (1896-1970), an American engineer and army general
ENRICO FERMI (1900-1954), an Italian American nuclear physicist
NIELS BOHR (1885-1962), a Danish physicist

## Energy on a Large Scale

The first evidence of uranium fission (the splitting of uranium atoms) was observed by German chemists Otto Hahn and Fritz Strassmann in Berlin at the end of 1938. When these scientists discovered radioactive barium impurities in neutron-irradiated uranium, they wrote to their colleague Lise Meitner in Sweden. She and her nephew, physicist Otto Robert Frisch, calculated the large release of energy that would be generated during the nuclear fission of certain elements. This result was reported to Niels Bohr in Copenhagen.

Meanwhile, similar fission energies were measured by Frédéric Joliot and his associates in Paris, who demonstrated the release of up to three additional neutrons during nuclear fission. It was recognized immediately that if neutron-induced fission released enough additional neutrons to cause at least one more such fission, a self-sustaining chain reaction would result, yielding energy on a large scale.

While visiting the United States from January to May of 1939, Bohr derived a theory of fission with John Wheeler of Princeton University. This theory led Bohr to predict that the common isotope uranium 238 (which constitutes 99.3 percent of naturally occurring uranium) would require fast neutrons for fission, but that the rarer uranium 235 would fission with neutrons of any energy. This meant that uranium 235 would be far more suitable for use in any sort of bomb. Uranium bombardment in a cyclotron led to the discovery of plutonium in 1940 and the discovery that plutonium 239 was fissionable—and thus potentially good bomb material. Uranium 238 was then used to "breed" (create) plutonium 239, which was then separated from the uranium by chemical methods.

During 1942, the Manhattan District of the Army Corps of Engineers was formed under General Leslie Richard Groves, an engineer and army general who contracted with E. I. du Pont de Nemours and Company to construct three secret atomic cities at a total cost of $2 billion. At Oak Ridge, Tennessee, twenty-five thousand workers built a 1,000-kilowatt reactor as a pilot plant. A second city of sixty thousand inhabitants was built at Hanford, Washington, where three huge reactors and remotely controlled plutonium-extraction plants were completed in early 1945.

## A Sustained and Awesome Roar

Studies of fast-neutron reactions for an atomic bomb were brought together in Chicago in June of 1942 under the leadership of J. Robert Oppenheimer. He soon became a personal adviser to Groves, who built for Oppenheimer a laboratory for the design and construction of the bomb at Los Alamos, New Mexico. In 1943, Oppenheimer gathered two hundred of the best scientists in what was by then being called the Manhattan Project to live and work in this third secret city.

Two bomb designs were developed. A gun-type bomb called "Little Boy" used 15 kilograms of uranium 235 in a 4,500-kilogram cylinder about 2 meters long and 0.5 meter in diameter, in which a uranium bullet could be fired into three uranium target rings to form a critical mass. An implosion-type bomb called "Fat Man" had a 5-kilogram spherical core of plutonium about the size of an orange, which could be squeezed inside a 2,300-kilogram sphere about 1.5 meters in diameter by properly shaped explosives to make the mass critical in the shorter time required for the faster plutonium fission process.

A flat scrub region 200 kilometers southeast of Alamogordo, called Trinity, was chosen for the test site, and observer bunkers were built about 10 kilometers from a 30-meter steel tower. On July 13, 1945, one of the plutonium bombs was assembled at the site; the next morning, it was raised to the top of the tower. Two days later, on July 16, after a short thunderstorm delay, the bomb was detonated at 5:30 A.M. The resulting implosion initiated a chain reaction of nearly 60 fission generations in about a microsecond. It produced an intense flash of light and a fireball that expanded to a diameter of about 600 meters in two seconds, rose to a height of more than 12 kilometers, and formed an ominous mushroom shape. Forty seconds later, an air blast hit the observer bunkers, followed by a sustained and awesome roar. Measurements confirmed that the explosion had the power of 18.6 kilotons of trinitrotoluene (TNT), nearly four times the predicted value.

*J. Robert Oppenheimer.*

National Archives

## Consequences

On March 9, 1945, 325 American B-29 bombers dropped 2,000 tons of incendiary bombs on Tokyo, resulting in 100,000 deaths from the fire storms that swept the city. Still, the Japanese military refused to surrender, and American military plans called for an invasion of Japan, with estimates of up to a half million American casualties, plus as many as 2 million Japanese casualties. On August 6, 1945, after authorization by President Harry S. Truman, the B-29 *Enola Gay* dropped the uranium Little Boy bomb on Hiroshima at 8:15 A.M. On August 9, the remaining plutonium Fat Man bomb was dropped on Nagasaki. Approximately 100,000 people died at Hiroshima (out of a population of 400,000), and about 50,000 more died at Nagasaki. Japan offered to surrender on August 10, and after a brief attempt by some army officers to rebel, an official announcement by Emperor Hirohito was broadcast on August 15.

The development of the thermonuclear fusion bomb, in which hydrogen isotopes could be fused together by the force of a fission explosion to produce helium nuclei and almost unlimited energy, had been proposed early in the Manhattan Project by physicist Edward Teller. Little effort was invested in the hydrogen bomb until after the surprise explosion of a Soviet atomic bomb in September, 1949, which had been built with information stolen from the Manhattan Project. After three years of development under

**999**

Teller's guidance, the first successful H-bomb was exploded on November 1, 1952, obliterating the Elugelab atoll in the Marshall Islands of the South Pacific. The arms race then accelerated until each side had stockpiles of thousands of H-bombs.

The Manhattan Project opened a Pandora's box of nuclear weapons that would plague succeeding generations, but it contributed more than merely weapons. About 19 percent of the electrical energy in the United States is generated by about 110 nuclear reactors producing more than 100,000 megawatts of power. More than 400 reactors in thirty countries provide 300,000 megawatts of the world's power. Reactors have made possible the widespread use of radioisotopes in medical diagnosis and therapy. Many of the techniques for producing and using these isotopes were developed by the hundreds of nuclear physicists who switched to the field of radiation biophysics after the war, ensuring that the benefits of their wartime efforts would reach the public.

*Joseph L. Spradley*

# Potsdam Conference Decides Germany's Fate

> *Conflicts between the Soviet Union and the Western Allies led to compromises over the fate of Eastern Europe and Germany after World War II.*

**What:** International relations
**When:** July 17-August 2, 1945
**Where:** Potsdam, Germany
**Who:**
HARRY S. TRUMAN (1884-1972), president of the United States from 1945 to 1953
JOSEPH STALIN (1879-1953), dictator of the Soviet Union from 1924 to 1953
SIR WINSTON CHURCHILL (1874-1965), prime minister of Great Britain from 1940 to 1945
CLEMENT ATTLEE (1883-1967), prime minister of Great Britain from 1945 to 1951

## The Conference Is Planned

At the close of the Yalta Conference in February, 1945, the leaders of the United States, Great Britain, and the Soviet Union agreed that they would meet again after World War II was ended in Europe. They realized that they needed to discuss a number of problems, including peace treaties with the former Axis nations. Great Britain and the United States also wanted to make sure that the Soviet Union stuck to its promise to allow free governments to be established in Eastern Europe.

After Germany surrendered in May, 1945, the Allied leaders began preparing for another conference. Winston Churchill, prime minister of Great Britain, was anxious that the meeting take place as soon as possible: first, to keep the Soviets from establishing firmer control over Eastern Europe, and second, to ensure that he himself had a chance to participate. Elections were scheduled in Great Britain in July, and he realized that he might be voted out of office.

The concerns of Harry S. Truman, however, conflicted with Churchill's. Truman had just become president of the United States in April, after Franklin D. Roosevelt died. Needing time to become familiar with the problems, Truman wanted the conference to be delayed. It was finally agreed that the conference would open on July 17 in Potsdam, Germany.

Two other problems had to be solved. One had to do with the provisional president of France, Charles de Gaulle, who had led the Free French Forces during the war. By prior agreement among the "Big Three"—the United States, Great Britain, and the Soviet Union—France's new government had been given responsibility for one of Germany's occupied zones after the war. Though this arrangement brought France back into the group of Allied Powers, the Allied leaders had found de Gaulle difficult to work with. In the end, they decided not to invite him to Potsdam.

The other problem concerned the occupation armies in Germany. As the Western and Soviet armies had moved into Germany, both of them had occupied areas that were not part of their assigned occupational zones. For example, there were no American troops in Berlin, though that city was supposed to be divided four ways; the American military commanders wanted soldiers there to protect President Truman on his trip to Potsdam. The Soviets refused to allow American troops into Berlin, however, until the Western soldiers left the Soviet occupational zone.

Churchill was suspicious of the Soviets, fearing that they intended to push the Western Allies out of Germany and that they would not abide by their bargain. Truman ordered the Western forces to retreat from the Soviet zone, however, and the Soviets did allow Americans into Berlin.

National Archives

*From left to right, Winston Churchill, Harry S. Truman, and Joseph Stalin, at the Potsdam Conference.*

### Bargaining at Potsdam

The Potsdam Conference began on July 17, 1945, and lasted for two weeks. In the middle of the conference, British election returns showed that Churchill had been defeated. The Labour Party leader, Clement Attlee, became the new prime minister; since Attlee was already attending the conference, however, there was no change in British policy.

Matters to be discussed at Potsdam included peace treaties, Eastern Europe, and Germany. Remembering the hasty decisions and mistakes made at the Paris Peace Conference of 1919, Truman suggested that the Council of Foreign Ministers be given as long as necessary to write peace treaties for the defeated Axis nations. The British and Soviet leaders agreed, and the foreign ministers soon began their discussions. Two years passed before the final treaties were drawn up, but those carefully written treaties were quickly accepted.

Eastern Europe was another matter. Stalin wanted the Western nations to form a diplomatic relationship with the new procommunist governments there, but Truman and Churchill refused. They were especially concerned about Poland. At Yalta the Allies had agreed that free elections would be held there "as soon as possible." Since that time, however, the Communist-controlled Warsaw government had taken over Poland and had made no plans for elections. When members of the anticommunist Polish government returned from London, where they had lived in exile during the war, they had been arrested.

In May, Truman had sent Harry Hopkins, special assistant to the president, to Moscow to try to iron out the problem. Hopkins had worked out a compromise between the London and Warsaw Poles: They would form a coalition government, and there would be free elections.

At Potsdam, this agreement was ratified, though one requirement was added: The Allied press would be admitted to Poland and allowed to report on the elections. Meanwhile, the Western Powers would grant diplomatic recognition to the interim Polish government.

When the Soviets had liberated Poland, they had turned over all German lands east of the Oder and western Neisse rivers to the Poles, along with the southern half of East Prussia. Stalin asked that the Allies confirm these boundaries as permanent, but Truman and Churchill accepted them only for the time being, until a final peace treaty with Germany could settle all border questions.

What to do about Germany was the major question at Potsdam. The Soviets wanted Germany to be greatly weakened and to be made to pay large reparations for its war crimes. Though the Western Allies agreed that the Germans should not be allowed to build up a strong military, they did not want to destroy Germany, for then the Soviet Union would have even more opportunities to extend its power into Europe.

In the end, the Allies agreed that Nazi institutions would be abolished and that Germany would not be allowed to manufacture weapons. The Control Council in Berlin would make decisions for Germany's economic future, as long as its members could come to unanimous agreement. Within each zone, the Allied military commander would have sovereign authority. Yet on a local level the Germans would be encouraged to govern themselves, and the German economy would be geared toward peace.

As reparations, it was agreed that each Allied power could take what it wanted from its own oc-cupational zone. The British agreed to trade some industrial machinery from the Ruhr area, which lay in their zone, for food products from the Soviet zone. This food would be given to refugees who had fled from the Soviet-controlled areas in the last days of the war.

The Potsdam Conference produced a few other decisions. Austria was divided into occupational zones like Germany's. The Allies condemned Spain for having supported the Axis Powers during the war, and Spain was forbidden from membership in the United Nations. Since World War II was still raging in the Pacific, a declaration from the conference called upon Japan to surrender unconditionally or face total destruction.

## Consequences

Whether the American and British leaders allowed the Soviet Union too much say in the decisions at Potsdam has been argued by many historians. Some say that the Americans, in particular, sold out to the Soviets, and that their weakness led to the Cold War that divided the democratic West from the Communist East for several decades. Others have argued the opposite: that the Americans felt secure in the knowledge that they owned the atomic bomb and therefore were able to bully the Soviets at Potsdam. A middle-of-the-road view is that the American negotiators did the best they could at Potsdam and that the Cold War would have come about even if they had acted differently.

*José M. Sánchez*

# Labour Party Leads Postwar Britain

*Fearing economic problems after the end of World War II in Europe, British voters replaced Winston Churchill's Conservative Party leadership with the Labour Party's program of socialist reforms.*

**What:** Political reform
**When:** July 26, 1945
**Where:** Great Britain
**Who:**
GEORGE VI (1895-1952), king of Great Britain and Northern Ireland from 1936 to 1952
SIR WINSTON CHURCHILL (1874-1965), prime minister of Great Britain from 1940 to 1945
CLEMENT ATTLEE (1883-1967), prime minister of Great Britain from 1945 to 1951

## The 1945 Election

Beginning in 1931, Great Britain was headed by a National Coalition government, made up of Conservatives, Liberals, and a few members of the Labour Party. The Great Depression had brought serious economic problems to Great Britain, and the coalition government was an attempt to solve those problems. In fact, however, the country was ruled by the Conservatives, since they held a majority in Parliament after the elections of 1931 and 1936.

Conservative leader Stanley Baldwin headed the coalition as prime minister from 1935 to 1937. After Baldwin resigned, King George VI asked Neville Chamberlain, another Conservative, to lead the nation. Chamberlain's appeasement policy toward Nazi Germany in the late 1930's led him to resign in 1940, when German chancellor Adolf Hitler ordered German troops to overrun Western Europe.

Once Chamberlain had resigned, King George asked Winston Churchill to form a government. Churchill was the leader of Conservatives who had criticized the policy of appeasement, and he reorganized the British cabinet, bringing in Labourites and Conservatives who had also opposed appeasement.

A tough, aggressive leader, Churchill played a major role in planning for Germany's defeat, which finally came about in May, 1945. By that time Parliament had been in session for nine years. Soon after the German surrender, Churchill agreed to hold a general election. The king dissolved Parliament on June 15.

To prepare the way for the election, which was scheduled for July 5, the coalition cabinet resigned, and Churchill appointed a new cabinet made up entirely of Conservatives. The main contest in the election campaign was between the Conservatives and the Labourites. Each party claimed to be the best qualified to solve Great Britain's desperate economic problems. The country had a national debt of more than three billion pounds, and personal property, ships, and factories destroyed in the war needed to be replaced.

The Labour Party said that the government should plan the country's economy: Many industries should be "nationalized" (taken over by the government), and those left under private ownership should be regulated by the government. Though the Conservatives agreed that the government should keep some economic controls, they argued that the Labourite program of nationalization would infringe on private enterprise. Both parties promised to prevent unemployment and continue the social services.

The Labour Party, headed by Clement Attlee, won the election by a strong majority: 392 Parliament seats, as opposed to 189 kept by the Conservatives. Attlee officially replaced Churchill as prime minister on July 26, 1945, the date on which the vote was counted. Attlee also became Great Britain's chief representative at the Potsdam Conference, which was then meeting to decide the fate of Germany.

## Labourite Reforms

The Labour Party remained in power from 1945 through most of 1951. During these years, with the help of loans from the United States, it tried to deal with Great Britain's social and economic problems through its socialistic reforms.

In October, 1945, the House of Commons allowed Attlee to extend the government's wartime emergency powers for five years. Gradually, then, the government nationalized the Bank of England, overseas telegraph and radio services, airlines, coal mines, railroads, trucking, canals and docks, electric companies and plants, and the iron and steel industries.

Businesses left under private ownership were required to meet strict government standards, mostly because the government needed to balance the country's exports and imports. Wartime rationing of food, clothing, and fuel was extended. The Attlee government also set price supports for milk, livestock, eggs, and potatoes, so that farmers were guaranteed a return on their investment. In return, farmers had to meet government standards for efficient production.

To improve the people's standard of living, the Labourites expanded social services, as they had promised. Between May and July, 1948, Parliament passed the National Insurance Act and the National Health Service Act. Through the National Insurance Act, a greater number of Britons would benefit from government-controlled health, old age, and unemployment insurance. The second bill provided to everyone free medical services and supplies, hospital care, and nursing aid.

Attlee also tried to redistribute Great Britain's wealth through a progressive tax: Taxes on low incomes were reduced, while the rate of taxation of higher incomes was raised. This reform and others were opposed in the House of Lords; in response, the Attlee government persuaded Parliament to limit the veto power of the Lords.

When its term expired in February, 1950, the Labour government campaigned for reelection with the promise to continue reforms. In turn, the Conservatives promised to supply more housing and to end further nationalization; they claimed that government control had damaged Great Britain's businesses. At the same time, they pledged not to bring back unemployment or to reduce social services.

Many people who had supported the Labour Party in 1945 now cast their votes for the Conservatives, who captured 297 seats. The Labour Party, however, kept control of the government with 315 seats. During the next two years, the Labourites barely managed to keep this slight majority.

## Consequences

A split in the Labour Party in mid-1951 finally brought Attlee to call for another election in October of that year. This time the Conservatives won a slim majority, and Churchill returned to 10 Downing Street, the prime minister's residence.

The Conservatives immediately denationalized the iron and steel industries. Their biggest job, however, was to find a solution to the country's huge deficit in the balance of payments—a problem that had existed since the end of the war. Churchill began cutting back government spending in an attempt to bring an end to the crisis. This pattern of spending followed by cutbacks has remained in British politics since that time. Nevertheless, many programs begun by the Labourites, such as the National Health Service, have continued to survive.

*Edward P. Keleher*

# United States Drops Atomic Bomb on Hiroshima

> *The dropping of an atomic bomb on Hiroshima, Japan, brought a quick end to World War II in the Pacific but began the era of nuclear weaponry.*

**What:** War; Weapons technology
**When:** August 6, 1945
**Where:** Hiroshima, Japan
**Who:**
HARRY S. TRUMAN (1884-1972), president of the United States from 1945 to 1953
HENRY LEWIS STIMSON (1867-1950), U.S. secretary of war from 1940 to 1945
HIROHITO (1901-1989), emperor of Japan from 1926 to 1989

## Planning for the Bomb

Development of the atomic bomb began in 1939, after a small group of scientists persuaded the U.S. government that building such a weapon was possible and warned that Germany was already experimenting with atomic energy. The research program that began in October, 1939, developed into the two-billion-dollar Manhattan Project, which aimed to produce a bomb before the Germans did so. Political and military leaders in the United States had little doubt that once such a weapon was developed, it would be used.

Before the first bomb was developed and tested, however, Germany surrendered. Only Japan remained at war with the Allies. Early in 1945, as the first bomb was being completed, some scientists began to have doubts about using it. They wondered whether it would cause such great destruction that people around the world would be horrified, and guilt would be heaped upon American leaders who had approved its use. Would dropping the bomb be worth this terrible cost?

The bomb could be used indirectly, however, as a warning. The United States might give a demonstration of the new weapon's power, dropping it on a deserted island in the presence of United Nations witnesses, who could then warn the Japanese of its destructive capability. Another possibility was to drop the bomb only on a military target in Japan, after giving a warning. Or the United States could refuse to drop it at all.

*The United States dropped a second atomic bomb on Nagasaki on August 9.*

While scientists mulled over these choices, military officials prepared to use the bomb. By the end of 1944, they had selected possible targets in Japan, and a squadron of B-29 bombers had begun training.

Two weeks following President Franklin D. Roosevelt's death in 1945, Secretary of War Henry L. Stimson met with the new president, Harry S. Truman, informed him about the bomb—which had been developed in great secrecy—and predicted that it would be ready for use in about four months. Truman then appointed a special Interim Committee, headed by Stimson, to consider how the bomb might be used.

On June 1, 1945, the committee recommended to the president that the bomb be used against Japan as soon as possible, that it be dropped on a military target, and that it be used without prior warning. By early July, when Truman left for the Potsdam Conference to discuss postwar settlements with Great Britain and the Soviet Union, he had decided to use the bomb once it was perfected.

## A Fearsome Weapon

On July 16, the first atomic bomb was successfully tested in the Trinity Flats near Alamogordo, New Mexico. The United States now had its weapon, although the war in the Pacific had already weakened Japan so much that it was close to surrendering.

As early as September, 1944, the Japanese had begun trying to learn how the Allies would handle a surrender. Just before the Potsdam Conference, the Japanese ambassador in Moscow asked the Soviet government to help bring about an end to the war. Japan could not accept unconditional surrender, but they might be willing to surrender if they were allowed to keep the position of the emperor in the Japanese system of government.

The Truman administration faced difficult questions. Bringing Japan to total defeat and unconditional surrender might require a long, expensive invasion. According to a promise at the Yalta Conference in February, 1945, the Soviets would be entering the war in the Pacific in early August. Although in the spring American leaders had been eager for this help from the Soviets, by summer it seemed less important. In fact, Truman hoped to defeat Japan before the Soviet Union could enter the war and gain any control over the postwar settlement with Japan.

On July 26, Truman and the other Allied leaders called upon Japan to surrender unconditionally or suffer "the utter devastation of the Japanese homeland." The Potsdam Declaration did not mention the atomic bomb, nor did it offer any possibility of bargaining over terms of peace.

**1007**

Waiting for a reply to the peace appeal they had made to the Soviet government, Japan's leaders chose not to respond to the declaration. Within Japan, the government called the Potsdam Declaration "unworthy of public notice."

In the Mariana Islands, two bombs had been made ready, and the B-29 crews were standing by. Truman ordered the U.S. Air Corps to drop the bombs.

At 2:45 A.M. on August 6, 1945, the *Enola Gay* took off from the island of Tinian. Just after 8:15 A.M. it released an atomic bomb from an altitude of about 31,600 feet over Hiroshima, Japan. The bomb exploded with terrible fury over the center of the city, immediately killing more than eighty thousand people and maiming thousands more. The searing heat from the explosion set the city afire and completely destroyed it.

Two days later, on August 8, the Soviet Union declared war on Japan. On August 9, at about 11:00 A.M., the United States dropped a second atomic bomb on Nagasaki, Japan. This explosion killed more than forty thousand people.

## Consequences

The destruction of Hiroshima was a shock to the Japanese, but the Soviet Union's declaration of war was devastating, since it removed all hope of Soviet help in ending the war. Also, Japanese nationalists had hoped to bring the Kwangtung Army home from its post in Manchuria so that it could help fight off the expected Allied invasion; now it was necessary to keep that force in Manchuria to fend off a Soviet invasion.

Throughout the day and into the night, the Japanese Supreme War Council grimly discussed the options. At 2:00 A.M. on August 10, the prime minister asked the emperor to decide Japan's future. Speaking softly, Emperor Hirohito told his ministers that he wanted the war to be brought to an end. The same day, Japan announced that it would surrender unconditionally.

The dropping of the atomic bomb has remained controversial over the years. While some argue that it actually prevented the greater loss of life that would have resulted from an invasion, others speculate that Japan would have surrendered quickly in any case.

As the Cold War began and the Soviet Union joined the United States in racing to build even more terrifying weapons, the debate continued: Should such destructive bombs be built at all? If they were built, what would be the best way to make sure that they were never used again?

*Burl L. Noggle*

# General MacArthur Oversees Japan's Reconstruction

> *Through a military occupation of Japan, U.S. general Douglas A. MacArthur brought about deep changes in Japanese government and society.*

**What:** Political reform; Economics
**When:** September 2, 1945-April 28, 1952
**Where:** Japan
**Who:**
Douglas A. MacArthur (1880-1964), supreme commander for the Allied Powers in Japan from 1945 to 1951

## The Goal: Rebuilding Japan

World War II officially ended in the Pacific on September 2, 1945. Within Japan, involvement in the war had led to disaster. More than two million servicemen had been killed or wounded since 1937, cities had been reduced to rubble, factories had been heavily damaged and lacked raw materials, and the transportation system had broken down. Japan's people were underfed, discouraged, and fearful.

The Allies' Potsdam Declaration of July 26, 1945, had required unconditional surrender. In the words of Emperor Hirohito, broadcast on August 15, Japan had to "bear the unbearable." Having failed in war, the Japanese people were open to change.

The seven-year military occupation that followed was the first time in history that Japan had been invaded. The American role in reshaping Japanese society was unusual. Expecting a harsh military rule, the Japanese found the occupation to be constructive; expecting treachery and hatred, the occupying Americans found openness and cooperation.

The occupation's leadership was provided by General Douglas A. MacArthur, Supreme Commander for the Allied Powers (SCAP). MacArthur was the ideal person to head the occupation. He was well versed in history, he had a dramatic public image, and he was full of confidence that the Allies could create a democratic Japan. SCAP, the acronym for MacArthur's title, soon came to stand for the occupation of Japan in general.

MacArthur's power in Japan was nearly absolute. The thirteen-nation Far Eastern Commission, based in Washington, D.C., provided only general guidelines, and the Allied Council in Tokyo, which included representatives from the United States, the British Commonwealth, China, and the Soviet Union, had little influence. As a Republican who was popular in Harry S. Truman's Democratic administration, MacArthur could bring about revolutionary change in Japan without being labeled a radical.

In the occupation of Germany, Soviet policy clashed with that of the Western Allies, and a military government was in control. In Japan, however, MacArthur did not allow the Soviets to have any influence, and SCAP worked through a rebuilt Japanese government. Within Japan's prefectures (local areas), military teams were assigned to make sure that SCAP reforms were carried out.

Traditional Japanese society was based on an alliance between the military and the *zaibatsu* (large business conglomerates). The people were divided into strict social classes, headed by the emperor, who in the state Shintō religion was viewed as a god.

The Americans believed that political, economic, and social reforms were essential if Japan was to become peaceful and democratic. SCAP reforms were designed to destroy Japan's "feudal" system and its values. The goal was to rebuild society completely.

SCAP policies had support within Japan. Politicians, intellectuals, and some leftists were glad

*General Douglas MacArthur (seated) at the surrender of the Japanese in Tokyo Bay, September 2, 1945.*

to see a return to the liberal trends of the 1920's. Because of this widespread support, most SCAP reforms took root and survived beyond the occupation.

## The Process of Reforms

MacArthur demanded that Japan's constitution be changed, but a draft presented to SCAP in February, 1946, seemed only a slight revision of the 1889 Meiji Constitution. Under the leadership of General Courtney Whitney, the government section of SCAP quickly produced another draft, which was presented to the Japanese on February 19. With minor revisions, this document was approved by the Diet (parliament) in November, 1946.

The new constitution, which went into effect May 3, 1947, was clearly an American document, though it claimed to represent "the freely expressed will of the Japanese people." The emperor became a "symbol of the state" whose du-

ties mostly involved government ceremonies; on January 1, 1946, Emperor Hirohito had already renounced all claims to divinity. Under the new constitution, Japan could no longer maintain military forces or resort to war. The Privy Council, the military services, and advisers to the emperor were all abolished, and human rights were guaranteed.

A British-style cabinet, chosen by the majority party or coalition in the Diet, became the new center of political power. Diet members would be chosen by an electorate that included all persons over twenty years of age—including women, who had previously had no right to vote. Local governments were given increased power over taxation, police, and schools.

Japan's overseas empire was taken away, and 6.5 million overseas Japanese were brought home. Paramilitary and extreme nationalist organizations were disbanded, and Shintō lost its position as the state religion. Political prisoners

**1010**

were released, and education was reformed; for example, nationalistic statements were removed from textbooks.

Trials for war crimes were held, and seven Japanese leaders were executed in December, 1948. About 200,000 former military officers, high officials, and business leaders were banned from participating in public life.

There were also reforms in agriculture and industry, aimed at redistributing wealth in Japan. *Zaibatsu* holding companies were banned, and the fortunes of *zaibatsu* families were seized. Conglomerates were separated into individual industries and banks, and laws forbidding monopolies were passed.

In the villages, 46 percent of the people were tenants rather than owners of the land that they worked; this system had contributed to poverty and repression. Because of strong SCAP pressure, a land-reform measure was passed by the Diet in October, 1946. The government bought all land held by absentee landlords and all other large landlord estates; this land was sold to farmer tenants. Tenancy dropped to 10 percent.

MacArthur favored a strong labor movement, for he believed it to be essential for democracy. By 1949, the unions boasted 6.5 million members. In 1947, however, SCAP policy toward un-

ions changed when MacArthur forbade government railway employees to strike.

In 1951, U.S. president Harry S. Truman dismissed MacArthur because of disagreements over the Korean War, which had begun in 1950. General Matthew Ridgway directed the last year of the occupation. An American-Japanese peace treaty was signed in San Francisco on September 8, 1951, ending the occupation on April 28, 1952.

## Consequences

Though the war had been bitter and there were vast differences between American and Japanese culture, MacArthur's occupation was remarkably successful. Some of the reforms were reversed in Japan after 1952, but the constitution and many other SCAP changes have had a lasting impact.

The destruction of Japan's military worked in the long run to the country's economic advantage, enabling the Japanese to focus their resources on business and manufacturing. In the second half of the twentieth century, Japan rose to become one of the world's major economic powers.

*Richard Rice*

# Nuremberg Trials Convict War Criminals

*After a trial of nine months, an international court found more than twenty Nazi leaders guilty of war crimes.*

**What:** International law
**When:** November 20, 1945-August 31, 1946
**Where:** Nuremberg, Germany
**Who:**
HERMANN GÖRING (1893-1946), former commander in chief of the Luftwaffe (German air force)
RUDOLF HESS (1881-1973), former deputy to Adolf Hitler
JOACHIM VON RIBBENTROP (1893-1946), former German minister for foreign affairs
ALFRED ROSENBERG (1893-1946), former director of the Foreign Policy Office of the Nazi Party
HJALMAR SCHACHT (1877-1970), former German minister of the National Economy
KARL DÖNITZ (1891-1980), former commander of the German navy
MARTIN BORMANN (1900-1945?), former third deputy to Hitler

## The Accused

As early as 1942, the Allies learned from various sources in Nazi-occupied Europe that the Germans were carrying out many atrocities against Jews, prisoners of war, and other groups. On October 7 of that year, the United States and Great Britain established the War Crimes Commission, which made lists of German war criminals. In the Moscow Declaration of November 1, 1943, the United States, Great Britain, and the Soviet Union affirmed that they intended to bring German war criminals to justice.

After the defeat of Nazi Germany in May, 1945, American, British, Soviet, and French leaders met at the London Conference to make more detailed plans for the trials. They made a list of twenty-four Nazi leaders and six Nazi organizations that should be brought to trial as soon as possible. The final list did not include Chancellor Adolf Hitler, Propaganda Minister Joseph Goebbels, or Gestapo chief Heinrich Himmler, for they had all committed suicide.

Among the individuals named were the following: Hermann Göring, who had commanded the Luftwaffe and served as president of the Council for War Economy; Rudolf Hess, who had been one of Hitler's deputies; Joachim von Ribbentrop, former minister of foreign affairs in Germany; Robert Ley, who had headed the Labor Front; Field Marshal Wilhelm Keitel, former Supreme Commander of the Armed Forces of Germany; Ernst Kaltenbrunner, who had headed the SS (Schützstauffeln); Hans Frank, an administrator of Polish territory in 1939; Wilhelm Frick, former minister of the interior in Germany; Julius Streicher, who had been managing editor of *Der Stürmer,* a Nazi publication; Walter Emmanuel Funk, who had organized business contacts for the Nazi Party; Gustav Krupp, a leading German businessman who had helped finance the Nazis; Erich Raeder, who had commanded the German navy from 1935 to 1943; Karl Dönitz, who had succeeded Raeder as navy commander; Baldur von Schirach, the long-time leader of the Nazi youth organization; Fritz Sauckel, who had overseen the use of labor in occupied territories; Alfred Jodl, former German army chief of staff; Franz von Papen, former vice chancellor of Germany and German ambassador to Turkey; Arthur Seyss-Inquart, who had served as high commissioner of the Netherlands; Albert Speer, who had been minister for armament and war production; Konstantin von Neurath, who had been minister of foreign affairs from 1932 to 1938; and Hans Fritzsche, a former official in the Ministry of Propaganda.

The six organizations named were the Reich Cabinet, the leadership corps of the Nazi Party, the SS and the SD (Sicherheitsdienst), the Gestapo, the SA (Sturm Abteilung), and the general staff and High Command of the German army.

Four counts were included in the final indictment of October 6, 1945: (1) participation in a common plan or conspiracy to commit crimes against peace, war crimes, and crimes against humanity; (2) commission of crimes against peace; (3) commission of war crimes; and (4) commission of crimes against humanity.

### The Trials

The Nuremberg Trials began on November 20, 1945, with the reading of the long indictment document in the Palace of Justice in Nuremberg.

The International Military Tribunal consisted of four members and four alternates, two judges from each of the four Allied Powers.

Only twenty-one of the twenty-four Nazis who had been indicted were seated in the dock. Martin Bormann had not been captured; there were rumors that he had committed suicide alongside Hitler. (According to other rumors years later, he had escaped and moved to South America, where he took a false name.) Gustav Krupp was excused because of poor health, and Robert Ley had committed suicide.

The trials lasted for more than nine months. During this time, the prosecution used many captured German documents to describe details of the Nazis' crimes. There was no good defense against this overwhelming evidence.

*Accused German war criminals at the Nuremberg trials: Front row, left to right, Hermann Göring, Rudolf Hess, Joachim von Ribbentrop, and Wilhelm Keitel; back row, left to right, Karl Dönitz, Erich Raeder, Baldur von Schirach, and Fritz Sauckel.*

## Consequences

The trials ended on August 31, 1946; on October 1, the tribunal read its judgments. Papen, Fritzsche, and Hjalmar Schacht were acquitted, while nineteen others were found guilty under one or more of the indictment counts.

Hess, Funk, and Raeder were sentenced to life imprisonment, while Dönitz, Speer, Schirach, and Neurath received shorter prison sentences. The others were sentenced to death by hanging. These executions were carried out on the night of October 15-16, but Göring escaped the noose by swallowing a vial of poison the day before.

The tribunal also found the Nazi Party leaders, the SS, the SD, and the Gestapo guilty. Later, the occupying Allies also held trials of lower-level Nazi officials who had been captured.

Some have criticized the Nuremberg Trials, saying that the International Military Tribunal was not a true court of law, since it did not have the authority of a world government to back it. These critics of the trials view them as a simple act of vengeance—the winners punishing the losers.

Others insist that the trials were legal, however, because the Nazis had committed enormous crimes by any nation's standards. Though there are no international laws that specifically forbid slaughtering and torturing millions of people, such acts are so horrible that they do not need to be formally prohibited. A world state does not have to exist before war criminals can be brought to justice.

*Johnpeter Horst Grill*

# Congress Passes Employment Act of 1946

*With the Employment Act of 1946, the American government officially accepted responsibility for stabilizing the national economy.*

**What:** Economics; Labor
**When:** February 20, 1946
**Where:** Washington, D.C.
**Who:**
HARRY S. TRUMAN (1884-1972), president of the United States from 1945 to 1953
JAMES E. MURRAY (1876-1961), senator from Montana from 1934 to 1961
JOHN MAYNARD KEYNES (1883-1946), a British economist

## Fear of Hard Times

During the last months of World War II, Americans began looking ahead anxiously to the nation's postwar economy. Their greatest worry was that another Great Depression would follow the war. Once the war was over, the nation would face the task of bringing eleven million members of the armed services back into peacetime occupations; the whole economy would have to be reconverted from the business of war to the business of peace. As soon as possible, war-weary Americans hoped to get rid of price controls and rationing, cut taxes, and turn industry back to producing consumer goods, which had been scarce during the war.

Yet the memory of the Depression and its breadlines still haunted adults. They suspected that it had been World War II, rather than the New Deal programs of President Franklin D. Roosevelt, that had solved the problem of unemployment. Maybe the sudden drop in defense spending would plunge the United States back into economic depression.

By early 1945, many economists were predicting that eight to ten million men would be unemployed when the returning troops were released from service. During the war, however, production and income had risen, and Americans had become used to the new prosperity. They were determined to settle for nothing less.

In 1944, the Democratic campaign platform guaranteed "full employment" after the war, while the Republicans made almost the same promise. In a campaign speech, Roosevelt set a postwar goal of "close to sixty million jobs."

## The Employment Act

On January 22, 1945, Senator James E. Murray of Montana, joined by several cosponsors, introduced a full employment bill in Congress. As chairman of the War Contracts Subcommittee of the Military Affairs Committee, Murray was especially concerned about making the shift to a peacetime economy. His bill stated that each American had the "right to a job" and obligated the federal government to make sure that everyone "able to work and seeking work" could actually find job opportunities. In times when business was slow, the government would have to take action to improve economic conditions; in a desperate situation, it might spend money to create programs that would employ new workers.

Soon after Japan surrendered, ending World War II, President Harry S. Truman, who had succeeded Roosevelt in April, 1945, urged that Congress pass the Murray bill. Yet the bill was already undergoing major changes. Conservatives in Congress cut out all mention of the federal guarantee of the "right to a job" and of "full employment." They also cut the specific mention of public works programs, substituting vague statements: The government would use "all practicable means . . . to promote maximum employment, production, and purchasing power."

As it stood when it was finally passed, the Employment Act of 1946 did not require the government to take action against depressions. It did not contain a statement of economic philosophy. It did, however, set up a Council of Economic Ad-

**1015**

visers, made up of three skilled economists; this council would help the president write an annual report on the state of the economy. In Congress, a Joint Committee on the Economic Report would review the president's report and make its own recommendations.

The Employment Act passed the Senate easily and was passed in the House by a 320-84 vote. Many supporters of the original Murray bill complained that it had been watered down, but at least the new law, in its vagueness, did not contradict the original bill. More important, liberals and conservatives had agreed that the federal government should assume responsibility for stabilizing the economy. Truman signed the bill into law on February 20, 1946.

The Employment Act actually expressed in law what Americans had learned during the Great Depression and World War II. It also grew out of the thinking of the British economist John Maynard Keynes, who in 1936 had published *The General Theory of Employment, Interest, and Money.* In this book Keynes attacked laissez-faire economics—the idea that prosperity would automatically grow if the government did not interfere with the free market. Instead, Keynes argued that a nation could actually pull itself out of a depression if the government stimulated the economy by going into debt for public works and other programs. Similarly, Keynes said, a government could slow down economic "booms"— rapid economic growth and rising prices—by raising taxes, raising interest rates, and keeping a budget surplus.

The first real test of Keynes's theory in the United States had come in the defense effort of World War II. The War Manpower Commission and the Office of War Mobilization had been successful in organizing labor and industry to help win the war. Many Americans quickly accepted the idea that government action could help bring about economic stability, even after the war was over.

## Consequences

Since World War II, every American president has used the basic ideas of the Employment Act to keep the nation from falling into a dangerous boom-and-bust cycle. Just after the war, Truman faced a confusing situation that was just opposite of what experts had predicted: shortages instead of surplus and inflation instead of depression and deflation. In August, 1946, employment did rise to sixty million jobs; unemployment never became a serious problem. Unexpectedly, Americans found their economy taking off on one of the longest boom periods of their history.

The main task of the Council of Economic Advisers has not been to prevent depressions, but to control inflation while still encouraging economic growth. During the twentieth century the United States experienced the Great Depression and several brief recessions. In response to the recessions, both Democrats and Republicans followed Keynes's advice: cutting taxes, lowering interest rates, and increasing spending on public works.

*Donald Holley*

# Perón Creates Populist Alliance in Argentina

*A coalition of labor and the military elected Juan Perón president of Argentina, ending sixteen years of conservative rule.*

**What:** National politics; Civil rights and liberties; Labor

**When:** February 24, 1946

**Where:** Argentina

**Who:**

ROBERTO ORTIZ (1886-1942), president of Argentina from 1938 to 1940

RAMÓN S. CASTILLO (1873-1944), president of Argentina from 1940 to 1943

JUAN PERÓN (1895-1974), president of Argentina from 1946 to 1955 and from 1973 to 1974

EVA "EVITA" DUARTE DE PERÓN (1919-1952), a radio personality, Perón's second wife

## Decade of Infamy

Argentina's last constitutionally elected president before World War II, Hipólito Irigoyen, was overthrown by a military coup in September, 1930. For the next sixteen years, conservative political leaders ruled the country by fraud and fear. The Argentine people had few political rights, and elections were rigged so that the government's candidates always won.

In the early part of the twentieth century, most workers in Argentina had been immigrants who worked in meat-packing plants or in the seaports. During the Depression of the 1930's, however, Argentina began expanding its industry, since its exports of beef and other agricultural products decreased. As more factories were built, more Argentines moved to the cities to take jobs in industry.

But during the 1930's, which came to be known as the "Decade of Infamy," labor laws were ignored. Many workers believed that the government was not upholding their rights to safe working conditions and impartial mediation when there were conflicts between workers and management.

When Roberto Ortiz was elected president in 1937, many people hoped that there would be better relations between labor and government. The *concordancia*, as the ruling coalition was called, allowed little political opposition, but President Ortiz made an effort to begin political reforms by refusing to recognize dishonest elections in Buenos Aires, the Argentinian capital.

Ortiz had to resign in 1938 because of poor health, however, and Vice President Ramón Castillo took power. Castillo soon showed that he was not interested in the people's right to choose their leaders or in the rights of labor.

New elections were scheduled for 1943, and the Argentinian people were concerned mostly about one issue. Could they bring an end to government by fraud? Castillo had picked a powerful conservative politician to succeed him, but the people were tired of having no real voice in the government.

The military stepped in on June 4, 1943, when a secret fraternity known as the Group of United Officers (GOU) set up a new government with General Pedro Ramírez as president. Under the leadership of Juan Perón, an assistant to General Edelmino Farrell, the government quickly took control of the press and radio. At a rally for earthquake victims in San Juan in 1944, Perón met a radio personality known as Evita (Eva Duarte), who eventually became his wife.

## Perón's Rise

Soon Perón became head of the Department of Labor and began responding to the complaints of workers and poor people. Because of his influence, an independent Secretariat of Labor and Social Welfare was created, and he used it to grant many benefits and rights to Argentina's poorer classes.

Workers' income had fallen during the 1930's, and shantytowns had sprung up around the city of Buenos Aires. Labor leaders, workers, and the poor came to see Perón as their representative in the government. In 1943 he made sure that workers gained a 40 percent increase in wages and the legal right to organize and negotiate. By the next year, workers were getting paid holidays, the government had helped to arrange more than two hundred agreements between workers and employers, and 2 million workers had gained benefits.

When General Farrell became president of Argentina, Perón became vice president. Liberals, however, suspected Perón of sympathizing with the Nazis and Fascists in World War II, and some military leaders were offended by his alliance with Eva Duarte. Farrell forced Perón to resign in October, 1945.

Perón was taken to an island in the Río de la Plata. When workers in Buenos Aires heard of his arrest, however, they called strikes and began a march to the center of the capital on October 17. Because many of them took off their shirts and waved them as banners, middle-class Argentines began calling Perón's supporters "the shirtless ones." The march was so huge that President Farrell was forced to bring Perón to address the people. Perón stood on a balcony, happily receiving their cheers.

Now Perón began preparing for the upcoming presidential election; he was supported by the new Labor Party as well as a group from the middle-class Radical Party. Because he called for religious instruction in the schools, the Catholic Church gave him indirect support as well. In February, 1946, in one of the most open and honest elections ever held in Argentina, Perón received nearly 54 percent of the vote. He was inaugurated as president in June of that year.

### Consequences

Under Perón, workers' wages continued to rise. In 1947, the government took over a private

*Juan Perón (left) and his wife, Eva.*

charity which was eventually renamed the María Eva Duarte de Perón Social Aid Foundation and was headed by the first lady. This foundation built schools, gave funding to hospitals, and helped poor people and orphans. At Evita's insistence, women were given the vote in 1947. Because of these reforms, Perón and Evita were adored by millions of Argentines.

Perón brought a sense of dignity to the workers of Argentina. He wanted Argentina's economy to be dominated not by wealthy landowners but by modern industries. In March, 1949, a new national constitution was proclaimed. It guaranteed workers the right to work, fair pay, good working conditions, dignity, and health.

The new constitution also gave a president the right to succeed himself, and Perón was reelected in 1951. During the early 1950's, however, his government became more and more authoritarian. Some said that his political opponents were being arrested and even tortured. For a time the government closed *La Prensa*, a respected newspaper.

A military coup brought an end to Perón's government in 1955, but many working-class Argentines hoped for his return to power. He was again elected president of Argentina in 1973, only to die in office in July, 1974.

*James A. Baer*

**1018**

# Churchill Delivers "Iron Curtain" Speech

---

*In a speech delivered in Missouri, Sir Winston Churchill fired the first verbal shot of the Cold War.*

---

**What:** International relations
**When:** March 5, 1946
**Where:** Fulton, Missouri
**Who:**
Sir Winston Churchill (1874-1965),
   former prime minister of Great Britain
Harry S. Truman (1884-1972), president
   of the United States from 1945 to
   1953
James Francis Byrnes (1879-1972), U.S.
   secretary of state from 1945 to 1947

## Message to the President

The phrase "Iron Curtain" was invented by Sir Winston Churchill, who served as Great Britain's prime minister from 1940 to 1945. On May 12, 1945—exactly one month after the death of U.S. president Franklin D. Roosevelt—Churchill sent a message to Roosevelt's successor, Harry S. Truman. Germany had just surrendered, and in his message Churchill tried to persuade Truman to ignore the German occupation zones the Allies had set up at the Quebec Conference in August, 1943.

Concerned about the Soviet army's presence in Eastern Europe, Churchill also urged Truman to keep holding the British and American positions in Yugoslavia, Austria, Czechoslovakia, northern Germany, and Denmark. Churchill ended his message by advising Truman not to move his armies until the leaders of Great Britain, the United States, and the Soviet Union had met.

Although Nazi Germany had been defeated, the United States was still fighting Japan in the Pacific. Truman was suspicious of the Soviet Union, yet some of his advisers encouraged him to keep cooperating with the Soviets. If he was too tough, the Soviets might become equally tough and push the British and Americans out of

Berlin and Vienna. Also, the atomic bomb had not yet been tested, and American leaders believed that they needed the Soviets' help to defeat Japan. This gave Truman another reason to try to work with Joseph Stalin, the Soviet dictator.

So it was that the first "Iron Curtain" message—a secret wartime dispatch—was rejected in Washington. By the time Churchill used the phrase in public a year later, however, the situation had changed.

## The Speech at Fulton

Churchill was visiting Washington, D.C., in the winter of 1946 when he was invited to deliver the John Findley Green Foundation Lecture at Westminster College in Fulton, Missouri. Because President Truman agreed to preside at the lecture, it became an official government event.

Up to that time, Truman's foreign policy had been based on confidence in the United Nations and faith in the Soviet Union's willingness to cooperate. These attitudes were in harmony with traditional American isolationism, yet the State Department was actually finding that the policy was not working very well, especially in Poland and Iran. Secretary of State James F. Byrnes reviewed the notes of Churchill's lecture and approved of them—a signal that the Truman administration was getting ready to change its foreign policy.

Churchill's speech, "The Sinews of Peace," was delivered on March 5, 1946, before an audience of forty thousand people. He opened by urgently reminding the American people that victory in war had left them at the height of power, giving them a special responsibility for peace. He asked all nations to cooperate with the United Nations and to make it more effective by establishing an "International Armed Force." Yet the secrets of atomic weaponry, which still belonged only to the United States, Great Britain, and Canada, must remain in safe hands until "the essential brotherhood of men is truly embodied and expressed in a world organization."

Encouraging the American people toward peace, Churchill warned them, however, that they should beware of tyranny; together with all liberty-loving people, they should proclaim "in fearless tones the great principles of freedom and the rights of man." Britons, Americans, and other English-speaking peoples should, Churchill said, join in a permanent alliance for mutual defense.

Next, Churchill expressed admiration for the Soviet people and his wartime comrade, Joseph Stalin. As if he were talking directly to the Soviet leaders, he acknowledged their right to have secure western frontiers so that they would be safe from any future German aggression. British and American leaders were determined, Churchill said, to create a lasting friendship with the

Soviets in spite of "the many differences and rebuffs."

Nevertheless, Churchill believed that it was his duty "not to misstate the facts" about the current situation in Europe. He then made his famous warning: "From Stettin in the Baltic to Trieste in the Adriatic, an iron curtain has descended across the Continent. Behind that line lie all the capitals of the ancient states of Central and Eastern Europe. Warsaw, Berlin, Prague, Vienna, Budapest, Belgrade, Bucharest, and Sophia . . . lie in what I must call the Soviet sphere and all are subject in one form or another, not only to Soviet influence but to a very high and increasing measure of control from Moscow."

Churchill was sure that while Soviet leaders did not want war, they did want to keep expanding their power and spreading Communist influence. Believing that the Soviets admired strength

Hulton Archive

*Winston Churchill (left) and President Harry S. Truman leave for Fulton, Missouri, where Churchill delivered his famous speech.*

and had only contempt for military weakness, Churchill argued that the West should create military and moral unity to oppose the "Iron Curtain." The first step was to complete a defense pact between Great Britain and the United States.

## Consequences

Americans were anxious to keep the peace after World War II, and many of them found Churchill's speech shocking because he argued so strongly that the Soviet Union was acting aggressively in Eastern Europe. There were reactions abroad as well. Eight days after the speech, the Soviet newspaper *Pravda* angrily condemned the speech for creating conflict among the Allied governments.

Though Truman and Byrnes basically agreed with Churchill's interpretation of world problems, they continued to try to cooperate with the Soviet Union. By 1947, however, the Soviet Union was extending its political influence in Central and Eastern Europe and refusing to cooperate in the joint Allied occupation of Berlin. The Soviets had also become involved in a guerrilla war in Greece and were threatening to build military bases in Turkey.

When Great Britain announced that it was withdrawing from Greece, Truman responded in March, 1947, with the Truman Doctrine, promising military and economic help for Greece and Turkey. A major change in American foreign policy had occurred: The United States had committed itself to "containing" Communism. This process continued in 1948 with the Marshall Plan and in 1949 with the North Atlantic Treaty Organization. By that time, "Iron Curtain" had become a common phrase in discussions about international relations.

*Christopher J. Kauffmann*

# Philippines Republic Regains Its Independence

*The Philippine Islands, which came under American rule in 1898 and were occupied by Japanese military forces from 1942 to 1945, were declared an independent republic by an agreement between the governments of the United States and the Philippines.*

**What:** Political independence; International relations

**When:** July 4, 1946

**Where:** Manila, Philippines

**Who:**

SERGIO OSMEÑA (1878-1961), Quezon's successor as president of the Philippine Commonwealth

MANUEL ROXAS (1892-1948), president of the modern Republic of the Philippines

DOUGLAS MACARTHUR (1880-1964), U.S. general

## Independence Regained

At about 9:00 A.M. on July 4, 1946, Paul V. McNutt, U.S. high commissioner of the Philippine Commonwealth, lowered the U.S. flag in Manila, capital city of the Philippines. Before a cheering crowd of Filipinos, new Philippine president Manuel Roxas raised the flag of his own nation. Although the Republic of the Philippines would continue to have close ties to the United States, the Southeast Asian island nation was once again independent.

The Philippines became a republic for the second time, or the third, if the short-lived republic established by the Japanese during their military occupation of the islands is counted. The Philippines was colonized by Spain in the 1500's. During the late 1800's, Filipinos rose up against Spanish rule, and by 1898, Philippine troops had driven the Spanish out of most of the countryside. However, in 1898, war broke out between Spain and the United States, which was trying to drive Spain out of Cuba.

After Spain surrendered to U.S. forces, the Spanish agreed to grant Cuba independence and to cede the Philippines, Guam, and Puerto Rico to the United States for $20 million. Many Filipinos saw this as a betrayal by their U.S. allies. On January 23, 1899, an assembly in the town of Malolos declared the Philippines an independent republic with the young general Emilio Aguinaldo as its president. The following February, fighting broke out between U.S. and Philippine forces in Manila, and Aguinaldo declared war against the United States.

### The Americans and the Japanese

The struggle between the United States and the Philippine independence fighters was a long and bloody one. No one knows how many Filipinos died in the course of this war, but estimates usually run from 200,000 to 500,000 people. By 1902, the United States had defeated Aguinaldo's troops. Aguinaldo was captured and swore allegiance to the U.S. government.

Having established themselves in the Philippines by military conquest, the Americans proceeded to create what they saw as a democratic government. Political offices were filled by popular election. Two major political parties competed. The Nacionalista party favored independence for the Philippines, and the Federalista party favored becoming a state in the United States. However, the Americans made few attempts to overcome one of the biggest problems in establishing a truly democratic system in their new colony. Ownership of land and wealth was highly unequal in the Philippines, and most of those who took part in the government came from the wealthy families. Because these wealthy families were closely connected to the Americans, foreign domination and economic inequality were intertwined in Philippine society.

In March, 1934, the U.S. Congress passed the Tydings-McDuffie Act, establishing a commonwealth in the Philippines. The Commonwealth of the Philippines was to be a self-governing region with its own constitution that would become fully independent after a ten-year period. On September 17, 1935, Nacionalista leader Manuel Luis Quezon was elected president of the Philippine Commonwealth, and his fellow Nacionalista Sergio Osmeña was elected vice president.

Japan attacked the U.S. naval base at Pearl Harbor on December 7, 1941, drawing the United States into World War II. By April, 1942, the Japanese had defeated the combined U.S. and Philippine forces in the Bataan Peninsula and taken control of the country. In the beginning, some Filipinos, including the aging Aguinaldo, favored the Japanese occupation, believing that this would lead to independence from the United States. Japanese troops were widely unpopular, however, and their harsh treatment of Filipinos made them more unpopular. In October, 1943, the Japanese, hoping to rule through the Filipino elite, established a Philippine republic under President José Laurel. Laurel was widely seen as a Japanese puppet and his government won little popular support, but many Filipinos list Laurel, as well as Aguinaldo, among the country's presidents.

## Consequences

On July 29, 1944, the U.S. Congress passed a joint resolution declaring that a completely independent government would be established in the Philippines. However, critics have often argued that the new government continued to be dependent on the United States. In 1945, American forces retook the Philippines from the Japanese in fighting so intense that much of Manila was destroyed. General Douglas MacArthur, former military adviser to Quezon and supreme commander for the Pacific during the war, was the most powerful individual in the Philippines and played a large part in shaping the new republic. MacArthur instructed Osmeña, who had become president of the Commonwealth after the death of Quezon, to convene the Philippine legislature that had been elected in 1941 in order to prepare the country for independence. Roxas, a friend of MacArthur, became president of the Commonwealth senate and chair of the Committee on Appointments. These offices, combined with MacArthur's support, enabled Roxas to outmaneuver Osmeña in the election for president of the new republic.

Some historians and political commentators argue that the events of July 4, 1946, did not confer true independence on the Philippines. They argue that the Philippines continued to be economically and politically dependent on the United States, and they point out that the United States retained vast military bases in the Philippines. Nevertheless, the Philippines gradually developed economic relations with many other nations, including Japan, and in 1991, the Philippine legislature voted to remove the U.S. military bases from the islands.

*Carl L. Bankston III*

# Schaefer Seeds Clouds with Dry Ice

> *Working alone in the General Electric Research Laboratory in New York, Vincent Joseph Schaefer created a miniature snowstorm inside a deep freezer by tossing a handful of dry ice inside.*

**What:** Earth science
**When:** July 12, 1946
**Where:** General Electric Research Laboratory, near Schenectady, New York
**Who:**
VINCENT JOSEPH SCHAEFER (1906-1993), an American chemist and meteorologist
IRVING LANGMUIR (1881-1957), an American physicist and chemist who won the 1932 Nobel Prize in Chemistry
BERNARD VONNEGUT (1914-1997), an American physical chemist and meteorologist

## Praying for Rain

Beginning in 1943, an intense interest in the study of clouds developed into the practice of weather "modification." Working for the General Electric Research Laboratory, Nobel laureate Irving Langmuir and his assistant researcher and technician, Vincent Joseph Schaefer, began an intensive study of precipitation and its causes.

Past research and study had indicated two possible ways that clouds produce rain. The first possibility is called "coalescing," a process by which tiny droplets of water vapor in a cloud merge after bumping into one another and become heavier and fatter until they drop to earth. The second possibility is the "Bergeron process" of droplet growth, named after the Swedish meteorologist Tor Bergeron. Bergeron's process relates to supercooled clouds, or clouds that are at or below freezing temperatures and yet still contain both ice crystals and liquid water droplets. The size of the water droplets allows the droplets to remain liquid despite freezing temperatures; while small droplets can remain liquid only down to 4 degrees Celsius, larger droplets may not freeze until reaching −15 degrees Celsius. Precipitation occurs when the ice crystals become heavy enough to fall. If the temperature at some point below the cloud is warm enough, it will melt the ice crystals before they reach the earth, producing rain. If the temperature remains at the freezing point, the ice crystals retain their form and fall as snow.

Schaefer used a deep-freezing unit in order to observe water droplets in pure cloud form. In order to observe the droplets better, Schaefer lined the chest with black velvet and concentrated a beam of light inside. The first agent he introduced inside the supercooled freezer was his own breath. When that failed to form the desired ice crystals, he proceeded to try other agents. His hope was to form ice crystals that would then cause the moisture in the surrounding air to condense into more ice crystals, which would produce a miniature snowfall.

He eventually achieved success when he tossed a handful of dry ice inside and was rewarded with the long-awaited snow. The freezer was set at the freezing point of water, 0 degrees Celsius, but not all the particles were ice crystals, so when the dry ice was introduced all the stray water droplets froze instantly, producing ice crystals, or snowflakes.

## Planting the First Seeds

On November 13, 1946, Schaefer took to the air over Mount Greylock with several pounds of dry ice in order to repeat the experiment in nature. After he had finished sprinkling, or seeding, a supercooled cloud, he instructed the pilot to fly underneath the cloud he had just seeded. Schaefer was greeted by the sight of snow. By the time it reached the ground, it had melted into the first-ever human-made rainfall.

Independently of Schaefer and Langmuir, another General Electric scientist, Bernard Vonnegut, was also seeking a way to cause rain. He found that silver iodide crystals, which have the same size and shape as ice crystals, could "fool" water droplets into condensing on them. When a certain chemical mixture containing silver iodide is heated on a special burner called a "generator," silver iodide crystals appear in the smoke of the mixture. Vonnegut's discovery allowed seeding to occur in a way very different from seeding with dry ice, but with the same result. Using Vonnegut's process, the seeding is done from the ground. The generators are placed outside and the chemicals are mixed. As the smoke wafts upward, it carries the newly formed silver iodide crystals into the clouds.

The results of the scientific experiments by Langmuir, Vonnegut, and Schaefer were alternately hailed and rejected as legitimate. Critics argue that the process of seeding is too complex and would have to require more than just the addition of dry ice or silver nitrate in order to produce rain. One of the major problems surrounding the question of weather modification by cloud seeding is the scarcity of knowledge about the earth's atmosphere. A journey begun over fifty years ago is still a long way from being completed.

## Consequences

Although the actual statistical and other proofs needed to support cloud seeding are lacking, the discovery in 1946 by the General Electric employees set off a wave of interest and demand for information that far surpassed the interest generated by the discovery of nuclear fission shortly before. The possibility of ending drought and, in the process, hunger excited many people. The discovery also prompted both legitimate and false "rainmakers" who used the information gathered by Schaefer, Langmuir, and Vonnegut to set up cloud-seeding businesses. Weather modification, in its current stage of development, cannot be used to end worldwide drought. It does, however, have beneficial results in some cases on the crops of smaller farms that have been affected by drought.

In order to understand the advances made in weather modification, new instruments are needed to record accurately the results of further experimentation. The storm of interest— both favorable and nonfavorable—generated by the discoveries of Schaefer, Langmuir, and Vonnegut has had and will continue to have far-reaching effects on many aspects of society.

*Earl G. Hoover*

# Paris Peace Conference Makes Treaties with Axis Powers

*After World War II, representatives of Allied nations met in Paris to approve treaties of peace with Italy, Finland, Romania, Bulgaria, and Hungary.*

**What:** International relations
**When:** July 29, 1946-February 10, 1947
**Where:** Paris
**Who:**
JAMES FRANCIS BYRNES (1879-1972), U.S. secretary of state from 1945 to 1947
ERNEST BEVIN (1884-1951), British foreign secretary from 1945 to 1951
VYACHESLAV MIKHAILOVICH MOLOTOV (1890-1986), Soviet commissar for foreign affairs from 1939 to 1949

## The Nations Gather

In the summer of 1945, shortly after Nazi Germany had been defeated, the Big Three wartime allies—the United States, the Soviet Union, and Great Britain—met in the Potsdam Conference to consider many problems related to the end of the war. At Potsdam they agreed to establish a Council of Foreign Ministers whose task would be to write the peace treaties: first with Italy, Romania, Bulgaria, Hungary, and Finland, and eventually with Germany.

Between December, 1945, and July, 1946, the Council of Ministers discussed the terms of the treaties to be presented to a peace conference. The council included James F. Byrnes, U.S. secretary of state, and foreign ministers Vyacheslav M. Molotov of the Soviet Union, Ernest Bevin of Great Britain, and Georges Bidault of France. Bidault was not as involved, however, as the other three.

As the foreign ministers continued their talks, hostility was arising between the Western Powers and the Soviet Union; each side feared that the other was trying to gain an unfair advantage. By midsummer, after many disagreements, the Council of Foreign Ministers had written almost complete drafts of the peace treaties for each of the five minor Axis Powers. They never succeeded in preparing a draft treaty for Germany, because the Soviet Union could not come to an agreement with the United States and Great Britain over the future of that country.

The peace conference, which was to meet in Paris, would have the task of accepting, rejecting, or amending the articles of the draft treaties. So it was that the Second Paris Peace Conference, like the first (in 1919, after World War I), was dominated by a "Big Four."

The Second Paris Peace Conference formally opened on July 29, 1946. Twenty-one Allied nations, including the Big Four, were represented. Representatives of the five enemy nations were invited to present their views in speeches before the conference. Most of the work of the conference, however, had to do with examining the draft treaties very closely, and this was handled by various committees.

By October 7, the committees had finished their reports and were ready to present them to the full conference. In the week that followed, the five treaties were approved, and the conference adjourned on October 14. The Big Four had decided ahead of time, however, that the Council of Foreign Ministers must meet again after the conference to approve any changes that had been made in the treaties.

## The Treaties

Of the five treaties, the longest, and perhaps the most important, was with Italy. Italy was forced to pay $100 million in reparations (war damages) to the Soviet Union and $260 million to be divided among Greece, Yugoslavia, Albania, and Ethiopia. Limits were put on the Italian armed forces, and Italy lost all of its African empire, which had included Libya, Ethiopia, Eritrea, and Italian Somaliland.

The Italian treaty also guaranteed that Albania, which Italy had tried to dominate, would keep its independence. The treaty did not resolve the question of Trieste and its surrounding area, which had been claimed by both Yugoslavia and Italy. A compromise solution to this dispute was not found until 1954.

The peace treaties with Finland, Hungary, Romania, and Bulgaria were similar to Italy's. All four states were forced to put strict limits on their armed forces and to make reparation payments to various neighboring nations, including the Soviet Union and other Eastern European countries.

Finland had to give the Soviet Union certain territories, the most valuable of which was the Petsamo district, which contained mineral de-

posits and provided an outlet to the Arctic Sea. Hungary returned to the borders it had had in January, 1938, except for a small area taken by Czechoslovakia. Romania gave up Bessarabia and northern Bukovina to the Soviet Union, which had seized those areas in June, 1940. Also, Romania was forced to accept that the southern Dobruja area would remain with Bulgaria.

The final versions of the five peace treaties were signed in Paris on February 10, 1947, by representatives of the Axis states and representatives of all the Allied states that had participated in the peace conference, except for the United States. Secretary Byrnes had already signed the treaties in Washington, D.C., on January 20, the day before he left office.

## Consequences

In some ways, the Paris Peace Conference of 1946-1947 was based on the peace settlements af-

*The opening of the Paris Peace Conference.*

ter World War I. The 1946-1947 treaties reaffirmed the general territorial lines set for the Eastern European nations in the 1919 conference. This time, however, not much was said about the "self-determination of peoples"—a major concern at the earlier conference.

Soviet influence in Eastern Europe was growing rapidly, even as the final documents were being prepared for signatures. Romania, Hungary, and Bulgaria were becoming Soviet satellites; though the United States and Great Britain soon protested actions in those nations that violated human-rights sections of the peace treaties, the protests had no effect.

From the Western point of view, the settlements with Italy and Finland were more positive. The Finland treaty helped to strengthen Finnish independence and national sovereignty, protecting it against aggression from its Soviet neighbor. The Italian treaty helped the Italians to develop a partnership with the West and to resolve its long-standing territorial dispute with Yugoslavia.

Peace with Japan was handled as a completely separate matter, and a formal treaty with Germany was never written at all. The Allies' settlement with Austria came only in 1955 and had remarkably easy terms. By that time the East-West conflict had grown into the Cold War; Great Britain and the United States found it more important to build alliances in Central Europe than to punish Austria and Germany for their roles in World War II.

*Paul D. Mageli*

# Physicists Build Synchrocyclotron

*Theoretical developments by Edwin Mattison McMillan and Vladimir Iosifovich Veksler led to the construction of the synchrocyclotron, a powerful particle accelerator that performed better than its predecessor, the cyclotron.*

**What:** Physics
**When:** November, 1946
**Where:** Berkeley, California
**Who:**

EDWIN MATTISON MCMILLAN (1907-1991), an American physicist who won the Nobel Prize in Chemistry in 1951

VLADIMIR IOSIFOVICH VEKSLER (1907-1966), a Soviet physicist

ERNEST ORLANDO LAWRENCE (1901-1958), an American physicist

HANS ALBRECHT BETHE (1906-    ), a German American physicist

## The First Cyclotron

The synchrocyclotron is a large electromagnetic apparatus designed to accelerate atomic and subatomic particles at high energies. Therefore, it falls under the broad class of scientific devices known as "particle accelerators." By the early 1920's, the experimental work of physicists such as Ernest Rutherford and George Gamow demanded that an artificial means be developed to generate streams of atomic and subatomic particles at energies much greater than those occurring naturally. This requirement led Ernest Orlando Lawrence to develop the cyclotron, the prototype for most modern accelerators. The synchrocyclotron was developed in response to the limitations of the early cyclotron.

In September, 1930, Lawrence announced the basic principles behind the cyclotron. Ionized—that is, electrically charged—particles are admitted into the central section of a circular metal drum. Once inside the drum, the particles are exposed to an electric field alternating within a constant magnetic field. The combined action of the electric and magnetic fields accelerates the particles into a circular path, or orbit. This increases the particles' energy and orbital radii. This process continues until the particles reach the desired energy and velocity and are extracted from the machine for use in experiments ranging from particle-to-particle collisions to the synthesis of radioactive elements.

Although Lawrence was interested in the practical applications of his invention in medicine and biology, the cyclotron also was applied to a variety of experiments in a subfield of physics called "high-energy physics." Among the earliest applications were studies of the subatomic, or nuclear, structure of matter. The energetic particles generated by the cyclotron made possible the very type of experiment that Rutherford and Gamow had attempted earlier. These experiments, which bombarded lithium targets with streams of highly energetic accelerated protons, attempted to probe the inner structure of matter.

Although funding for scientific research on a large scale was scarce before World War II (1939-1945), Lawrence nevertheless conceived of a 467-centimeter cyclotron that would generate particles with energies approaching 100 million electronvolts. By the end of the war, increases in the public and private funding of scientific research and a demand for higher-energy particles created a situation in which this plan looked as if it would become reality, were it not for an inherent limit in the physics of cyclotron operation.

## Overcoming the Problem of Mass

In 1937, Hans Albrecht Bethe discovered a severe theoretical limitation to the energies that could be produced in a cyclotron. Physicist Albert Einstein's special theory of relativity had demonstrated that as any mass particle gains velocity relative to the speed of light, its mass increases. Bethe showed that this increase in mass

**1029**

Lawrence Radiation Laboratory; Courtesy AIP Emilio Segrè Visual Archives

*Edwin Mattison McMillan at the controls of the synchrocyclotron in 1948.*

would eventually slow the rotation of each particle. Therefore, as the rotation of each particle slows and the frequency of the alternating electric field remains constant, particle velocity will decrease eventually. This factor set an upper limit on the energies that any cyclotron could produce.

Edwin Mattison McMillan, a colleague of Lawrence at Berkeley, proposed a solution to Bethe's problem in 1945. Simultaneously and independently, Vladimir Iosifovich Veksler of the Soviet Union proposed the same solution. They suggested that the frequency of the alternating electric field be slowed to meet the decreasing rotational frequencies of the accelerating particles—in essence, "synchronizing" the electric field with the moving particles. The result was the synchrocyclotron.

Prior to World War II, Lawrence and his colleagues had obtained the massive electromagnet

for the new 100-million-electronvolt cyclotron. This 467-centimeter magnet would become the heart of the new Berkeley synchrocyclotron. After initial tests proved successful, the Berkeley team decided that it would be reasonable to convert the cyclotron magnet for use in a new synchrocyclotron. The apparatus was operational in November of 1946.

These high energies combined with economic factors to make the synchrocyclotron a major achievement for the Berkeley Radiation Laboratory. The synchrocyclotron required less voltage to produce higher energies than the cyclotron because the obstacles cited by Bethe were virtually nonexistent. In essence, the energies produced by synchrocyclotrons are limited only by the economics of building them. These factors led to the planning and construction of other synchrocyclotrons in the United States and Europe. In 1957, the Berkeley apparatus was redesigned in order to achieve energies of 720 million electronvolts, at that time the record for cyclotrons of any kind.

## Consequences

Previously, scientists had had to rely on natural sources for highly energetic subatomic and atomic particles with which to experiment. In the mid-1920's, the American physicist Robert Andrews Millikan began his experimental work in cosmic rays, which are one natural source of energetic particles called "mesons." Mesons are charged particles that have a mass more than two hundred times that of the electron and are therefore of great benefit in high-energy physics experiments. In February of 1949, McMillan announced the first synthetically produced mesons using the synchrocyclotron.

McMillan's theoretical development led not only to the development of the synchrocyclotron but also to the development of the electron synchrotron, the proton synchrotron, the microtron, and the linear accelerator. Both proton and electron synchrotrons have been used successfully to produce precise beams of muons and pi-mesons, or pions (a type of meson).

The increased use of accelerator apparatus ushered in a new era of physics research, which has become dominated increasingly large accelerators and, subsequently, larger teams of scientists and engineers required to run individual experiments. More sophisticated machines have generated energies in excess of 2 trillion electronvolts at the United States' Fermi National Accelerator Laboratory, or Fermilab, in Illinois. Part of the huge Tevatron apparatus at Fermilab, which generates these particles, is a proton synchrotron, a direct descendant of McMillan and Lawrence's early efforts.

*David Wason Hollar, Jr.*

# Gabor Invents Holography

*Dennis Gabor created a lensless system of three-dimensional photography that is one of the most important developments in twentieth century optical science.*

**What:** Physics; Photography
**When:** 1947
**Where:** Rugby, England
**Who:**

DENNIS GABOR (1900-1979), a Hungarian-born inventor and physicist who was awarded the 1971 Nobel Prize in Physics

EMMETT LEITH (1927-    ), a radar researcher who, with Juris Upatnieks, produced the first laser holograms

JURIS UPATNIEKS (1936-    ), a radar researcher who, with Emmett Leith, produced the first laser holograms

## Easter Inspiration

The development of photography in the early 1900's made possible the recording of events and information in ways unknown before the twentieth century: the photographing of star clusters, the recording of the emission spectra of heated elements, the storing of data in the form of small recorded images (for example, microfilm), and the photographing of microscopic specimens, among other things. Because of its vast importance to the scientist, the science of photography has developed steadily.

An understanding of the photographic and holographic processes requires some knowledge of the wave behavior of light. Light is an electromagnetic wave that, like a water wave, has an amplitude and a phase. The amplitude corresponds to the wave height, while the phase indicates which part of the wave is passing a given point at a given time. A cork floating in a pond bobs up and down as waves pass under it. The position of the cork at any time depends on both amplitude and phase: The phase determines on which part of the wave the cork is floating at any given time,

and the amplitude determines how high or low the cork can be moved. Waves from more than one source arriving at the cork combine in ways that depend on their relative phases. If the waves meet in the same phase, they add and produce a large amplitude; if they arrive out of phase, they subtract and produce a small amplitude. The total amplitude, or intensity, depends on the phases of the combining waves.

Dennis Gabor, the inventor of holography, was intrigued by the way in which the photographic image of an object was stored by a photographic plate but was unable to devote any consistent research effort to the question until the 1940's. At that time, Gabor was involved in the development of the electron microscope. On Easter morning in 1947, as Gabor was pondering the problem of how to improve the electron microscope, the solution came to him. He would attempt to take a poor electron picture and then correct it optically. The process would require coherent electron beams—that is, electron waves with a definite phase.

This two-stage method was inspired by the work of Lawrence Bragg. Bragg had formed the image of a crystal lattice by diffracting the photographic X-ray diffraction pattern of the original lattice. This double diffraction process is the basis of the holographic process. Bragg's method was limited because of his inability to record the phase information of the X-ray photograph. Therefore, he could study only those crystals for which the phase relationship of the reflected waves could be predicted.

## Waiting for the Laser

Gabor devised a way of capturing the phase information after he realized that adding coherent background to the wave reflected from an object would make it possible to produce an interference pattern on the photographic plate. When

**1032**

the phases of the two waves are identical, a maximum intensity will be recorded; when they are out of phase, a minimum intensity is recorded. Therefore, what is recorded in a hologram is not an image of the object but rather the interference pattern of the two coherent waves. This pattern looks like a collection of swirls and blank spots. The hologram (or photograph) is then illuminated by the reference beam, and part of the transmitted light is a replica of the original object wave. When viewing this object wave, one sees an exact replica of the original object.

The major impediment at the time in making holograms using any form of radiation was a lack of coherent sources. For example, the coherence of the mercury lamp used by Gabor and his assistant Ivor Williams was so short that they were able to make holograms of only about a centimeter in diameter. The early results were rather poor in terms of image quality and also had a double image. For this reason, there was little interest in holography, and the subject lay almost untouched for more than ten years.

Interest in the field was rekindled after the laser (*light amplification by stimulated emission of radiation*) was developed in 1962. Emmett Leith and Juris Upatnieks, who were conducting radar research at the University of Michigan, published the first laser holographs in 1963. The laser was an intense light source with a very long coherence length. Its monochromatic nature improved the resolution of the images greatly. Also, there was no longer any restriction on the size of the object to be photographed.

The availability of the laser allowed Leith and Upatnieks to propose another improvement in holographic technique. Before 1964, holograms were made of only thin transparent objects. A small region of the hologram bore a one-to-one correspondence to a region of the object. Only a small portion of the image could be viewed at one time without the aid of additional optical components. Illuminating the transparency diffusely allowed the whole image to be seen at one time. This development also made it possible to record holograms of diffusely reflected three-dimensional objects. Gabor had seen from the beginning that this should make it possible to create three-dimensional images.

After the early 1960's, the field of holography developed very quickly. Because holography is different from conventional photography, the two techniques often complement each other. Gabor saw his idea blossom into a very important technique in optical science.

## Consequences

The development of the laser and the publication of the first laser holograms in 1963 caused a blossoming of the new technique in many fields. Soon, techniques were developed that allowed holograms to be viewed with white light. It also became possible for holograms to reconstruct multicolored images. Holographic methods have been used to map terrain with radar waves and to conduct surveillance in the fields of forestry, agriculture, and meteorology.

At the beginning of the twenty-first century, holography was a multimillion-dollar industry, finding applications in advertising, as an art form, and in security devices on credit cards, as well as in scientific fields. An alternate form of holography, also suggested by Gabor, uses sound waves. Acoustical imaging is useful whenever the medium around the object to be viewed is opaque to light rays—for example, in medical diagnosis. Holography has affected many areas of science, technology, and culture.

*Grace A. Banks*

# Libby Uses Carbon 14 to Date Ancient Objects

---

*The radioactive decay of carbon 14 has been used to measure the age of archaeological objects going back ten thousand years or more.*

---

**What:** Archaeology
**When:** 1947-1949
**Where:** Chicago, Illinois
**Who:**
WILLARD FRANK LIBBY (1908-1980), an
   American chemist who won the 1960
   Nobel Prize in Chemistry
CHARLES WESLEY FERGUSON (1922-1986),
   a scientist who demonstrated that
   carbon 14 dates before 1500 B.C.E.
   needed to be corrected

## One in a Trillion

Carbon dioxide in the earth's atmosphere contains a mixture of three carbon isotopes (isotopes are atoms of the same element that contain different numbers of neutrons), which occur in the following percentages: about 99 percent carbon 12, about 1 percent carbon 13, and approximately one atom in a trillion of radioactive carbon 14. Plants absorb carbon dioxide from the atmosphere during photosynthesis, and then animals eat the plants, so all living plants and animals contain a small amount of radioactive carbon.

When a plant or animal dies, its radioactivity slowly decreases as the radioactive carbon 14 decays. The time it takes for half of any radioactive substance to decay is known as its "half-life." The half-life for carbon 14 is known to be about fifty-seven hundred years. The carbon 14 activity will drop to one-half after one half-life, one-fourth after two half-lives, one-eighth after three half-lives, and so forth. After ten or twenty half-lives, the activity becomes too low to be measurable. Coal and oil, which were formed from organic matter millions of years ago, have long since lost any carbon 14 activity. Wood samples from an Egyptian tomb or charcoal from a prehistoric fireplace a few thousand years ago, however, can be dated with good reliability from the leftover radioactivity.

In the 1940's, the properties of radioactive elements were still being discovered and were just beginning to be used to solve practical problems. Scientists still did not know the half-life of carbon 14, and archaeologists still depended mainly on historical evidence to determine the ages of ancient objects.

In early 1947, Willard Frank Libby started a crucial experiment in testing for radioactive carbon. He decided to test samples of methane gas from two different sources. One group of samples came from the sewage disposal plant at Baltimore, Maryland, which was rich in fresh organic matter. The other sample of methane came from an oil refinery, which should have contained only ancient carbon from fossils whose radioactivity should have completely decayed. The experimental results confirmed Libby's suspicions: The methane from fresh sewage was radioactive, but the methane from oil was not. Evidently, radioactive carbon was present in fresh organic material, but it decays away eventually.

## Tree-Ring Dating

In order to establish the validity of radiocarbon dating, Libby analyzed known samples of varying ages. These included tree-ring samples from 1072 C.E. and 575 C.E. and one redwood from 979 B.C.E., as well as artifacts from Egyptian tombs going back to about 3000 B.C.E. In 1949, he published an article in the journal *Science* that contained a graph comparing the historical ages and the measured radiocarbon ages of eleven objects. The results were accurate within 10 percent, which meant that the general method was sound.

The first archaeological object analyzed by carbon dating, obtained from the Metropolitan Museum of Art in New York, was a piece of cypress wood from the tomb of King Djoser of Egypt. Based on historical evidence, the age of this piece of wood was about forty-six hundred years. A small sample of carbon obtained from this wood was deposited on the inside of Libby's radiation counter, giving a count rate that was about 40 percent lower than that of modern organic carbon. The resulting age of the wood calculated from its residual radioactivity was about thirty-eight hundred years, a difference of eight hundred years. Considering that this was the first object to be analyzed, even such a rough agreement with the historic age was considered to be encouraging.

The validity of radiocarbon dating depends on an important assumption—namely, that the abundance of carbon 14 in nature has been constant for many thousands of years. If carbon 14 was less abundant at some point in history, organic samples from that era would have started with less radioactivity. When analyzed today, their reduced activity would make them appear to be older than they really are.

Charles Wesley Ferguson from the Tree-Ring Research Laboratory at the University of Arizona tackled this problem. He measured the age of bristlecone pine trees both by counting the rings and by using carbon 14 methods. He found that carbon 14 dates before 1500 B.C.E. needed to be corrected. The results show that radiocarbon dates are older than tree-ring counting dates by as much as several hundred years for the oldest samples. He knew that the number of tree rings had given him the correct age of the pines, because trees accumulate one ring of growth for every year of life. Apparently, the carbon 14 content in the atmosphere has not been constant. Fortunately, tree-ring counting gives reliable dates that can be used to correct radiocarbon measurements back to about 6000 B.C.E.

## Consequences

Some interesting samples were dated by Libby's group. The Dead Sea Scrolls had been found in a cave by an Arab shepherd in 1947, but some Bible scholars at first questioned whether they were genuine. The linen wrapping from the Book of Isaiah was tested for carbon 14, giving a date of 100 B.C.E., which helped to establish its authenticity. Human hair from an Egyptian tomb was determined to be nearly five thousand years old. Well-preserved sandals from a cave in eastern Oregon were determined to be ninety-three hundred years old. A charcoal sample from a prehistoric site in western South Dakota was found to be about seven thousand years old.

The Shroud of Turin, located in Turin, Italy, has been a controversial object for many years. It is a linen cloth, more than four meters long, which shows the image of a man's body, both front and back. Some people think it may have been the burial shroud of Jesus Christ after his crucifixion. A team of scientists in 1978 was permitted to study the shroud, using infrared photography, analysis of possible blood stains, microscopic examination of the linen fibers, and other methods. The results were ambiguous. A carbon 14 test was not permitted because it would have required cutting a piece about the size of a handkerchief from the shroud.

The Nobel Foundation

*Willard Frank Libby.*

A new method of measuring carbon 14 was developed in the late 1980's. It is called "accelerator mass spectrometry," or AMS. Unlike Libby's method, it does not count the radioactivity of carbon. Instead, a mass spectrometer directly measures the ratio of carbon 14 to ordinary carbon. The main advantage of this method is that the sample size needed for analysis is about a thousand times smaller than before. The archbishop of Turin permitted three laboratories with the appropriate AMS apparatus to test the shroud material. The results agreed that the material was from the fourteenth century, not from the time of Christ. The figure on the shroud may be a watercolor painting on linen.

Since Libby's pioneering experiments in the late 1940's, carbon 14 dating has established itself as a reliable dating technique for archaeologists and cultural historians. Further improvements are expected to increase precision, to make it possible to use smaller samples, and to extend the effective time range of the method back to fifty thousand years or earlier.

*Hans G. Graetzer*

# Truman Presents His Doctrine of Containment

President Harry S. Truman announced that by giving help to countries threatened by the Soviet Union, the United States would try to restrain Soviet expansion.

**What:** International relations
**When:** March 12, 1947
**Where:** Washington, D.C.
**Who:**

HARRY S. TRUMAN (1884-1972), president of the United States from 1945 to 1953

SIR WINSTON CHURCHILL (1874-1965), prime minister of Great Britain from 1940 to 1945

JAMES FRANCIS BYRNES (1879-1972), U.S. secretary of state from 1945 to 1947

GEORGE CATLETT MARSHALL (1880-1959), U.S. secretary of state from 1947 to 1949

GEORGE F. KENNAN (1904-    ), director of the Policy Planning Staff, U.S. State Department, 1947

## Groping for a Policy

Soon after World War II ended, the United States was faced with the need to find a new approach to international relations. President Franklin D. Roosevelt had hoped for a postwar peace based on cooperation between the United States and the Soviet Union, but that idea was not working. The Red Army had occupied most of Eastern and Central Europe, and the Soviets were making it clear that they would not tolerate independent governments in those areas. Soviet dictator Joseph Stalin imposed Communist governments on Poland, Hungary, Bulgaria, and Romania, and protests from the United States and Great Britain seemed to fall on deaf ears.

The Soviet Union was also trying to expand into areas where it had no military control, including Greece, Turkey, and Iran. Having kept their occupation forces in Iran even though the war was over, the Soviets demanded exclusive oil and mineral rights there. After American and British leaders implied that they might give military aid to Iran, however, the Soviets began pulling their troops out in March, 1946.

In 1945 and 1946, the Soviet Union demanded that Turkey hand over some of its border territory. The Soviets also wanted Turkey to allow joint Soviet-Turkish control of the Dardanelles, along with Soviet navy and army bases in that area. After the second Soviet message to Turkey, however, the United States sent a strong naval fleet into the Mediterranean, and a week later British and American leaders rejected Soviet demands on Turkey. Meanwhile, Greece was succeeding in fighting off Communist guerrillas only with large amounts of military and economic help from Great Britain.

How to respond to the Soviet Union was debated hotly by Americans. One position in the debate was expressed by Winston Churchill, former British prime minister, in a speech at Fulton, Missouri, in early 1946. With Truman on the platform, Churchill painted the Soviet Union as an aggressive nation that could only be kept in line by a strong counterforce. The "Iron Curtain" that had fallen across Eastern Europe, Churchill warned, could be kept from extending itself only if the British and American governments cooperated to preserve the independence of Europe.

Another view was given by U.S. commerce secretary Henry A. Wallace, who insisted that only American-Soviet cooperation could prevent another war. He argued that the Soviet desire to control its border regions was understandable and reasonable—not much different from the United States' own actions to bring about stability in the Western Hemisphere.

The State Department tried to find a middle ground between these views. Secretary of State James F. Byrnes urged the Soviet Union to be more cooperative. Meanwhile, he said, the United States should have a policy of "firmness and patience," waiting for the Soviets to see that negotiation was the best way of solving problems.

Yet many Americans, including President Harry S. Truman, believed that the United States kept trying to be reasonable while the Soviet Union was unwilling to change. By 1947, the Truman administration had taken the position that the Soviet Union's Communist principles made traditional diplomacy impossible.

## Truman's Doctrine

In February, 1947, Great Britain informed the U.S. State Department that the British government could not continue to help the government of Greece. Like all of Western Europe, Great Britain was suffering from serious economic problems. As the British retreated, the Americans stepped forward. Within the next few weeks, Truman decided that the independence of Greece and the economic recovery of Europe were essential for American security.

On March 12, 1947, the president came before a joint session of Congress and presented what became known as the Truman Doctrine. Explaining the desperate situation in both Greece and Turkey, he called upon the American people to "help free peoples to maintain their free institutions and their national integrity against aggressive movements." Totalitarian aggression in these countries, he said, was a direct threat to the security of the United States.

In response, Congress appropriated $400 million for economic aid to Greece and Turkey. The president was also given authority to send civilian and military advisers to help both nations keep their independence.

If Greece and Turkey deserved aid, it seemed clear that the United States should help other European nations whose situations were equally desperate. Large numbers of British factories had been destroyed in the war, and it had few manufactured goods to export. Germany was in ruins, practically unable to feed its population. In France and Italy, the Communist Party had much support within the working class and was

trying to bring about radical changes in the French and Italian governments.

At Harvard University in June, 1947, the new U.S. secretary of state, George C. Marshall, outlined the American response to these European problems. The Marshall Plan would be based on cooperation among the European nations to solve their economic problems, and the United States would supply large amounts of aid. The American commitment to rebuilding Europe and keeping the Soviet Union at bay would be expressed in cold, hard cash.

The ideas behind the new American policy were explained in an unsigned article in the July, 1947, issue of *Foreign Affairs*. It was later revealed that the article's author was George F. Kennan, a high-ranking member of the State Department. According to Kennan, the growing hostility between the United States and the Soviet Union was simply a logical result of certain basic Soviet beliefs, for Communist teaching demanded war against capitalist states.

In response, Kennan said, the United States should adopt a policy of "long-term, patient, but firm and vigilant containment." The American government would have to be consistent in standing against Soviet expansion. Through its steadfast commitment to containment, Kennan argued, the United States might be able to bring about changes in the Soviet system.

## Consequences

Actually, the Soviet Union received the American offer of economic help along with other European nations, but it refused, preferring to keep its economic secrets and its economic control over the nations of Eastern Europe. The Marshall Plan proved to be a great success: The United States spent about $12 billion to support the recovery of Western and Central Europe, and by 1952 the nations receiving aid had doubled their prewar production.

Although many Americans criticized the policy of containment, it was the official U.S. Cold War policy for several decades. Not until the 1980's, when the Soviets began releasing control of Eastern Europe and other satellite states, did the United States and the Soviet Union enter into a new period of cooperation.

*George Q. Flynn*

**1038**

# Marshall Plan Helps Rebuild Europe

> *U.S. secretary of state George C. Marshall proposed a cooperative plan for rebuilding the economy of Western Europe with the help of American dollars.*

**What:** Economics; International relations
**When:** March 12, 1947
**Where:** Washington, D.C.
**Who:**

GEORGE CATLETT MARSHALL (1880-1959), U.S. secretary of state from 1947 to 1949

GEORGE F. KENNAN (1904-　　　　), director of Policy Planning Staff, U.S. State Department, 1947

ERNEST BEVIN (1884-1951), British foreign secretary from 1945 to 1951

GEORGES BIDAULT (1899-1983), French foreign minister from 1944 to 1947

VYACHESLAV MIKHAILOVICH MOLOTOV (1890-1986), Soviet commissar for foreign affairs from 1939 to 1949

## Marshall's Proposal

For two years after World War II ended in Europe, the nations of Western Europe struggled to bring their economic productivity up to what it had been before the war. Though they had received more than $11 billion worth of American aid, it was a losing battle. The war had had a huge impact on the world's economic structure. Manufacturing, transportation, and agriculture in Europe were far behind prewar levels; raw materials from Germany and Eastern Europe and food imports from Asia were not available at all. The Soviet Union had taken control of much of Eastern Europe, and the Western European nations had lost many of their overseas colonies.

There was some hope in 1946 that things would improve, but the harsh winter of 1946-1947 quickly shattered this optimism. There were shortages of fuel, food, and machinery even in Great Britain, which had once been called "the shopkeeper of the world."

Matters got worse in 1947. Productivity was low, inflation was high, and there was a shortage of dollars. Workers were underfed and poorly trained, and factories had been slow to reconvert to peacetime industries. Because food and consumer goods were scarce, prices soared.

Western Europe was short of dollars because it depended on American goods and could not produce enough exports to balance its imports. Without dollars, the nations of Western Europe could not make investments to increase production, stop inflation, and balance their trade.

The economic depression brought social unrest and political instability along with it. If these problems continued, it was quite possible that new totalitarian governments would arise to replace Nazism and fascism. The Communist parties of France and Italy were growing rapidly.

U.S. secretary of state George C. Marshall believed that his country should help lead Europe toward a long-range plan for solving its economic problems. After presenting his thoughts to George Kennan's Policy Planning Staff in the State Department, Marshall presented an outline of his plan in the Harvard Commencement Address of June 5, 1947.

Marshall expressed his ideas quietly, emphasizing their practical usefulness. Focusing on Europe's low productivity, its shortages of raw materials, fuel, and modern machinery, and its need for dollars, Marshall warned Americans that a European depression would mean political instability. Europe's economic problems were a threat to peace, he said, and would mean trouble for the American economy as well.

"Our policy is directed not against any country or doctrine," Marshall said, "but against hunger, poverty, desperation, and chaos." A plan for European recovery should be drawn up by representatives of willing European nations, he said, and American aid would support it.

## Program for Recovery

British and French leaders responded quickly. On June 23, 1947, British foreign secretary Ernest Bevin joined French foreign minister Georges Bidault, and together they met with Vyacheslav M. Molotov, Soviet commissar for Foreign Affairs. The two Western Powers proposed that a coordinating committee be set up to make a plan for recovery and oversee various other committees that would evaluate the needs.

Molotov rejected this proposal, insisting that such a committee would interfere in the internal affairs of sovereign states. He offered a substitute plan: After the United States had specified how much money it was willing to grant, each participating state could list its needs. This plan did not fit with Marshall's proposal for European cooperation, however, and Bevin and Bidault would not accept it.

On July 2, Molotov left the meetings. The next day, Bidault and Bevin invited twenty-two other European nations to a meeting in Paris. Because of Soviet dominance in Eastern Europe, Czechoslovakia was the only nation from that area to accept, and because of Soviet pressure the Czechs later withdrew.

The Soviets had decided to respond to the Marshall Plan with their own actions: The old Comintern—the international Communist organization—was resurrected under the new name "Cominform." A series of agreements for supplying grain aid and for trade among the Soviet satellite nations was called the Molotov Plan.

Meanwhile, representatives of sixteen nations met as the Committee of European Economic Cooperation (CEEC) in Paris on July 22. The delegates came from Austria, Belgium, Denmark, France, Greece, Iceland, Ireland, Italy, Luxembourg, the Netherlands, Norway, Portugal, Sweden, Switzerland, Turkey, and the United Kingdom. On September 23, the conference's report outlined a four-year plan for recovery for these nations, plus West Germany.

Projects for self-help and mutual help would increase productivity in farms and factories, so

*General George Catlett Marshall.*

Library of Congress

that exports could rise and imports could fall. The participating nations promised to stabilize their currencies. There would be an organization to bring about cooperation in resources, productivity, transportation, trade, and labor supply. A series of power plants would be built in Germany, the Alps, and Italy. The CEEC Report predicted that more than $57 billion worth of imports of food, fuel, machinery, iron, and steel would be needed.

Through the summer and fall of 1947, the Truman administration examined the Marshall Plan more closely. The Foreign Aid Act of December, 1947, granted $522 million to France, Italy, and Austria, and in 1948 the Marshall Plan became law in the Foreign Assistance Act of 1948. It set up the Economic Cooperation Administration (ECA), which would coordinate the first year's aid: about $4.3 billion, plus another $1 billion in loans.

The ECA worked closely with the CEEC, which in March, 1948, became the Organization for European Economic Cooperation (OEEC). About $8.6 billion in aid was processed by the OEEC in three years: $2.2 billion to the United Kingdom, $1.7 billion to France, $1.7 billion to Germany, $9.18 million to Italy, $3.53 million to Greece, $6.18 million to the Netherlands, and $1.6 billion to others.

## Consequences

By 1940, productivity in Western Europe was 45 percent higher than in 1947, and exports during the same period had increased by 91 percent.

Farm production in 1950-1951 was above the average prewar level.

The Marshall Plan proved to be not only a solution for Europe's immediate economic problems after the war but also the basis for continuing economic cooperation among the European nations. There were political benefits, too: As West Germany became economically strong, it served as a bulwark against Soviet expansion, just as American and British leaders had hoped.

*Stanley Archer*

# Dead Sea Scrolls Are Found

> *The discovery of the Dead Sea Scrolls helped investigators to understand the Old Testament Scriptures, the early growth of the Christian church, and the nature of Judaism.*

**What:** Archaeology
**When:** Spring, 1947
**Where:** Qumran, Palestine
**Who:**
ELIEZER L. SUKENIK (1889-1953), a professor of archaeology
MAR ATHANASIUS YESHUE SAMUEL (1907-      ), an archbishop in the Syrian Orthodox church
MUHAMMAD ADH-DHIB, one of the Bedouin shepherds who found the first scrolls

## While Shepherds Watched Their Flocks

In the spring of 1947, Muhammad adh-Dhib and another young Bedouin (a member of a group of nomadic Arabs) of the Ta'amireh tribe were climbing after some sheep that had wandered among cliffs. They found a cave, and without much thought, one of them threw a rock inside. They were surprised by the sound of breaking pottery. Some days later they returned, hoping to find hidden treasure. The floor of the cave was covered with rubble, but along one wall were several narrow jars. They looked into one and tore the cover from another, but they found nothing. Another contained dirt. Finally, in one they found three old leather scrolls, which they could not read. Their hopes of finding treasure faded.

The young shepherds did not know that the scrolls were the ancient Book of Isaiah in Hebrew, a commentary on the biblical Book of Habakkuk, and a work called *The Manual of Discipline*, a set of guidelines written by an ancient religious sect. A few weeks later, one of the young men returned with another Bedouin to find four more scrolls. These included a second scroll of Isaiah, a damaged but fascinating narrative called "Genesis Apocryphon," a book of thanks-giving psalms, and a work titled *The War of the Sons of Light Against the Sons of Darkness.*

In early 1947, two of the Bedouin brought the first three scrolls and two of the jars to Bethlehem to sell. They contacted two agents who agreed to handle the sale of the scrolls. The scrolls were described to Mar Athanasius Yeshue Samuel, the archbishop of St. Mark's Monastery in Jerusalem, who agreed to buy them all even though he was unable to read them.

Meanwhile, Eliezer L. Sukenik, a professor of archaeology at the Hebrew University, had already managed to buy three of the seven scrolls. Sukenik recognized the Hebrew on the pieces of leather as being a type used between 100 B.C.E. and 100 C.E. He wrote in his diary that this was one of the greatest finds ever made in Palestine.

In late January, 1948, Sukenik asked to buy the rest of the scrolls from Archbishop Samuel. Not sure how much the scrolls were worth, Samuel called on John Trever at the American School of Oriental Research. Trever excitedly sent photographs to William Foxwell Albright of The Johns Hopkins University. Albright wrote back: "incredible . . . there can happily not be the slightest doubt about the genuineness."

## The Ultimate Jigsaw Puzzle

With thoughts of making additional profits, Bedouin tribesmen began to comb the hills, and in 1952 they found a second cave at Murabbaat. By 1956, the Bedouin and teams of archaeologists had found eleven caves containing about eight hundred scrolls, all apparently part of an ancient library. About one-third of the scrolls were biblical; all the books of the Bible except the Book of Esther were represented at least in part. Other scrolls included commentaries on the books of the Bible, religious writings, a marriage contract, and letters by Simeon ben Kozibah (also known as Bar Kokhba), the leader of the second revolt against the Romans. One copper scroll told of hidden treasure. The manuscripts were in Aramaic, Hebrew, and Greek.

Many scrolls were damaged and incomplete. The Bedouin were not always careful when handling them, and some of the scrolls seemed to have been deliberately destroyed in ancient times. Cave 4, the main library, contained fifteen thousand postage-stamp-size scraps of some seven hundred different writings; one archaeologist called them pieces of "the ultimate in jigsaw puzzles."

Modern technology has proven useful in studying the scrolls. For example, the Genesis Apocryphon scroll looked as if coffee had been spilled over it. Nevertheless, when the scroll was heated, letters became visible when photographed with a new infrared film. Noah's words after the Flood appeared: "We gathered together and went . . . to see the Lord of Heaven . . . who saved us from ruin."

## Consequences

The Hebrew Bible, or Old Testament, is the foundation of both Judaism and Christianity. Although it was formed in the ancient Mideast, it has shaped modern Western thought. The Bible has been called "the most quoted, the most published, the most translated, and the most influential book in the history of mankind."

Before the discovery of the Dead Sea Scrolls, biblical scholars had to base their researches on ninth century versions of the Hebrew Scriptures and on the Septuagint, a Greek translation made in the third century B.C.E. The scrolls at Qumran thus allowed investigators to see a thousand years beyond the previous Hebrew texts. The scrolls are extremely important to the understanding of the text of the Hebrew, or Old Testament, Scriptures, the early growth of the Christian church, and the nature of ancient Judaism.

When the discovery of the scrolls was announced, many people wondered if the scrolls might contain information that would change religious doctrine. The scrolls, though, do not

differ in any important way from the other versions of the Scriptures.

Despite their importance to numerous researchers, the scrolls were long monopolized by a handful of scholars who had been granted access by the Israeli government. Over the years, the restricted access to the scrolls and the slow publication of their contents led to many complaints. In September, 1991, the Huntington Library in San Marino, California, which had come into possession of photographs of most of the scrolls, declared that it would make them available to all researchers. Despite threats of legal action by the Israeli government, the Huntington went ahead with its plans, and by November, the Israeli government itself lifted its own restrictions on access to the originals.

*Paul R. Boehlke*

**1043**

# Congress Passes National Security Act

The National Security Act of 1947 was an effort to create greater efficiency and flexibility in the American armed services.

**What:** National politics; Military
**When:** July 26, 1947
**Where:** Washington, D.C.
**Who:**
HARRY S. TRUMAN (1884-1972), president of the United States from 1945 to 1953

JAMES VINCENT FORRESTAL (1892-1949), U.S. secretary of the Navy from 1944 to 1947, and secretary of defense from 1947 to 1949

LOUIS A. JOHNSON (1891-1966), secretary of defense from 1949 to 1950

GEORGE CATLETT MARSHALL (1880-1959), secretary of defense from 1950 to 1951

## Reorganizing the Armed Forces

As the United States government strengthened its commitment to European security through the Truman Doctrine and the Marshall Plan, it became more and more clear that work needed to be done to make the American military more efficient. During World War II, major weaknesses in the American defense system had been revealed. Pearl Harbor had been a great disaster, and there had been many cases of duplicated effort and wasteful competition among the army, navy, and air force. Before the war there had been little cooperation between the military and the diplomatic corps, but the new Cold War situation made that kind of cooperation essential.

Many American government leaders, including President Harry S. Truman, believed that the United States needed a more efficient system of defense. Truman had brought one reorganization plan to Congress in 1945, but it took time to settle many differences of opinion among the three branches of the armed forces.

Navy leaders feared that the new defense establishment would be dominated by the army. In particular, they feared that the marines might be abolished, or transferred to the army. There was also the matter of the navy's new air capability. Having become convinced during World War II that aircraft carriers were extremely valuable, the navy wanted to build supercarriers that would be able to carry the new jet airplanes. If the army dominated American defense, the admirals argued, more money might be spent on land-based, long-distance bombers.

During 1946 and 1947, Truman worked to bring together the army, represented by Secretary of War Robert P. Patterson, and the navy, represented by Secretary of the Navy James V. Forrestal. Although Forrestal was quite sympathetic to the admirals' concerns, he was especially helpful to Truman in working for a compromise.

These meetings brought about agreements that were eventually expressed in the National Security Act of 1947. The act created a Department of Defense (called the National Military Establishment until 1949), headed by a cabinet secretary. The Department of the Army, the Department of the Navy, and a new Department of the Air Force were made into separate agencies within the Department of Defense.

The Joint Chiefs of Staff were given legal recognition as a committee with representatives of each of the three services. The Joint Chiefs would coordinate military action, prepare defense plans, and make strategy recommendations to another new agency, the National Security Council (NSC).

The NSC would be chaired by the president of the United States and would also include the secretary of state, the secretary of defense, the secretaries of each of the three services, and the chair-

**1044**

person of the board of another new agency, the National Security Resources Board.

Finally, the act created the Central Intelligence Agency to gather information having to do with national security.

## The Rivalries Continue

This reorganization plan had barely gotten off the ground when serious problems arose. Some of these problems were simply carryovers from the traditional competition between the army and the navy. Though Forrestal did outstanding work as the first secretary of defense, the new system did not seem to have an effect on rivalries between the services.

Soon the three services were spending much time and money on separate publicity and lobbying campaigns to gain more money for their defense goals. The navy tried to "sell" its supercarrier program, while the air force argued that the new B-36 bomber was the best defense investment.

Secretary Forrestal tried to mediate, but the problems seemed to be too great for one person to control. Still, in early 1949 he was able to report that the new defense system had already saved the American taxpayers more than $56.5 million.

Because of ill health, Forrestal had to resign on March 3, 1949. The new secretary of defense appointed by President Truman was Louis A. Johnson of West Virginia. Johnson was a rather argumentative person who had not held such a high government post before. Soon he became

part of the interservice rivalry by favoring the air force. The building of the navy's new aircraft carriers was stopped, and large amounts of money went into expanding the strength of the air force.

Although this approach saved money, some critics claimed that it weakened national defense. State Department officials joined in the growing criticism of Johnson because they did not like his way of making defense decisions independently.

In September, 1950, Truman responded to these criticisms by turning to General George C. Marshall, former secretary of state, to take over the Department of Defense. At the same time, Marshall's World War II colleague General Omar Bradley was appointed chairman of the Joint Chiefs of Staff. Together, Marshall and Bradley succeeded in making things run more smoothly in the Department of Defense.

## Consequences

Congress remained involved in the process of reorganizing defense in the years following the passage of the National Security Act. In 1949 the Hoover Commission, which Truman had appointed to investigate the workings of the federal government, reported that changes were needed in the Department of Defense. In August, 1949, a new bill was passed giving the secretary of defense more power in dealing with the individual services. Nine years later, in July, 1958, the Defense Department Reorganization Act made more changes to help bring an end to the "Battle of the Pentagon."

# India Is Partitioned

> Before India could become independent from the British Empire, conflicts between Hindu and Muslim communities made it necessary to divide the Asian subcontinent into the nations of India and Pakistan.

**What:** Political reform
**When:** August 15, 1947
**Where:** Southern Asia
**Who:**
MOHAMMED ALI JINNAH (1876-1948), president of the All-India Muslim League from 1935 to 1947, and the founder of Pakistan
JAWAHARLAL NEHRU (1889-1964), leader of the Indian National Congress
MOHANDAS KARAMCHAND GANDHI (1869-1948), leader of the Indian National Congress
LORD LOUIS MOUNTBATTEN (1900-1979), viceroy of India in 1947, and governor-general of India from 1947 to 1948

## Rise of the Muslim League

After five years in England, Mohammed Ali Jinnah returned to his native India in 1935 to lead the All-India Muslim League into the important 1937 elections. This mission, however, was a failure; the League failed to prove itself as the main voice of the Muslim community. In the two areas in which Muslims were a majority, important politicians such as Fazl al-Haq of Bengal and Sir Sikander Hayat Khan of the Punjab rejected Jinnah's leadership and cooperated with non-Muslim parties.

Yet Jinnah did not give up. After reorganizing itself, the League began an intense rivalry with the Indian National Congress, disputing the Congress's claim to represent all Indians in their struggle against British rule. Congress ministries that held power in six Indian provinces were attacked by Jinnah, who accused them of wrongdoing and especially of discrimination against Muslims. Across India, Muslims began to be fearful that once India became independent, it would be dominated by the Congress, and Muslims would suffer.

When World War II broke out, India's viceroy, Lord Victor Linlithgow, brought India into the war without consulting with native Indian leaders. Congress ministries across the country resigned in protest, creating a political vacuum. The effect was to strengthen Jinnah, for the imperial government began to seek his support and brought him into British councils alongside Jawaharlal Nehru and Mohandas K. (Mahatma) Gandhi, leaders of the Congress. Because of this recognition, Jinnah gained greater respect among Muslims as well.

Having gained the bargaining power he had sought, Jinnah decided that Muslims across India needed to be brought together under a clear political goal. At the Lahore meeting of the League in early 1940, Jinnah stated that the Muslims were a nation by any definition of the word "nation," and that they deserved "their homelands, their territory, and their state." The delegates at this meeting passed a resolution declaring that the final goal of the League would be the creation of "Pakistan" (meaning "land of the pure"), an independent Muslim state.

## Independence and Partition

No group would be forced to join a future Indian state, the British government and the Congress responded separately. The Cripps Mission of 1942, sent by Great Britain, gave the League further encouragement: The mission's leader, Sir Richard Cripps, stated that the British would accept an independent Muslim state if the people of the two Muslim majority areas supported the idea.

At the end of World War II, India looked forward to its first provincial elections since 1937. Appealing to the religious feelings of the Muslim people, the League platform centered on the call for the creation of Pakistan. This strategy was an enormous success: The League won 439 of 494 seats reserved for Muslims in the provincial assemblies, and all 30 Muslim seats in the Central Legislative Assembly.

Some time later, the British announced a plan for giving India its independence; to meet Muslim demands, the new nation would be set off as a very loose confederation of Indian states. Though both the Congress and the League approved the plan at first, they soon fell to squabbling over the Congress's claim that it, too, had the right to represent Muslims.

In response, Jinnah declared August 16, 1946, to be "Direct Action Day." The result was the "Great Calcutta Killing"—widespread rioting in which more than four thousand people were killed. Thousands more were killed as rioting spread to neighboring areas.

In October, 1946, the Muslim League decided to join Congress leaders in India's Executive Council—but only to make it difficult for the council to do its work. Two months later, the League refused to take part in an assembly called by the British to write a new constitution for India. Meanwhile, riots and killings spread to the Punjab, where Sikhs complicated matters by insisting that they, too, had a right to self-rule.

## PARTITION OF INDIA, 1947

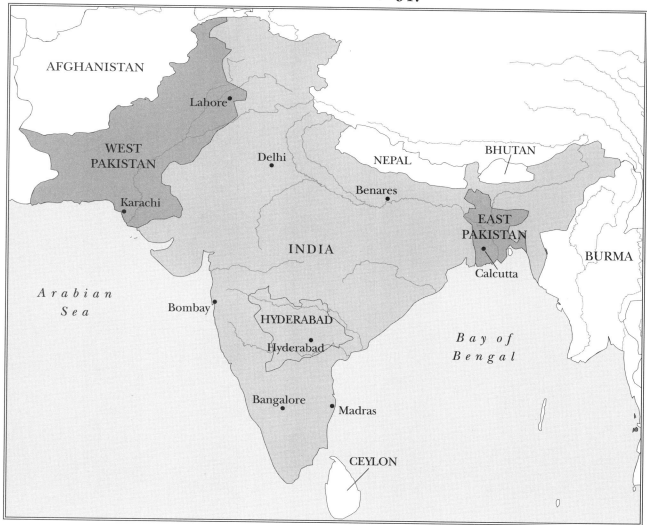

In a last-ditch effort to solve the problems, the new prime minister of Great Britain, Clement Attlee, announced that his country intended to transfer power in India by June, 1948. Lord Louis Mountbatten was chosen as India's viceroy and was asked to prepare a plan for the transfer. These actions, however, failed to bring about a compromise.

Finally, in March, 1947, Congress leaders realized that if they did not give in to the League's demands, India would explode into civil war. In May, they recommended that the Punjab be partitioned. The goal of the Congress now became to limit the new Muslim state to as small an area as possible—a "maimed, mutilated, and moth-eaten Pakistan," as Gandhi said.

Lord Mountbatten, too, had decided that partition was the only solution. On June 3, he announced that the British cabinet had approved a plan to transfer power in India to two separate Hindu and Muslim states. The British would leave India in mid-August, 1947.

At midnight on August 14, 1947, Mohammed Ali Jinnah became the first governor-general of independent Pakistan, while Lord Mountbatten took the same office in India.

## Consequences

Less than ten years before, many Indian and British leaders had hoped to see the "jewel" in the British imperial crown transformed into a peaceful, unified Indian state. That hope was now shattered, and even partition did not end the bloodshed. More than 200,000 people were killed as Hindus fled their homes in Pakistan and Muslims left theirs in India.

As wars and threats of wars continued for many years between India and Pakistan, millions of impoverished Indian and Pakistani people continued to suffer. Pakistan itself was partitioned in 1971, when the nation of Bangladesh was created out of Eastern Pakistan.

*George Q. Flynn*

# Bell's X-1 Rocket Plane Breaks Sound Barrier

> United States Air Force captain Chuck Yeager flew the Bell X-1, a rocket-propelled airplane, past the speed of sound, a crucial step in aerospace research.

**What:** Space and aviation
**When:** October 14, 1947
**Where:** Edwards Air Force Base, California
**Who:**

CHARLES E. "CHUCK" YEAGER (1923-    ), a military test pilot

EZRA KOTCHER (1903-1990), an American aeronautical engineer

THEODORE VON KÁRMÁN (1881-1963), a Hungarian-born aerodynamicist

JOHN STACK (1906-1972), an American aeronautical engineer

## The Sound Barrier

Pilots do not need to worry about air turbulence produced by their aircraft at low speeds. However, as an airplane nears the speed of sound (about 1,266 kilometers per hour at sea level, referred to as Mach 1), dangerous turbulence builds up, especially at the tips of propellers and along the wings. Because of the turbulence, until the mid-1930's, most scientists thought that the speed of sound was a barrier that pilots could not pass without destroying their planes and themselves.

By the late 1930's, engineers were designing military aircraft that, when diving, could fly so fast that turbulence stressed their propellers and wings. Research was badly needed to learn how to prevent damage from the stress. Some American aeronautical engineers, led by John Stack, believed that specially designed planes could eventually break into faster-than-sound, or supersonic, speeds. They called for the construction of experimental airplanes to perform the research, even though most scientists claimed that

no propeller-driven airplane could ever fly fast enough. Although the National Advisory Committee for Aeronautics (NACA) showed interest in supersonic research, the U.S. government did not set aside money for a program until 1943.

## Yeager Breaks the Barrier

In 1943, World War II was at its height, the Germans were known to be manufacturing jet aircraft, and the U.S. Army thought that it would soon need faster fighter planes. Army Air Force Scientific Advisory group director Theodore von Kármán, a longtime supporter of supersonic research, received approval to begin developing of a supersonic aircraft. He told Ezra Kotcher, a NACA engineer, to work on designs. Kotcher found that a rocket-propelled airplane, rather than a jet or propeller plane, would be the best research aircraft.

At the end of 1944, Bell Aircraft Corporation in Buffalo, New York, was hired to build the first model, the XS-1 (X for experimental and S for supersonic; the abbreviation later became simply X-1). The first X-1 was finished in late 1945. It looked like a long bullet with short, thin wings and a rudder.

Military pilots began testing the X-1 in mid-1947 at Edwards Air Force Base in California. In a series of carefully planned flights, Captain Charles E. "Chuck" Yeager took the rocket plane ever closer to the sound barrier and found that it handled well despite some turbulence problems.

Yeager was eager to be the first man to travel faster than sound, but there was good reason to believe that he would not survive the attempt: Less than a year earlier, a British experimental plane had disintegrated near supersonic speed. Finally, for his thirteenth test flight, Yeager's superiors told him to break the sound barrier.

Just after 10:00 A.M. on October 14, 1947, a B-29 bomber carried Yeager and the X-1 to about 6,100 meters altitude. (The X-1 could not take off on its own.) Everything went well, even though Yeager himself was in pain. The night before he had broken two ribs in a horse-riding accident. He did not tell his superiors about the injury for fear that they would not let him make the barrier-breaking flight.

At 10:26 A.M., as the B-29 was cruising at 420 kilometers per hour, the X-1 dropped away like a bomb. The rocket plane had no throttle to control its speed. Instead, it had four combustion chambers in the engine. The pilot could ignite any chamber alone or any combination to vary his speed. As the X-1 fell, Yeager test-fired all the chambers, and then with the thrust from two he began to climb, soon leaving behind the B-29 and the two fast "chase planes" assigned to follow him. Then he activated the other two chambers and zoomed upward at full thrust, while paying close attention to the plane's rudder and wings, which briefly lost effectiveness. At 10,700 meters, hurtling up at Mach 0.95, Yeager shut down two chambers and leveled off at 12,200 meters, the altitude for his record-setting run.

Yeager restarted a third chamber and watched as the Mach meter in the cockpit rose gradually to Mach 1.02. There it stuck, then jumped to Mach 1.06. Yeager had breached the sound barrier, with plenty to spare. The X-1 flew smoothly the whole time. The experiment a success, he turned off the rocket engine and glided toward the runway. The whole flight lasted only fourteen minutes.

## Consequences

In only a few moments, Yeager proved beyond doubt that supersonic flight was possible. The Air Force tried to keep the event secret, but news leaked out to the popular magazine *Aviation Week*, which announced it. The flight soon made headlines throughout the world.

The X-1's exceptionally thin wings, bullet-shaped fuselage, control system, cockpit design, and rocket engine were technological innovations upon which many later aircraft and rockets were based. Still, these technical matters, although they were important, had less immediate impact than the fact that Yeager had ended the myth of the sound barrier. Aircraft designers set out to mass-produce supersonic aircraft. In 1953, the United States unveiled its first supersonic jet fighter, the F-100 Super Sabre. The Soviet Union and other countries soon did likewise.

The X-1's achievement had more than military importance. Supersonic passenger planes were eventually built, and the X-1 made possible more powerful research aircraft such as the X-15, which flew to the brink of outer space. These in turn led to the first spacecraft, such as the space shuttle, vehicles that could exceed Mach 25 and escape Earth's gravity.

Yeager's 1947 flight was a crucial first step for all space programs, and the X-1 is the grandparent of all American spacecraft.

*Roger Smith*

# Shockley, Bardeen, and Brattain Invent Transistors

> *William Shockley and John Bardeen provided the theory, and Bardeen and Walter H. Brattain performed the experiments, that led to the discovery of the point-contact transistor.*

**What:** Energy
**When:** November-December, 1947
**Where:** Murray Hill, New Jersey
**Who:**
WILLIAM SHOCKLEY (1910-1989), a
    theoretical physicist
JOHN BARDEEN (1908-1991), a
    theoretician
WALTER H. BRATTAIN (1902-1987), an
    experimental physicist

## Vacuum Tube Limits

Vacuum tube technology was well established at the beginning of World War II (1939-1945). The tubes functioned by allowing current to flow only in one direction—from one electrode to another electrode, both of which were contained inside an evacuated tube. The current passing through the tube could also be controlled by inserting a metal grid between the two electrodes. By changing the voltage applied to the grid, the current between the electrodes could be increased. These two basic functions of vacuum tubes (rectification, or one-way flow, and amplification) had been established by 1906. In subsequent years, additional complicated tubes were designed to operate in more and more complicated circuitry.

Vacuum tubes had inherent limitations that were made very obvious by technological developments spurred by World War II. Vacuum tubes were bulky, required large amounts of power to operate, were not particularly rugged, and needed reliable cooling systems to prevent damage from overheating. Radar, developed during the war, required circuits that could operate over a range of frequencies, power, and voltages that vacuum tubes could not provide.

Although electronic computers did not play a major role in World War II, they did add impetus to the movement toward a replacement technology. The development of bigger and better computers was limited in part by the vast arrays of vacuum tubes and cooling systems required. Designers realized that significant advances in electronic computers would require circuits based on elements other than vacuum tubes.

## Negative Electrons and Positive Holes

Beginning in the summer of 1945, a group of scientists from Bell Telephone Laboratories set out to study semiconductors: substances that will, under certain conditions, carry (or conduct) an electric current. Headed by William Shockley, the group included Walter H. Brattain, John Bardeen, Gerald Pearson, and Robert Gibney. They studied the two types of semiconductors—the "p-type" and the "n-type"—so named because the carriers of the electric current were either positive or negative. Wartime work indicated that a very slight amount of phosphorus added to silicon or germanium made the semiconductor an n-type, because the phosphorus gave up one (negatively charged) electron per atom. Earlier work also had shown that other impurities take up electrons from the atoms in the semiconductor, leaving (positive) "holes" in the crystal structure where the electrons used to be. These holes can then carry current, and the semiconductor becomes a p-type semiconductor.

In the summer of 1945, Shockley theorized that amplification and rectification would be achieved by applying an external electric field to a block of quartz with a thin semiconducting film on one side and a metallic conductor on the other side. His calculations indicated that the current could be increased considerably by applying a voltage between the semiconducting film and the metallic conductor. Repeated experimental attempts to verify the amplification failed. Shockley had Bardeen check the calculations, which proved to be correct. By March, 1946, Bardeen had come up with a theoretical explanation for their inability to see this "field effect." He postulated that electrons were trapped on the surface of the semiconductor and were thus unable to contribute to the flow of the current.

By December, 1947, amplification had been achieved by replacing the quartz with two lines of gold foil spaced about half a micrometer apart. The researchers quickly realized that contacts in the form of a line were not necessary and replaced them with two closely spaced fine wires known as "cat's whiskers." The device was later named the "point-contact transistor" to indicate the type of connection made to the device and the fact that it transfers current from a low-resistance input to a high-resistance output.

The operation of the point-contact transistor was not well understood theoretically. Shockley suggested using a simpler system: a single crystal of semiconducting material with a p-type region sandwiched between two n-type regions. The junction between the regions would rectify the current, and Shockley's theory predicted that amplification would also occur. After 1951, the "junction transistor" dominated the field because it proved to be useful, versatile, and fairly easy to fabricate after physicists at Bell Labs were able to develop techniques to grow the crystals.

Aware of the significance of the transistor, the staff at Bell Labs spent six months drawing up patent applications and performing additional experiments before announcing the discovery at a press conference on July 1, 1948. A paper written jointly by Bardeen and Brattain appeared in a major scientific journal two weeks later.

## Consequences

The immediate impact of the transistor was negligible. Other scientists recognized its potential, but formidable technical obstacles hindered the transformation of the first point-contact tran-

sistor—a rather gawky-looking and crude device—into a practical, economical, and useful circuit element. Eventually, junction transistors would prove to be easier to mass-produce than point-contact transistors.

For years, transistors were difficult to manufacture reliably and even more difficult to keep operating within given specifications. In addition, transistors could not directly replace vacuum tubes in existing electronic devices: Different designs had to be produced to incorporate transistors into existing devices.

Transistors began to be used in telephone equipment in 1952 and appeared in hearing aids in 1953. The first hearing aids using transistors were actually more expensive than those using vacuum tubes, but the transistorized hearing aids used less power and saved their owners money on the cost of replacement batteries. Once established, the use of transistors in hearing aids dramatically changed the market. Before two years had passed, virtually all hearing aids used only transistors, and sales of hearing aids increased considerably.

The first transistorized radio appeared in 1954. In 1955, the International Business Machines Corporation marketed its model 7090 computer, which took up less space than, and needed only 5 percent of the power of, earlier vacuum tube computers. As of 1956, transistors had not yet conquered the consumer market, but Shockley, Bardeen, and Brattain were recognized for their research into semiconductors and for their discovery of the transistor effect by being awarded the Nobel Prize in Physics in 1956.

In 1958, ten years after the discovery of the point-contact transistor, transistor technology leaped forward with the development of the "planar technique" that made it possible to produce an integrated circuit—an entire electronic circuit on a single piece of semiconductor material. The commercial introduction of microprocessors ("computers on a chip" that incorporate highly complicated integrated circuits in a cheap, mass-producible form) appeared in 1971. Once microprocessors became available, microcomputers began to be built. The computer age in many ways owes its existence to the discovery of the transistor effects.

*Roger Sensenbaugh*

# Nationalists Take Power in South Africa

*Winning a majority of Parliament seats in the election of 1948, Daniel Malan and his National Party began to set up a system of laws to guarantee what they called "apartheid"—complete separation of the races.*

**What:** Political reform; Civil rights and liberties; Human rights
**When:** 1948
**Where:** South Africa
**Who:**
JAN CHRISTIAN SMUTS (1870-1950), prime minister of South Africa from 1919 to 1924 and from 1939 to 1948
DANIEL FRANÇOIS MALAN (1874-1959), prime minister of South Africa from 1948 to 1954

## A Diverse Land

According to historians, the first Africans arrived in South Africa in about the fifth century. Dutch settlers arrived at the southern tip of Africa centuries later, in 1652. These new South Africans, who farmed cattle, soon began to move north and east in search of pasture for their herds. It was not long before they came upon native tribespeople who also had a claim on the land. Many of the Khoi-khoi and San people were killed by white hunting parties, in wars, or by smallpox.

As the Dutch (who came to be called Boers) pushed east, African groups tried to move west. Then the British arrived, occupying the Cape of Good Hope region in 1806. Waves of British settlers brought liberal ideas about equality of the races. Slavery was abolished in 1834, and British missionaries criticized the way the Boers treated black Africans.

In response, the Boers decided to move on, and the Great Trek began. Around 1836, the Boer farmers went to Natal, where they fought a series of bloody wars with the Zulu. When the British began appearing in Natal, the Boers moved on again, this time to the Transvaal (which they called the South African Republic) and the Or-ange Free State. The governments they set up vowed, "No equality in church or state."

British whites dominated the other two states of South Africa—the Cape Colony and Natal. The Cape, Natal, the South African Republic, and the Orange Free State existed side by side for a number of years without uniting into one nation. When diamonds and gold were discovered in South Africa, however, British colonists began flooding the country. The Boers (who were now calling themselves Afrikaners) hoped to free themselves from British rule, but they were defeated and humiliated in the Anglo-Boer War of 1899-1902.

An Afrikaner war hero, Jan Christian Smuts, began working to make one nation out of the four colonies, and the Union of South Africa was created in 1910. Blacks and "coloreds" (people of mixed race) who met certain qualifications were allowed to vote in the Cape, but Natal had made a rule that excluded almost all nonwhites. Transvaal and the Orange Free State did not allow any nonwhites to vote.

## Building Apartheid

From the beginning, it was the Afrikaners who had the top posts in the government, and laws based on discrimination were also there from the start. The first law prohibiting marriage between whites and black Africans had come as early as 1685. In 1913, the Natives Land Act forced thousands of Africans off farms in the Transvaal and the Orange Free State; these people were moved onto "reserves"—lands that were to be for Africans only. The Mines and Works Act set aside certain jobs for certain races.

The National Party was formed in response to the needs of poor Afrikaners. By the 1930's, about 300,000 Afrikaners (more than 17 percent of the Afrikaner population) were living in poverty. Many of these Afrikaners resented the blacks, because it seemed that the blacks were taking jobs that whites should have. They also hated and feared the British, who had elbowed their way into South Africa and forced it to become part of the British Empire.

Having helped to found the League of Nations after World War I, Smuts became prime minister of South Africa in 1919. During World War II Smuts helped keep South Africa on the side of the Allies, even though other Afrikaner leaders felt quite sympathetic to the ideas of Adolf Hitler. After the war, Smuts was involved in forming the United Nations.

As the elections of 1948 approached, Smuts and the United Party faced the challenge of Daniel François Malan and his "purified" National Party. Malan's greatest concern was not "keeping the blacks in their place" but raising the Afrikaners to victory over the British.

Malan, who had been trained as a clergyman, was certain that God had called the Afrikaner people for a special purpose. With the help of God, they would break out of the British Empire, form a pure Christian republic, and rule South Africa without anyone's interference.

In 1946, Smuts's government appointed a special commission to look into the problems of race

relations in South Africa. This commission reported in 1948 that "the idea of total segregation is utterly impracticable." In response, the National Party set up its own committee to establish a race policy before the election. Its report stated that South Africa's goal must be "total apartheid [apartness] between whites and natives."

The Nationalists knew that businesses in South Africa depended heavily on cheap black labor, so they did not say that all blacks must be removed at once. Instead, they decided that apartheid would have two stages. For a time, blacks would be allowed to work in white areas, but eventually they would be phased out. No one knew exactly how manufacturing and mining could continue without black workers, so they avoided talking about specific plans.

In the elections, Smuts's United Party and its allies won 547,437 votes, or 50.9 percent, while Malan's National Party and its allies took only 443,278 votes (41.2 percent). Because of the way electoral districts were set up, however, the Nationalists won seventy-nine seats in Parliament, while the United Party ended up with only seventy-one. The Nationalists were in control, and Malan became the new prime minister.

## Consequences

Malan quickly restricted immigration from Great Britain, but tried to encourage people from Germany and the Netherlands to move to South Africa. He also pushed many English speakers out of prominent positions in the government.

In response to the lower-class Afrikaners who were deeply worried about competition from blacks, the Nationalists then began creating a complicated system of racial laws—the Population Registration Act, the Mixed Marriages Act, and the Group Areas Act. These laws classified people according to race and restricted whom they could marry and where they could live. Colored people lost the vote. Separate (and inferior) education was set up for black children, and all blacks were assigned to "homelands"—rural reservations—even if they had lived in the city all their lives. This system of segregation by law survived in South Africa until the 1980's, when the government finally began responding to pressure from the black community and the world.

*Ruth Goring Stewart*

Library of Congress

*Jan Christian Smuts.*

# Gamow Proposes "Big Bang" Theory of Universe

> *George Gamow proposed that the universe was created by the explosion of a hot, dense primordial fireball—an idea that became known as the "big bang" theory.*

**What:** Astronomy
**When:** 1948
**Where:** Washington, D.C.
**Who:**
GEORGE GAMOW (1904-1968), a Russian American nuclear physicist and cosmologist
RALPH ASHER ALPHER (1921-    ), an American physicist
ROBERT C. HERMAN (1914-1997), an American physicist
GEORGES LEMAÎTRE (1894-1966), a Belgian Jesuit priest, astronomer, and cosmologist
FRED HOYLE (1915-    ), an English astronomer

## The Big Bang

After World War II, George Gamow, a professor at George Washington University in Washington, D.C., began to prove that reversing the process of galactic expansion pointed to a time when all matter was confined to an extremely small space (perhaps thirty times the sun's diameter) at a temperature of trillions of degrees. He proposed that the density of the radiation that existed at that time was greater than the density of the matter, a condition that caused the explosion leading to the formation of the present universe. His theory was known as the "big bang" theory.

Gamow's big bang theory was based on the American astronomer Edwin Powell Hubble's discovery in 1929 that the universe is expanding. In Russia, the mathematician and physicist Aleksandr A. Friedmann had used this discovery to find a dynamic solution to German American physicist Albert Einstein's general theory of rela-

tivity, published in 1916, which implied that the universe was expanding. Georges Lemaître, unaware of Friedmann's solution, proposed in 1927 that the universe originated from a "cosmic egg."

The primary motive for Gamow's proposal was his desire to explain how the heavier chemical elements could be formed in the relative abundances in which they are found in the universe. Hydrogen and helium are presumed to constitute approximately 99 percent of the matter in the universe. The other 1 percent consists of the heavier elements, which decline in abundance through the periodic table until zinc is reached. At this point (roughly halfway down the periodic table) the abundance flattens out and approximately the same amount of all the remaining elements occurs.

Gamow reasoned that, because of this pattern, the interiors of stars could not be the birthplaces of the heavy elements. He proposed that they had been formed in the first thirty minutes of the initial explosion, before the temperature had cooled too much. He believed that he could explain how deuterium (heavy hydrogen) could have been formed in the big bang, and he was convinced that it could be destroyed only in stellar interiors. He also believed that helium was too abundant to have been formed in stars and had to have resulted from the initial explosion. This portion of his initial ideas has stood the test of time.

## Heavy Element Formation

Gamow enlisted the aid of two physicists to calculate the mathematics involved in heavy element formation: Ralph Asher Alpher, a doctoral candidate at The Johns Hopkins University, and Robert Herman, who was selected because of his expertise in operating the early computers used by the Bureau of Standards. A major element of their theory came from a surprising source. During World War II, scientists at Brookhaven National Laboratory had measured the neutron-capture characteristics of several atoms and had found that capture increased during the first half of the periodic table and then flattened out—the inverse of the pattern of abundance of the elements. Since the differences between elements are based on the number of neutrons and protons in the nucleus of atoms, Alpher proposed that neutron capture explained how elements formed during the big bang.

Because the predictions of his theory produced some odd results, he conceded eventually that the heavy elements were not created in the initial big bang. Although his assumption that the dynamics of gravity caused the condensation of the cooling gases into galaxies and stars presented further difficulties, Gamow believed that the outline of his theory was firm enough to present it publicly in 1948. It was popularly presented in 1952 in the book *The Creation of the Universe.*

Although Gamow's presentation of the big bang was accepted quickly by many astronomers as a proper interpretation of the astronomical evidence, specific proof of the theory was slower in coming. Alpher and Herman pointed out in 1948 that the level of radiation in the universe had steadily declined since the big bang to a level that they estimated to be 5 Kelvins (above absolute zero). They thought that it might still be detectable, not as light but perhaps as a low-level microwave radiation.

Arno Penzias and Robert Wilson of Bell Laboratories discovered the microwave radiation in their efforts to study sources of background radiation causing static in radio transmission. The discovery of this background radiation (which was actually 3 Kelvins) provided a major confirmation of the hot big bang theory. The big bang theory explained the expansion of the universe better than other theories could, and it has gradually become the accepted view of the origin of the universe.

**Consequences**

Because of the problem with heavy elements, there was some early neglect of the success of Gamow's theory in explaining the buildup of helium and the abundance of hydrogen and helium compared with the rest of the elements. Gamow's team was also successful in identifying the process of heavy element formation through neutron capture. They merely had the wrong location—the big bang rather than the interiors of massive stars.

The clear implication of the big bang theory is that the universe had a hot beginning and will die a cold and isolated death as the galaxies become farther apart, with the individual stars eventually burning out as a result of an insufficient rate of birth of new stars. Some cosmologists have proposed a coming collapse of the universe (an idea Gamow described in 1952), and others have suggested a cycle of big bangs and collapses.

The principal difficulty of his theory eventually forced Gamow to accept Fred Hoyle's explanation of heavy element formation in the interior of stars. It is now the widely accepted view that heavy elements were created from fundamental particles during the radiative life of massive stars and dispersed into space through supernova explosions. The prediction of microwave radiation was a brilliant insight that initially was inappropriately neglected but that ultimately established the big bang as the most reasonable explanation of the creation of the universe.

*Ivan L. Zabilka*

# Lyons Constructs First Atomic Clock

*Harold Lyons and his colleagues developed a clock that surpassed mechanical clocks in long-term stability, precision, and accuracy.*

**What:** Metrology
**When:** 1948
**Where:** Washington, D.C.
**Who:**
HAROLD LYONS (1913-1984), an American physicist

## Time Measurement

The accurate measurement of basic quantities, such as length, electrical charge, and temperature, is the foundation of science. The results of such measurements dictate whether a scientific theory is valid or must be modified or even rejected. Many experimental quantities change over time, but time cannot be measured directly. It must be measured by the occurrence of an oscillation or rotation, such as the twenty-four-hour rotation of the earth. For centuries, the rising of the sun was sufficient as a timekeeper, but the need for more precision and accuracy increased as human knowledge grew.

Progress in science can be measured by how accurately time has been measured at any given point. In 1713, the British government, after the disastrous sinking of a British fleet in 1707 because of a miscalculation of longitude, offered a reward of 20,000 pounds for the invention of a ship's chronometer (a very accurate clock). Latitude is determined by the altitude of the sun above the southern horizon at noon local time, but the determination of longitude requires an accurate clock set at Greenwich, England, time. The difference between the ship's clock and the local sun time gives the ship's longitude. This permits the accurate charting of new lands, such as those that were being explored in the eighteenth century. John Harrison, an English instrument maker, eventually built a chronometer that was accurate within one minute after five months at sea. He received his reward from Parliament in 1765.

## Atomic Clocks Provide Greater Stability

A clock contains four parts: energy to keep the clock operating, an oscillator, an oscillation counter, and a display. A grandfather clock has weights that fall slowly, providing energy that powers the clock's gears. The pendulum, a weight on the end of a rod, swings back and forth (oscillates) with a regular beat. The length of the rod determines the pendulum's period of oscillation. The pendulum is attached to gears that count the oscillations and drive the display hands.

There are limits to a mechanical clock's accuracy and stability. The length of the rod changes as the temperature changes, so the period of oscillation changes. Friction in the gears changes as they wear out. Making the clock smaller increases its accuracy, precision, and stability. Accuracy is how close the clock is to telling the actual time. Stability indicates how the accuracy changes over time, while precision is the number of accurate decimal places in the display. A grandfather clock, for example, might be accurate to ten seconds per day and precise to a second, while having a stability of minutes per week.

Applying an electrical signal to a quartz crystal will make the crystal oscillate at its natural vibration frequency, which depends on its size, its shape, and the way in which it was cut from the larger crystal. Since the faster a clock's oscillator vibrates, the more precise the clock, a crystal-based clock is more precise than a large pendulum clock. By keeping the crystal under constant temperature, the clock is kept accurate, but it eventually loses its stability and slowly wears out.

In 1948, Harold Lyons and his colleagues at the National Bureau of Standards (NBS) constructed the first atomic clock, which used the ammonia molecule as its oscillator. Such a clock is called an atomic clock because, when it operates, a nitrogen atom vibrates. The pyramid-shaped ammonia molecule is composed of a triangular base; there is a hydrogen atom at each corner and a nitrogen atom at the top of the pyramid. The nitrogen atom does not remain at the top; if it absorbs radio waves of the right energy and frequency, it passes through the base to produce an upside-down pyramid and then moves back to the top. This oscillation frequency occurs at 23,870 megacycles (1 megacycle equals 1 million cycles) per second.

Lyons's clock was actually a quartz-ammonia clock, since the signal from a quartz crystal produced radio waves of the crystal's frequency that were fed into an ammonia-filled tube. If the radio waves were at 23,870 megacycles, the ammonia molecules absorbed the waves; a detector sensed this, and it sent no correction signal to the crystal. If radio waves deviated from 23,870 megacycles, the ammonia did not absorb them, the detector sensed the unabsorbed radio waves, and a correction signal was sent to the crystal. The atomic clock's accuracy and precision were comparable to those of a quartz-based clock—one part in a hundred million—but the atomic clock was more stable because molecules do not wear out.

The atomic clock's accuracy was improved by using cesium 133 atoms as the source of oscillation. These atoms oscillate at 9,192,631,770 plus or minus 20 cycles per second. They are accurate to a billionth of a second per day and precise to nine decimal places. A cesium clock is stable for years. Future developments in atomic clocks may see accuracies of one part in a million billions.

## Consequences

The development of stable, very accurate atomic clocks has far-reaching implications for many areas of science. Global positioning satellites send signals to receivers on ships and airplanes. By timing the signals, the receiver's position is calculated to within several meters of its true location.

Chemists are interested in finding the speed of chemical reactions, and atomic clocks are used for this purpose. The atomic clock led to the development of the maser (an acronym for *m*icrowave *a*mplification by *s*timulated *e*mission of *r*adiation), which is used to amplify weak radio signals, and the maser led to the development of the laser, a light-frequency maser that has more uses than can be listed here.

Atomic clocks have been used to test Einstein's theories of relativity that state that time on a moving clock, as observed by a stationary observer, slows down, and that a clock slows down near a large mass (because of the effects of gravity). Under normal conditions of low velocities and low mass, the changes in time are very small, but atomic clocks are accurate and stable enough to detect even these small changes. In such experiments, three sets of clocks were used—one group remained on Earth, one was flown west around the earth on a jet, and the last set was flown east. By comparing the times of the in-flight sets with the stationary set, the predicted slowdowns of time were observed and the theories were verified.

*Stephen J. Shulik*

# Hoyle Advances Steady-State Theory of Universe

*Hermann Bondi, Thomas Gold, and Sir Fred Hoyle presented the steady-state theory, which views the universe as infinite, eternal, and unchanging.*

**What:** Astronomy
**When:** 1948
**Where:** Cambridge, England
**Who:**

SIR FRED HOYLE (1915-      ), an English astronomer and astrophysicist

THOMAS GOLD (1920-      ), an American astronomer

HERMANN BONDI (1919-      ), an Austrian English mathematician and cosmologist

GEORGE GAMOW (1904-1968), an American physicist

GEORGES LEMAÎTRE (1894-1966), a Belgian astronomer

EDWIN POWELL HUBBLE (1889-1953), an American astronomer

ALBERT EINSTEIN (1879-1955), an American physicist

## Explaining an Expanding Universe

By the first half of the twentieth century, it was known that the sun is a star, one of hundreds of billions of stars that form the Milky Way galaxy. It was known that there are billions of galaxies like the Milky Way and that the farthest ones were more than a billion light-years away. In his general theory of relativity, Albert Einstein presented a framework grand enough that theories of the universe could be formulated within it. Unfortunately, this framework allows one to formulate many theories without giving any indication as to which of the theories, if any, is correct.

In 1929, Edwin Powell Hubble showed that only models that allowed for the expansion of the universe could be correct. He did this by demonstrating that distant galaxies in all directions are receding from the Milky Way. He also proved that the more distant a galaxy is, the faster it is receding (Hubble's law). This implies that the universe is expanding. If the universe is expanding, it must have been smaller in the past.

It seems to be a straightforward matter to use Hubble's law to calculate when the universe began. If it is known how far away the parts are and how fast they are going, one should be able to calculate how long ago they were all together. Using Hubble's original data, 2 billion years was given as the age of the universe. This was somewhat embarrassing since the accepted age of the earth was far older. The problem lies in measuring the distances to galaxies, a difficult and uncertain process. Using other data, the age of the universe is calculated to be between 10 and 20 billion years old, which fits well with the accepted age of the earth as approximately 4.6 billion years.

In 1927, Georges Lemaître proposed an expanding universe based upon a prediction of general relativity. He supposed that all of space, matter, and energy had been crushed together and then exploded outward. It is this explosion that Sir Fred Hoyle later named the "big bang" theory. In 1948, American physicists George Gamow and Ralph Alpher realized that the big bang must have been incredibly hot. Building upon a theory proposed by the German physicist Hans Albrecht Bethe in 1938, they showed how elements could form in the cooling fireball. Hydrogen would be first to form, since its nucleus is a single proton or sometimes a proton combined with a neutron. Nuclear reactions would then convert about 25 percent of the matter into helium and a trace of lithium. There the process stops because the matter of the fireball is no longer dense enough for the reactions to continue.

**1059**

## An Alternative

In 1948, the big bang theory seemed unable to explain the existence of the heavy elements, and it gave an age for the universe that was less than that of Earth. A further difficulty was the problem of forming galaxies. It seemed unlikely that matter flung outward in the violence of the big bang would ever be able to come together again in clumps large enough to form galaxies. To overcome these difficulties, Hermann Bondi and Thomas Gold proposed the "steady-state theory" in 1948. The basis of this theory is that, when considered in a large enough volume, the universe will appear the same everywhere and at any time. This meant that the universe is infinitely old. Accepting the fact that the universe is expanding, they reached the astounding conclusion that matter must be continuously created uniformly throughout space. It was assumed that the new matter would come together to form new galaxies.

Since the most abundant element in the universe is hydrogen, Gold and Bondi proposed that the new mass appears as hydrogen. Using the best estimate for the rate of expansion of the universe, they calculated that if only one new hydrogen atom popped into existence in each volume of space the size of a living room over every few million years, then the average number of galaxies in a given volume of space would remain constant.

While there was no way to observe directly the creation of such a tiny amount of matter, there should still be observable consequences. In more than a billion years or so, enough matter should collect in the space between galaxies to form new galaxies, while old galaxies should increase in size as they gather in new matter. Later in 1948, Hoyle showed how continuous creation of matter might possibly fit into the formidable framework of general relativity.

## Consequences

The steady-state theory had both a cultural and a scientific impact. It had great philosophi-cal appeal for many because it proclaimed a universe of order, an infinite and eternal universe. The ready market for the many popular books written by Hoyle, Bondi, and others says something about the public's fascination with the steady-state theory. According to science historian Wolfgang Yourgrau, the introduction of the steady-state theory caused a tremendous sensation among cosmologists (those who study models of the universe). It stimulated much theoretical and empirical work as they sought either to prove or disprove the theory. Astronomers looking for the distribution of young and old galaxies predicted by the steady-state theory, however, did not find it. Instead, they found that all galaxies close enough to be so studied have at least some old stars. It appears that all galaxies formed at about the same time and that there are no intrinsically young galaxies.

Hoyle, who became the chief spokesman for the steady-state theory, worked on the problem of the origin of the heavy elements. He and others were able to show that heavy elements can form under the fantastic densities and temperatures that exist in the cores of stars and that these elements are flung back into space during supernova explosions. Eventually, heavy elements are incorporated into a new generation of stars and perhaps planets. The origin of heavy elements seemed to fit the steady-state theory, but this was also true of the big bang theory later on.

As evidence against the steady-state theory mounted, Hoyle and his companions eventually abandoned it. It had stimulated scientific endeavor and fired the imaginations of many people, but its day eventually passed. With the discovery of microwave radiation left behind by the primordial fireball of the big bang by radio astronomers Arno Penzias and Robert W. Wilson in 1965, the big bang theory reigned supreme.

*Charles W. Rogers*

# Piccard Invents Bathyscaphe

*Auguste Piccard invented a deep-sea submersible, called a bathyscaphe, that was capable of exploring the deepest trenches of the world's oceans.*

**What:** Earth science
**When:** 1948-1960
**Where:** Antwerp, Belgium, and Trieste, Italy
**Who:**
WILLIAM BEEBE (1877-1962), an American biologist and explorer
AUGUSTE PICCARD (1884-1962), a Swiss-born Belgian physicist
JACQUES PICCARD (1922-      ), a Swiss ocean engineer

## Early Exploration of the Deep Sea

The first human penetration of the deep ocean was made by William Beebe in 1934, when he descended 923 meters into the Atlantic Ocean near Bermuda. His diving chamber was a 1.5-meter steel ball that he named *Bathysphere*, from the Greek word *bathys* (deep) and the word *sphere*, for its shape. He found that a sphere resists pressure in all directions equally and is not easily crushed if it is constructed of thick steel. The bathysphere weighed 2.5 metric tons. It had no buoyancy and was lowered from a surface ship on a single 2.2-centimeter cable; a broken cable would have meant certain death for the bathysphere's passengers.

Numerous deep dives by Beebe and his engineer colleague, Otis Barton, were the first uses of submersibles for science. Through two small viewing ports, they were able to observe and photograph many deep-sea creatures in their natural habitats for the first time. They also made valuable observations on the behavior of light as the submersible descended, noting that the green surface water became pale blue at 100 meters, dark blue at 200 meters, and nearly black at 300 meters. A technique called "contour diving" was particularly dangerous. In this practice, the bathysphere was slowly towed close to the seafloor. On one such dive, the bathysphere narrowly missed crashing into a coral crag, but the explorers learned a great deal about the submarine geology of Bermuda and the biology of a coral-reef community. Beebe wrote several popular and scientific books about his adventures that did much to arouse interest in the ocean.

## Testing the Bathyscaphe

The next important phase in the exploration of the deep ocean was led by the Swiss physicist Auguste Piccard. In 1948, he launched a new type of deep-sea research craft that did not require a cable and that could return to the surface by means of its own buoyancy. He called the craft a bathyscaphe, which is Greek for "deep boat." Piccard began work on the bathyscaphe in 1937, supported by a grant from the Belgian National Scientific Research Fund. The German occupation of Belgium early in World War II cut the project short, but Piccard continued his work after the war. The finished bathyscaphe was named *FNRS 2*, for the initials of the Belgian fund that had sponsored the project. The vessel was ready for testing in the fall of 1948.

The first bathyscaphe, as well as later versions, consisted of two basic components: first, a heavy steel cabin to accommodate observers, which looked somewhat like an enlarged version of Beebe's bathysphere; and second, a light container called a float, filled with gasoline, that provided lifting power because it was lighter than water. Enough iron shot was stored in silos to cause the vessel to descend. When this ballast was released, the gasoline in the float gave the bathyscaphe sufficient buoyancy to return to the surface.

Piccard's bathyscaphe had a number of ingenious devices. Jacques-Yves Cousteau, inventor of the Aqualung six years earlier, contributed a me-

*Auguste Piccard.*

dive demonstrated the potential of the bathyscaphe. On the second dive, the vessel was severely damaged by waves, and further tests were suspended. A redesigned and rebuilt bathyscaphe, renamed *FNRS* 3 and operated by the French navy, descended to a depth of 4,049 meters off Dakar, Senegal, on the west coast of Africa in early 1954.

In August, 1953, Auguste Piccard, with his son Jacques, launched a greatly improved bathyscaphe, the *Trieste*, which they named for the Italian city in which it was built. In September of the same year, the *Trieste* successfully dived to 3,150 meters in the Mediterranean Sea. The Piccards glimpsed, for the first time, animals living on the seafloor at that depth. In 1958, the U.S. Navy purchased the *Trieste* and transported it to California, where it was equipped with a new cabin designed to enable the vessel to reach the seabed of the great oceanic trenches. Several successful descents were made in the Pacific by Jacques Piccard, and on January 23, 1960, Piccard, accompanied by Lieutenant Donald Walsh of the U.S. Navy, dived a record 10,916 meters to the bottom of the Mariana Trench near the island of Guam.

## Consequences

The oceans have always raised formidable barriers to humanity's curiosity and understanding. In 1960, two events demonstrated the ability of humans to travel underwater for prolonged periods and to observe the extreme depths of the ocean. The nuclear submarine *Triton* circumnavigated the world while submerged, and Jacques Piccard and Lieutenant Donald Walsh descended nearly 11 kilometers to the bottom of the ocean's greatest depression aboard the *Trieste*. After sinking for four hours and forty-eight minutes, the *Trieste* landed in the Challenger Deep of the Mariana Trench, the deepest known spot on the ocean floor. The explorers remained on the bottom for only twenty minutes, but they answered one of the biggest questions

chanical claw that was used to take samples of rocks, sediment, and bottom creatures. A seven-barreled harpoon gun, operated by water pressure, was attached to the sphere to capture specimens of giant squids or other large marine animals for study. The harpoons had electrical-shock heads to stun the "sea monsters," and if that did not work, the harpoon could give a lethal injection of strychnine poison. Inside the sphere were various instruments for measuring the deep-sea environment, including a Geiger counter for monitoring cosmic rays. The air-purification system could support two people for up to twenty-four hours. The bathyscaphe had a radar mast to broadcast its location as soon as it surfaced. This was essential because there was no way for the crew to open the sphere from the inside.

The *FNRS* 2 was first tested off the Cape Verde Islands with the assistance of the French navy. Although Piccard descended to only 25 meters, the about the sea: Can animals live in the immense cold and pressure of the deep trenches? Observations of red shrimp and flatfishes proved that the answer was yes.

The *Trieste* played another important role in undersea exploration when, in 1963, it located and photographed the wreckage of the nuclear submarine *Thresher*. The *Thresher* had mysteriously disappeared on a test dive off the New England coast, and the Navy had been unable to find a trace of the lost submarine using surface vessels equipped with sonar and remote-control cameras on cables. Only the *Trieste* could actually search the bottom. On its third dive, the bathyscaphe found a piece of the wreckage, and it eventually photographed a 3,000-meter trail of debris that led to *Thresher*'s hull, at a depth of 2.5 kilometers.

These exploits showed clearly that scientific submersibles could be used anywhere in the ocean. Piccard's work thus opened the last geographic frontier on Earth.

*Charles E. Herdendorf*

# Communists Take Over Czechoslovakia

*Though Czechoslovakia had gained a democratic government after World War II, Communists soon succeeded in pushing out noncommunists and taking control of the country.*

**What:** Political reform; Coups
**When:** February 20-25, 1948
**Where:** Prague, Czechoslovakia
**Who:**
EDVARD BENEŠ (1884-1948), president of Czechoslovakia from 1945 to 1948
KLEMENT GOTTWALD (1896-1953), Czech prime minister from 1946 to 1948, and president of Czechoslovakia from 1948 to 1953
JOSEPH STALIN (1879-1953), dictator of the Soviet Union from 1924 to 1953
JAN MASARYK (1886-1948), Czech foreign minister from 1940 to 1948

## A New Democracy

From 1620 until the end of World War I, the territories that make up modern Czechoslovakia were part of the Habsburg Empire. Dissatisfied with domination by the Germans and Hungarians, Tomáš Masaryk openly sided with the Allied Powers in World War I and demanded the creation of an independent Czech state. The Allied victory and the disintegration of the Habsburg Empire enabled this dream to become reality.

In 1924, Czech leaders signed a pact of mutual assistance with France. Czechoslovakia was a democracy, allowing all political parties to compete freely. In elections, the Communist Party won slightly more than 10 percent of the votes.

When Adolf Hitler took power in Germany in 1933, Czechoslovakia's future was endangered. Hitler wished to dominate all of Eastern Europe, and since there were three million German citizens living in the border areas of Czechoslovakia, he demanded that these lands be given to Germany. Since these areas were mountainous and heavily fortified, however, giving them up would leave Czechoslovakia defenseless.

Nevertheless, at the Munich Conference of September, 1938, France (with Great Britain's support) refused to honor its treaty with the Czechs and agreed to let Germany annex the border territories. Czechoslovakia's well-being was less important to France than avoiding war. Abandoned by its ally, the Prague government allowed Hitler to move in peacefully. Six months later, he occupied all the Czech lands and set up a Nazi-controlled government in Slovakia.

The dismemberment of Czechoslovakia made a lasting impression on its president, Edvard Beneš, who had succeeded Tomáš Masaryk. Even though Beneš moved to London during World War II to set up a Czech government-in-exile, he decided that the Czechs could not rely on the West to protect them. Instead, Czechoslovakia would have to make an alliance with the Soviet Union. As a result, Beneš became dependent upon Soviet dictator Joseph Stalin.

While Beneš was in London, the leaders of the Czech Communist Party stayed in Moscow. There they came under greater Soviet influence. Within Czechoslovakia, the Communist Party—with the Soviets' help—became deeply involved in the Resistance movement. As an advocate of social justice and freedom from German domination, the Party gained much new support among the Czech people.

In 1945, Beneš traveled to Moscow and was obligated to recognize the Communists' achievements within Czechoslovakia. He gave the Communists control of important government ministries such as the Interior, which controlled the police; Agriculture, which distributed the lands taken back from the Sudeten Germans; and Information. The new prime minister and minister of defense were both sympathetic to Communism.

Also, Beneš agreed to set up National Committees to run local government until elections could be held. Since Czechoslovakia was liberated by the Soviets, these National Committees became dominated by the Communists. Finally, the Trade Unions, which had been divided among the political parties before the war, were united—with a membership of one million—under Communist leadership.

In spite of these large gains for the Communists, a democratic government was set up in Czechoslovakia. Its survival depended both on the strength of noncommunist groups within the country and on the cooperation of the Allied Powers—the Soviet Union, Great Britain, and the United States.

## The Communist Coup

In November, 1945, the Red Army left Czechoslovakia, and free elections were held in 1946; the Communists won 38 percent of the vote. As a

result, Communist leader Klement Gottwald became prime minister in a new coalition cabinet.

Meanwhile, serious conflicts were developing between the Soviet Union and the Western Allies (Great Britain and the United States). In June, 1947, U.S. secretary of state George C. Marshall announced a plan to help the European economy recover from the effects of war. The Soviets interpreted this plan as an American way of buying friends in Europe, and they refused to let Czechoslovakia participate.

Cominform (the Communist Information Bureau) criticized Czech Communists for not socializing their country as quickly as the other Eastern European Communists. Elections in Czechoslovakia were set for May, 1948, and a poll showed that the Communists had lost about 10 percent of their support. The Communists decided that it was time to take action.

*Czech youths carry banners depicting Prime Minister Klement Gottwald (left) and Joseph Stalin in a parade in Prague.*

In September, 1947, booby-trapped boxes of perfume were sent by an important Communist official to noncommunist government ministers, including Foreign Minister Jan Masaryk, son of Tomáš Masaryk. None of these bombs exploded. In February, 1948, a large Trade Union Congress assembled to ask the government to nationalize more industry and take over more land.

Around the same time, the minister of interior suddenly replaced eight noncommunist police commanders with Communists. A majority of ministers in the cabinet voted to overturn this order. When the minister of interior refused and was supported by Gottwald and the Communist Party, the twelve noncommunist ministers in the cabinet resigned in protest. They hoped that President Beneš would be able to solve the problem, but he was unable to convince the Communists to back down.

By then Prague was filled with delegates to the Trade Union Congress, and arms were passed out to workers' militia groups. Communists took over the ministries whose leaders had resigned, and a number of arrests were made. On February 25, faced with the possibility of civil war or even a Soviet invasion, Beneš accepted the twelve resignations and approved a new government completely dominated by Communists.

On March 10, Foreign Minister Masaryk was killed when he jumped—or was pushed—from his office window. In the May elections, the list of the Communist-dominated National Front was victorious. Beneš resigned as president on June 7, and a week later Gottwald succeeded him as head of state. A Communist dictatorship then reigned in Czechoslovakia.

**Consequences**

The Communists who dominated Czechoslovakia proved repressive, following the lead of their Soviet teachers. Their coup had important effects on international relations as well. The Western Allies became more determined to stand against further Communist expansion; this determination was expressed in the Berlin Airlift and the formation of the North Atlantic Treaty Organization (NATO). Within other European countries where the Communist Party had been active, governments became more cautious about dealing with Communists.

Czechoslovakia would not free itself from Communist rule until the late 1980's, when a wave of popular uprisings toppled Communist governments across Eastern Europe. Eventually, Czechoslovakia would split into two democratic nations, the Czech Republic and Slovakia.

# Organization of American States Is Formed

*The republics of both American continents joined in an organization to promote regional security and cooperation.*

**What:** International relations
**When:** March-May, 1948
**Where:** Bogotá, Colombia
**Who:**
HARRY S. TRUMAN (1884-1972), president of the United States from 1945 to 1953
GEORGE CATLETT MARSHALL (1880-1959), U.S. secretary of state from 1947 to 1949
JORGE ELIÉCER GAITÁN (1902-1948), leader of Colombia's Liberal Party

## Alliances and Conferences

In the three years following World War II, the United States focused most of its diplomacy on problems in Europe and the Far East. Latin America was not forgotten, but the United States concentrated on the Communist threat to Western Europe and made military and economic aid to that region its highest priority. This shift in focus eventually led to problems with Latin America. In the years between 1945 and 1948, the United States was interested primarily in security within the Western Hemisphere, while the Latin American nations wanted economic help.

The principle of hemispheric defense had been confirmed by the Act of Havana in July, 1940. This agreement, signed by the twenty-one republics of the Pan-American Union, stated that the American republics, individually or as a group, should take charge of any European possession in the Western Hemisphere that was threatened by any act of aggression.

After Pearl Harbor, most Latin American nations had declared war on Germany and Japan. This wartime cooperation was continued by the Act of Chapultepec on March 3, 1945, which an-

nounced that aggression against any American nation was aggression against all of them.

Two years later, steps were taken to create a permanent defensive alliance. On June 3, 1947, President Harry S. Truman announced that the United States was willing to negotiate an inter-American defense pact. Twenty-one American nations came together at a conference in Rio de Janeiro, Brazil, in August, 1947. Secretary of State George C. Marshall spoke for the United States. Faced with demands for a "Marshall Plan" for Latin America like the economic-recovery plan designed for Europe, Marshall asked that such economic questions be postponed.

The Rio conference's main achievement was the September 2 signing of the Inter-American Treaty of Reciprocal Assistance. Members agreed to consult together if any of them was threatened by aggression, and eventually all twenty-one American republics ratified the treaty. This treaty did not, however, set up ways for the republics to come together and reach decisions.

## Setting Up the OAS

The questions left unanswered at the Rio conference were saved for the Ninth International Conference of American States, which met at Bogotá, Colombia, in the spring of 1948. On April 9, soon after the delegates arrived, Liberal Party leader Jorge Eliécer Gaitán was assassinated in downtown Bogotá. The violence provoked riots that soon spread out of control. For several days the visiting delegates were confined to their embassies and hotel rooms; by the time the Colombian army gained control of the situation, much of Bogotá lay in ruins. On April 14, the conference was able to resume its meetings in a boys' school near the U.S. Embassy.

Secretary of State Marshall, who again headed the American delegation, addressed the second

**1067**

plenary session. Describing the major economic, military, and humanitarian efforts the United States was making in other parts of the world, Marshall confessed that American resources were limited. He frankly admitted that a Marshall Plan for Latin America would not be possible. If Europe could be strengthened, however, he believed that economic stability would spread across the world.

The major achievement of the Bogotá conference was the creation of the Organization of American States (OAS), a new name given to the reorganized Pan-American Union, which had existed for sixty years. Where the earlier organization had been loose and flexible, the OAS Charter clearly stated rights and obligations of member states: inter-American relations would be governed by international law and order, and member states would have equal power; an act of aggression against one member would be an act of aggression against all; any controversy between states would be decided by peaceful means; The well-being of the American peoples depended on social justice, political democracy, economic welfare, and respect for one another's culture.

The charter established six organizations: the Inter-American Conference, the Meeting of Consultation of Ministers of Foreign Affairs, Specialized Conferences, the OAS Council, the Pan-American Union, and Specialized Organizations. The first three bodies would have responsibility for general political and economic issues. The Meeting of Consultation of Foreign Ministers would be an emergency assembly when urgent problems arose, such as the threat of war.

## Consequences

The OAS was not an independent organization; its authority was limited, for it depended upon the nations that had brought it into existence. Nevertheless, the formation of the OAS did strengthen cooperation among the republics of the Western Hemisphere. In later years, it became a means for joint political and military action.

*Roger P. Davis*

# Communists Lose Italian Elections

> *Because of a split among leftists and Western support for the Christian Democrats, Italy's Christian Democratic Party managed to defeat the Communist Party in the important elections of 1948.*

**What:** Political reform
**When:** April 18, 1948
**Where:** Italy
**Who:**
PALMIRO TOGLIATTI (1893-1964), leader of the Italian Communist Party
ALCIDE DE GASPERI (1881-1954), leader of the Christian Democrats, and prime minister of Italy from 1948 to 1953

## The Communists Gain Strength

The Communist Party emerged in 1945 as a major force in Italian politics. With its superb organization, it had survived Fascist repression and, in fact, had been active in the Resistance movement in northern Italy during the years of German occupation. Italy's defeat brought about economic problems, political chaos, and general discouragement—a situation that helped the Communists even further.

The Communist and Socialist Parties cooperated in the coalitions that ruled Italy between 1945 and 1948. Palmiro Togliatti, a Communist who had stayed in Moscow during the Fascist era in Italy, built up party membership lists by recruiting northern industrial workers and peasants from the Po River Valley and southern Italy. The party's armed struggle against Benito Mussolini and the Germans had left it with weapons and a tradition of taking militant action.

The Liberal and Conservative parties, which had opposed Communism before the war, had been disgraced by their failure to stop fascism. They were replaced by the new Christian Democratic Party, which had its origins in the Popular Party—a non-Marxist Catholic group that had been active in Italian politics for a short time after World War I. Under the strong leadership of Alcide De Gasperi, the Christian Democrats attracted Italians who favored Catholicism and social reform and opposed Communism.

Between December, 1945, and May, 1947, the Christian Democrats and Communist-Socialists governed Italy in a coalition, while unofficially opposing each other. At first, it appeared that the Communists might drive the Christian Democrats from power. If De Gasperi tried to unite with the Liberals or the new semi-fascist Uomo Qualunque Party, he would probably lose the left wing of his party. Yet if the Christian Democrats gave in too much to the Communists and Socialists, they could alienate the right wing of the party. De Gasperi was forced to walk a tightrope.

## The Christian Democrats Rise

A split on the Left was the first event that rescued De Gasperi from this dilemma. Pietro Nenni, leader of the Socialist Party, had linked his party's candidates to the Communists in the elections of November, 1946. As a result, Communist candidates gained votes that would otherwise have gone to socialists. Moderate Socialists, led by Giuseppe Saragat—who preferred to support the Americans in the developing Cold War between East and West—were disgusted with Nenni's election strategy. They used Nenni's failure as a reason for creating a new party, the Social Democrats.

Yet Togliatti failed to realize how important this split was, and he continued to behave as if the Christian Democrats could not rule without Communist support. In May, 1947, however, De Gasperi managed to form a one-party cabinet. Since the Christian Democrats were still a minority, Togliatti and De Gasperi realized that this cabinet was only temporary. Both of them began preparing for a fight in the 1948 elections.

De Gasperi and his party made several important arguments during their election campaign. Boasting of Italy's dramatic economic improvement under the guidance of Finance Minister Luigi Einaudi, De Gasperi added that this progress would continue only if American aid continued—and American aid would continue only if the Christian Democrats could build a strong government.

The Christian Democrats received help from overseas. On March 15, 1948, the U.S. State Department announced that all economic aid to Italy would stop if the Communist-Socialist ticket won the elections. Italian-Americans were encouraged to urge their relatives in Italy to vote Christian Democrat.

On March 20, 1948, French, British, and American leaders joined to give disputed territory around Trieste to Italy—an action that strengthened De Gasperi's argument for an alliance with the Western Powers. Leaders of Italy's Catholic Church ordered priests to refuse the sacraments to supporters of Togliatti and Nenni.

Finally, De Gasperi argued not only against the Communists but also against other anticommunist groups. A vote cast for any other anticommunist party, he said, was a vote wasted. Pointing to Czechoslovakia, which had just come under Communist control, he said that Italy would come to a similar fate if anticommunists did not unite behind his party. The results of this campaign went beyond De Gasperi's greatest hopes. The Christian Democrats won 12,713,300 votes, while the Communist-Socialist ticket gathered only 8,137,000.

De Gasperi then had an absolute majority in the Italian legislature, with the right to continue one-party rule. He realized, however, that such a policy might be dangerous; instead, he chose to form a coalition with Saragat's Social Democrats and two smaller parties. This coalition allowed him to base his government on 60 percent of the votes cast.

**Consequences**

The Italians had decisively voted against Communism. The Christian Democrats continued to dominate Italian politics for nearly a decade, and Togliatti's Communists had to take the frustrating role of an opposition party. Great Britain and the United States, concerned about the expansion of Soviet influence in Europe, saw the Italian election of 1948 as a major achievement.

*Leftist sympathizers demonstrate against the De Gasperi regime in downtown Rome.*

# Jewish State of Israel Is Born

*In the face of Arab hostility, Jews in Palestine proclaimed the establishment of an independent Jewish state, a homeland for Jews scattered across the world.*

**What:** Political independence
**When:** May 14, 1948
**Where:** Palestine
**Who:**
THEODOR HERZL (1860-1904), the founder of Zionism
DAVID BEN-GURION (1886-1973), prime minister of the State of Israel from 1948 to 1953
CHAIM WEIZMANN (1874-1952), president of the State of Israel from 1948 to 1952

## The Rise of Zionism

Since the destruction of the Jerusalem Temple and the scattering of the Jews in the first century C.E. (common era), the Jews' history was one of trauma and tragedy. The Jews were expelled from medieval Europe, and again from Spain in the fifteenth century. In most cases, when Jews settled in a new land they could not achieve political equality because of anti-Semitism, and they had no land they could call their own.

With the coming of the Enlightenment and the French Revolution, it seemed to many Western European Jews that they would at last be emancipated. Emancipation would solve the Jews' problems because they would stop being treated as a separate nation and would become simply citizens of a particular country: Britons or Germans who happened to be Jews. Judaism would become no more than a religion.

In the East, however, the Jews began to hope for a different solution. In Russia and Poland there was no hope of emancipation. Stubborn anti-Semitism followed Eastern European Jews everywhere they went; they were assaulted, their property was taken. In this atmosphere of constant oppression, the idea of Zion—a land to which Jews could return—was born in the writings of Theodor Herzl in 1896. Zion would be a land where Hebrew was the spoken language, a land built by Jewish laborers and farmers as well as Jewish professionals.

At first this country was not necessarily Palestine, though Palestine was attractive because devout Jews believed that it was the land promised by God to their ancestor Abraham. South America was suggested; a country in Africa was seriously considered. In the meantime, some Jews did settle in Palestine, and others bought Palestinian land on the chance that Zion would be established there.

At last, during World War I, it seemed possible to negotiate an agreement with the British for a Jewish homeland in Palestine. British foreign secretary Arthur Balfour issued the Balfour Declaration in 1917, promising his country's support: "His Majesty's Government views with favour the establishment in Palestine of a national home for the Jewish people." Balfour specified that "nothing shall be done which may prejudice the civil and religious rights of existing non-Jewish communities in Palestine."

## A Question of Survival

After the war, Great Britain was given responsibility to govern Palestine as a "mandate" under the League of Nations. Across the world, Jews began to hope that they would have a homeland at last. Then, in a series of "White Papers" (in 1922, 1930, and 1939), British leaders made it clear that they did not intend to follow through on the promises of the Balfour Declaration.

Great Britain's approach to the mandate taught Jews that they could not trust the British, or any other single nation, to guarantee the safety of Jews. In the 1930's, Jews around the world could see that the survival of European Jews depended on mass immigration to Palestine and to friendly nations. Yet Great Britain and the United States closed their doors. It was Germany and Japan's aggressiveness, not the Nazis' savage attacks on the Jews, that brought the Allies into war against the Axis Powers.

During World War II, Sir Winston Churchill's government was not friendly to Zionism, and the Labour government that came into power in Great Britain in 1945 had little interest in the Jewish problem. It seemed clear to many that Great Britain preferred to ally itself with the Arab cause. In fact, British leaders wished to give up their Palestinian mandate and did so in May of 1948, leaving the Palestinian Jews at the mercy of invading Arab armies.

Jews around the world realized that they could survive only with a Jewish homeland, a sovereign state where Judaism was not only a religion but also a nationality. Yet the British and the Arabs re-mained opposed. The United Nations partitioned Palestine between Arabs and Jews in 1947, but Arab and British leaders refused to accept this solution. Arabs raided Jewish areas to disrupt communications, and the British looked the other way.

In the midst of this struggle, Jews in Palestine proclaimed the birth of the independent Jewish state of Israel on May 14, 1948, David Ben-Gurion was appointed prime minister in a provisional government, and Chaim Weizmann, a long-time Zionist, became president.

## Consequences

The proclamation of an independent Israel did not make matters easier for Israeli Jews: The surrounding Arab states rushed to attack the new nation. The fighting was difficult and bitter, yet somehow the Israelis were able to build a national army and to arm it, to some extent, with modern weapons. By the end of 1948, Israeli forces had been successful in pushing through southern Palestine and even into Egypt. One Israeli force made its way to the Red Sea and took control of the port of Eilat.

There were continuing tragedies along the way. During a cease-fire, a United Nations mediator named Count Folke Bernadotte Af Wisborg was assassinated by Zionist terrorists. Finally, by July, 1949, the Israelis signed armistice agreements with the Arab states.

Throughout the twentieth century there were other wars between Israel and its neighbors. Leaders of Palestinian Arab groups continued to insist that Israel had no right to exist in Palestine, while Israeli leaders continued to claim land for Jewish settlers. Peace conference after peace conference attempted to cut a way through the tangled problems of Israeli rights versus Arab rights. In 1991, Arab and Israeli leaders met in Madrid, Spain, and Washington, D.C., to discuss issues that divided them. No conclusive agreements were reached, but the participants were applauded for their efforts and the talks set a precedent for future negotiations.

*Jonathan Mendilow*

*In the late 1950's, thousands of Israelis take to the streets of Tel Aviv to celebrate the anniversary of their nation's independence.*

# Hale Telescope Begins Operation at Mount Palomar

*George Ellery Hale received funding to construct a 508-centimeter reflecting telescope atop Southern California's Mount Palomar.*

**What:** Astronomy
**When:** June 3, 1948
**Where:** Mount Palomar, California
**Who:**
GEORGE ELLERY HALE (1868-1938), an American physicist and astronomer
ELIHU THOMSON (1853-1937), an American scientist
WALTER SYDNEY ADAMS (1876-1956), the director of Mount Wilson
RAYMOND FOSDICK (1883-1972), the president of the Rockefeller Foundation
ANDREW CARNEGIE (1835-1919), a benefactor of the Carnegie Institute

## A Talent for Fund-Raising

George Ellery Hale was born in Chicago on June 29, 1868. He pursued his education at the Massachusetts Institute of Technology and was graduated in 1890. During that same year, Hale founded the Kenwood Astrophysical Observatory near his Chicago home. Hale conducted research in solar spectroscopy (the analysis of the sun's light) and developed, in 1891, a device called the "spectroheliograph," an instrument that photographs the sun in a narrow wavelength band. Such photographs reveal the structure of the sun's surface.

In 1892, Hale became the first director of the Yerkes Observatory, Williams Bay, Wisconsin, and received funding for a 102-centimeter refracting telescope. He always envisioned working in an astronomy department with state-of-the-art observational equipment. Hale was one of the original founders, in 1895, of a professional publication, *Astrophysical Journal.* Several years later,

in 1899, he helped establish the American Astronomical Society.

Hale was not only an inventor and astronomer but also an excellent fund-raiser. He continued to dream of bigger and better astronomical equipment. Hale made presentations before members of the Carnegie Institute of Washington to advise them of the unusual opportunity of building an observatory in the nearly cloudless mountains of Southern California. In 1904, he received a grant from Andrew Carnegie of the Carnegie Institute to construct the Mount Wilson Observatory (near Los Angeles, California), where 152-centimeter (in 1908) and 254-centimeter (in 1917) reflecting telescopes would later be erected. Hale served as director of Mount Wilson Observatory until 1923.

Yet Hale continued to dream of a larger telescope. By 1928, he had written many articles describing the benefits of constructing a giant telescope. He sent copies of his articles to representatives of the Carnegie Institute and Rockefeller Foundation, hoping to arouse financial support.

Miraculously, Hale negotiated an unusual financial arrangement between these foundations for the construction and future operation of a 508-centimeter telescope. The staff from Mount Wilson Observatory would supply scientists to construct and operate the telescope, while the California Institute of Technology would decide how it would be used.

Raymond Fosdick, then president of the Rockefeller Foundation, appropriated $6 million for the construction of the new telescope. The California Institute of Technology provided an endowment for its operation. On October 29, 1928, Hale told the media of his construction plans.

**1073**

## New Challenges

Hale was sixty years old when he received the funds to start building. His health, however, was not good; his stressful lifestyle caused Hale to be hospitalized for a short time in a sanatorium. Afterward, he strove to take life at a slower pace but continued to assist with the plans for the telescope. He had to solve two problems: where to build the new observatory and how to cast a 508-centimeter mirror.

It was decided that the new observatory would be built on Mount Palomar, north of San Diego, California. The bigger problem concerned the huge mirror. No mirror of that size had ever been built. The ideal glass for it was thought to be quartz, which would expand or contract very little as a result of variations in temperature. Elihu Thomson, a scientist at the General Electric Company in Lynn, Massachusetts, was contracted to cast the mirror, but unfortunately, each time the quartz was heated, evaporation caused bubbles to form. Finally, members of the Observatory Council, including Walter Adams (director of Mount Wilson Observatory), asked the Corning Glass Works, in Corning, New York, to develop a 508-centimeter Pyrex disk. (Pyrex is a low-expansion borosilicate glass.) Although temperature variations affect Pyrex more than quartz, Pyrex is still easier to work with than ordinary glass. A special annealing oven was built, and the 508-centimeter mirror was poured successfully on December 2, 1934.

The 508-centimeter telescope was constructed to be a "prime focus" instrument. With such a telescope, the observer actually climbs into a cage within the telescope and observes light reflected from the mirror onto a photographic plate. A platform moves up and down so that the astronomer can step directly into the prime focus area. This telescope is one of the few that allows an observer to view from the prime focus, rather than from an eyepiece below the body of the instrument.

## Consequences

On February 21, 1938, Hale died, ten years before his dream of the 508-centimeter telescope would be realized. The occurrence of World War II (1939-1945) caused optical astronomy to come almost to a halt. Many scientists who had worked on Mount Palomar were contracted into the war effort. Finally, on June 3, 1948, the 508-centimeter telescope was officially dedicated and named the "Hale Telescope," a monument to the man whose vision and leadership had made it a reality.

One of the first observations made with the Hale Telescope led to a doubling of the distance scale between the Milky Way and the Andromeda galaxies, from one million light-years to two million light-years.

The importance of a detailed investigation of neighboring galaxies is in establishing a diagram of their distribution. In addition, a spectroscopic study of galaxies (that is, an analysis of the light given off by them) allows astronomers to determine whether these galaxies are receding from or moving toward the Milky Way and how fast they are doing so. In fact, observations made with the Hale Telescope suggested that the universe is expanding, lending support to the big bang theory. The Hale Telescope remained the largest in the world from 1948 to 1976, when a 600-centimeter reflector telescope was completed at Zelenchukskaya, in the Soviet Union. Problems with the Soviet telescope, however, rendered it less accurate than the Palomar telescope, which remained the premier optical telescope for the next four decades.

*Noreen A. Grice*

# Western Allies Mount Berlin Air Lift

*After the Soviets blockaded Berlin, American and British airplanes began flying in food and fuel for the city's 2.25 million people.*

**What:** International relations; Economics
**When:** June 24, 1948-May 12, 1949
**Where:** Berlin, Germany
**Who:**

HARRY S. TRUMAN (1884-1972), president of the United States from 1945 to 1953

LUCIUS D. CLAY (1897-1978), military governor of Germany (U.S. zone) from 1947 to 1949

CURTIS E. LEMAY (1906-1990), commander of U.S. Air Forces in Europe

WILLIAM H. TUNNER, commander of the Berlin Air Lift

## The Air Lift Takes Off

After World War II, the Allied Powers agreed that Berlin, which lay deep within the Soviet-governed zone of Germany, would be divided into four sections, to be occupied by the four main Allies: France, Great Britain, the United States, and the Soviet Union. As these nations tried to work together, however, there were more and more conflicts between the United States, Great Britain, and France on one hand and the Soviet Union on the other.

Early in 1948, the Four Power Allied Control Council for Germany met to make a decision about reforming Germany's currency. To help Germany recover from the war, the Allies needed to help stop inflation, rebuild the economy, and control the black market. As usual, however, the Soviets' plans conflicted with those of the Western Allies, and the Soviets left the meeting.

On March 30, 1948, General Lucius D. Clay, military governor of the American Zone, approved a currency reform that was adopted by all but the Soviets, who established their own currency and tried to force everyone in Berlin to use it. At the same time, Soviet soldiers began demanding the right to inspect Western military trains passing through the Soviet zone into Berlin. When Western leaders refused, Soviet troops stopped all railway and river traffic and created tight road blockades, so that almost no supplies could enter Berlin. Then the Allies closed their zones to Soviet traffic.

Eager to push the Western Allies out of Berlin, the Soviets stopped all traffic into the city on June 24, 1948. Even shipments of food for Berlin's civilian population were turned back. At first the Soviets claimed that road repairs were needed; finally, however, they admitted that they simply wanted the Western Allies out.

The Soviets offered to feed all the population of Berlin, but the Berliners themselves rejected this offer, and the French, British, and Americans in Berlin refused to leave their zones of the city. At the time of the blockade, these zones had coal supplies to last for forty-five days and enough food to feed the population of 2.25 million for thirty-six days.

Huge amounts of supplies would be needed, for the Western zones had been importing six thousand tons of coal each day, along with hundreds of tons of food. All of this would have to be brought in by air.

On June 21, Clay ordered the "maximum number of suitable aircraft" to transport supplies to Berlin, and Lieutenant General Curtis E. LeMay gave the U.S. Air Force the order to supply Berlin on June 26. Battered old C-47 "Gooney Birds," which had been used in the Normandy Invasion and other World War II efforts, took the first eighty tons of medicine and food into Berlin. Only 102 C-47's were available in all Europe, besides a few British planes and two C-54's. The French had no suitable planes, so they helped in organizing.

**1075**

*A U.S. cargo plane, part of the Berlin Air Lift, flies above a ruined building near the Tempelhof Airport, on July 9, 1948.*

British and American crews flew the planes. At first, the airlift could bring only seven hundred tons a day. Since about forty-five hundred tons of coal, food, and other supplies were needed per day, the effort seemed doomed to failure.

Back in the United States, President Harry S. Truman was being pressured to pull American forces out of Berlin. Keeping them there and flying in supplies seemed an impossible task. Yet his response was clear: "The United States is going to stay—period."

### The Battle Is Won

The U.S. Air Force put out a call all over the world for pilots and planes. On June 29, 1948, the Berlin Air Lift Task Force was established, and the effort began to grow. By the end of July, American planes were delivering fifteen hundred tons a day to Berlin, while another 750 tons

came on Royal Air Force planes. Meanwhile, the Berliners refused the Soviets' offers of extra food if they would give in.

Lieutenant General William H. Tunner, an airlift expert, formed the crews into a professional force, flying planes out of four airfields every three minutes. It was difficult flying. Air traffic moved in one direction on two outside routes into Berlin and back through the center route. A missed approach meant returning to base without unloading.

In August, 120,672 tons were flown into Berlin. "Operation Vittles," as the airlift was called, was enabling Berlin to survive. By fall, five thousand tons a day were reaching the city.

In November and December, the weather turned bad. Temperatures dropped, aircraft were covered with ice, fierce winter storms reduced visibility to almost nothing, and snow clogged the

**1076**

airfields. Planes flew as usual, but often they were unable to land. In Berlin, food became scarce, and some people burned furniture to heat their homes. The Soviets were sure that soon the West would give up.

Yet by January, 1949, the worst weather was over. By mid-February, the millionth ton of supplies was delivered; in April, 235,476 tons were delivered. The airlift had become routine, and the West had proved that it could keep the effort up as long as necessary. On May 12, 1949, the Soviet Union called off the blockade.

## Consequences

The battle for survival had been won. Yet the airlift did not stop until September 30 so that emergency supplies could be stockpiled. Between June 24, 1948, and May 12, 1949, more than 1.5 million tons of supplies had been flown in, and another million tons came before the airlift stopped. The effort had cost the Western Allies $200 million, thirty-five aircraft, and seventy-five lives.

The Berlin Air Lift became a symbol of Western strength and determination. It was an outstanding example of how military might could be used for peacetime purposes. Within Germany, the Western Allies gained much support, while the Soviets became even more unpopular. Soon a number of nations of Western Europe joined with the United States in the North Atlantic Treaty Organization (NATO), and West Germany was organized into the Federal Republic of Germany.

*Theodore A. Wilson*

**1077**

# Cominform Expels Yugoslavia

> *Resisting Soviet efforts to control Yugoslavia, prime minister Josip Tito was unable to keep his country from being thrown out of the international Communist organization.*

**What:** International relations
**When:** June 28, 1948
**Where:** Eastern Europe
**Who:**
JOSEPH STALIN (1879-1953), dictator of the Soviet Union from 1924 to 1953
JOSIP BROZ TITO (1892-1980), marshal and prime minister of Yugoslavia from 1945 to 1953, and president of Yugoslavia from 1953 to 1980
GEORGI DIMITROV (1882-1949), prime minister of Bulgaria from 1946 to 1949

## Questions of Loyalty

In 1945, Soviet dictator Joseph Stalin seemed to have no more loyal ally and devoted Communist supporter than Josip Broz Tito, the leader of Yugoslavia's resistance movement against Nazi Germany in World War II. With Stalin's approval, Marshal Tito had become the leader of Yugoslavia's Communist Party in the 1930's. After the war, Tito was prime minister of Yugoslavia, and he quickly began turning his country into a socialist state. As the Soviet Union began to drift away from its British and American wartime allies, Tito joined in the Soviet criticism of the West. There was even some talk of Yugoslavia's becoming a constituent republic of the Soviet Union.

Yet relations between Moscow and Belgrade were not as smooth as they appeared. For one thing, Stalin had not supported Tito's movement during the war as much as the Yugoslav leader had wished. For months the Soviet Union had allowed the Allies to back Tito's rival's instead. Second, Stalin had worked out an agreement with British prime minister Winston Churchill in 1944: After the war, Soviet influence in Yugoslavia would be shared fifty-fifty with the West.

When news of this bargain came out in the Western press, Tito felt betrayed by Stalin.

At the same time, Stalin did not trust Tito, though Tito was actually very loyal to Communist doctrine and to Moscow's leadership. What Stalin really wanted was to build a worldwide base of support, with Communist parties that would depend on him personally. Tito had led a popular movement and had risen to power without the Red Army's help; this gave him too much room for independence. He was not likely to bow under Stalin's tight control.

One of the tools Stalin planned to use to control his new allies in Eastern Europe was the Communist Information Bureau (Cominform), which was established in Belgrade in September, 1947. Members included Yugoslavia, Bulgaria, Poland, Hungary, Romania, Czechoslovakia, and the Soviet Union. Albania, whose Communist Party and government were linked to Yugoslavia's, was not a member, but Italy's Communist Party participated. The Soviet Union also set up economic agreements with the governments of these countries.

## Tito Resists Stalin

Trying to keep tabs on Tito, Stalin found people working with Tito who agreed to serve as Soviet agents. One of these was Streten Zhuyovich Tsrni, a member of the Yugoslav Communist Party Central Committee. Another Central Committee member, Andreya Hebrang, was likely Stalin's agent as well.

Meanwhile, Tito was becoming unhappy over Stalin's efforts to dominate his country's politics, economy, and culture. In January, 1948, Tito dismissed Hebrang from his position as chairman of the Economic Planning Commission. Stalin began retaliating, first by refusing a trade treaty that Tito wanted, then by recalling Soviet military and civilian advisers from Yugoslavia. The

**1078**

Yugoslav Communist Party's Central Committee, however, decided to hold fast against Soviet pressure.

Throughout the spring of 1948, Belgrade and Moscow exchanged argumentative messages. This correspondence failed to solve the problem; in fact, it had the opposite effect. Moscow accused Belgrade of practicing an impure form of Communism; Tito proclaimed Yugoslavia's loyalty and accused Stalin of interfering in his country's internal affairs.

Moscow also forced the other countries of the Cominform to speak out against Tito's "deviation." Though all these nations obeyed the order, one leader seems to have been reluctant: Georgi Dimitrov, prime minister of Bulgaria. Because of his famous defiance of the Nazis at the Reichstag fire trial in 1933, Dimitrov was a hero among international Communists. Dimitrov had been Tito's partner in proposing a Balkan confederation. Out loud, Stalin agreed to this plan, but in fact he mistrusted it, fearing that it would keep the Soviet Union from achieving complete control over Eastern Europe.

On June 28, the Cominform officially expelled Yugoslavia for its "unsocialist behavior," especially the arrests of Hebrang and Zhuyovich-Tsrni.

## Consequences

Tito was not ready to crawl back, asking for mercy. At the Yugoslav Community Party Con-

gress that summer, he brought the dispute into the open. In response, the Soviets began punishing Yugoslavia in several ways.

The Cominform countries began to break off their trade agreements with Yugoslavia. The first country to do so was Albania, whose leader, Enver Hoxha, allied himself with Stalin in order to overthrow Yugoslav control of his party and country. Yugoslavia and Albania were already involved in territorial disputes, and the break between them became even more hostile and long-lived than the break between Belgrade and Moscow.

Also, the Cominform countries began to give special help to Yugoslav Communists who did not support Tito. These persons were invited to conferences and praised as martyrs of "true" socialism. There were minor military conflicts on Yugoslavia's borders with Albania and Bulgaria as well, but though the country prepared for a major Soviet invasion, none ever came.

Tito dealt with the ostracism in several ways: turning away from the Soviet Union for support and developing a more "liberal" kind of socialism within Yugoslavia. Many Yugoslav people respected him for his willingness to stand up to Stalin. After Stalin died in 1953, Tito continued his independent path and became a leader of the "nonaligned bloc"—a group of nations in Asia, Africa, and Latin America that did not wish to make close alliances with either the Soviet Union or the West.

# Truman Orders End of Segregation in U.S. Military

President Harry S. Truman issued an executive order mandating the desegregation of the U.S. military on July 26, 1948.

**What:** Civil rights and liberties; Military
**When:** July 26, 1948
**Where:** United States
**Who:**
HARRY S. TRUMAN (1884-1972), president of the United States
A. PHILIP RANDOLPH (1889-1979), labor leader and civil rights activist
STROM THURMOND (1902- ), governor of South Carolina and Dixiecrat presidential candidate

## Civil Rights Struggle

African Americans have participated in every war in which the United States has been involved. During the American Revolution, for example, about five thousand black soldiers served in the Continental army. However, prejudice kept black soldiers from receiving the recognition and opportunity they deserved. The tradition of separating the races in every aspect of society also influenced the military, in which whites and blacks served in separate units. Segregation and the official treatment of African Americans as second-class citizens in the military continued through World War II.

Harry S. Truman, who became president in 1945 after Franklin D. Roosevelt died in office, became involved in the civil rights struggle for a number of reasons. He viewed the time to be right for "the improvement of our democratic processes." He also saw the need to fight segregation to improve the international reputation of the United States. Politics was another motivating factor: A civil rights stance could win African American votes. One area in which Truman saw

the need for change was the military. Reports of African American World War II veterans being physically attacked by racists in the South upset Truman. He believed that desegregation and equal opportunity should be part of the military. Civil rights activists also influenced his decision to desegregate the armed forces.

African American labor and civil rights leader A. Philip Randolph led the struggle for a desegregated military. Randolph had been involved in trying to win more rights for African Americans for many years. In 1941, he had protested against the exclusion of blacks in defense industry jobs and influenced President Roosevelt to pass the Fair Employment Practices Act, which limited the amount of discrimination in defense industry jobs. However, his attempts to gain desegregation of the military failed. After World War II, Randolph intensified the struggle against discrimination in the military. He founded the Committee Against Jim Crow in Military Service and Training. Civil rights activists met with President Truman during the spring of 1948, urging him to ban military segregation. After little came out of these meetings, Randolph turned to civil disobedience to protest military segregation. He stated that African Americans should refuse to serve in a segregated military and that they should not be forced to serve in a segregated military in which they were treated as second-class citizens.

## Executive Order

Truman knew legislation banning segregation in the military would not pass through Congress because of the opposition of the South, so he issued an executive order. On July 26, 1948, Truman signed Executive Order 9981 calling for

**1080**

"equality of treatment and opportunity for all persons in the armed services." The order also established the President's Committee on Equality of Treatment and Opportunity in the Armed Forces, chaired by Charles Fahy, to implement the decision. The military gradually ended segregation, and by 1954, the U.S. military had completely integrated.

Not everyone agreed with Truman's pro-civil rights stance. A number of southerners left the Democratic Party because of Truman's policies. These dissatisfied Democrats started a third party, the Dixiecrat Party, and supported South Carolina governor Strom Thurmond for president in 1948 on a platform of continued segregation. One of Truman's proposals opposed by the Dixiecrats was the desegregation of the U.S. military. Thurmond denounced military desegregation as "un-American."

## Consequences

The executive order to desegregate the military had a great impact on the struggle for civil rights in United States. The armed forces represented a large and well-respected segment of the United States. Some military leaders disagreed with the decision. U.S. Army general Omar Bradley stated that the military should not be used for "social experiments" and that the Army would only desegregate when the rest of society did. However, the Army and other branches did desegregate. The military proved to be an example for the rest of the country. If the military pushed for the end of segregation, the rest of society would have to recognize that desegregation could occur. The military desegregated before many areas of civilian life did and proved that integration could be successful.

The integrated military had an effect on people's opinions concerning civil rights, and the military's policy influenced some civilian businesses in the South to desegregate. Integrated military bases often existed within segregated southern populations. In the early 1960's, the Commission on Equal Opportunity set up by President John F. Kennedy, stated that ci-

vilian establishments refusing to desegregate should be declared off limits for military personnel. Military personnel spent a lot of money in civilian establishments. If establishments wanted to make money from off-duty military personnel, owners had to desegregate. A significant number of establishments chose to desegregate to maintain profits.

The military also influenced many individuals. Nearly every American male spent at least a short period in the armed forces. After 1948, the military introduced them to the idea of an integrated society. Therefore, those leaving the military found it easier to accept the notion of integration.

The executive order also had an impact on the election of 1948 and American politics. Many people predicted that Truman would lose the election, especially with the defection of some

*A. Philip Randolph in 1937.*

Schomburg Center, The New York Public Library

southern Democrats. However, Truman won a narrow victory over Republican candidate Thomas Dewey. Many African Americans voted for Truman because of his pro-civil rights stance, including the desegregation of the military. Black voters helped Truman win the electoral votes of California, Illinois, and Ohio. The electoral votes of these states were enough to reelect Truman president in 1948. For many years, the majority of African Americans continued to vote Democratic, and Truman's actions had a tremendous impact on this trend.

President Truman made a courageous stand when publicly supporting the struggle for civil rights. A number of civil rights activists such as Randolph worked tirelessly to gain civil rights and knew that changes in the armed forces would be a huge step. The desegregation of the military sped up integration. When such an important organization as the U.S. military desegregated, this had a major influence on the rest of the country's views toward segregation.

*Marty Kuhlman*

# World Council of Churches Is Formed

> *In order to bring together Christians of various denominations for action and reflection, an Amsterdam conference organized 145 church bodies into the World Council of Churches.*

**What:** Religion
**When:** August 23, 1948
**Where:** Amsterdam
**Who:**

JOHN RALEIGH MOTT (1865-1955), a
   Methodist YMCA leader, and
   chairman of the International
   Missionary Council from 1921 to 1942

CHARLES HENRY BRENT (1862-1929),
   bishop of the Protestant Episcopal
   Church of the United States, and
   founder of the World Conference on
   Faith and Order

NATHAN SÖDERBLOM (1866-1931),
   Lutheran archbishop of Uppsala, and
   founder of the Universal Christian
   Conference on Life and Work

## Coming Together

At the beginning of the twentieth century, deep concern over divisions within Christianity and the damage that such divisions caused to missionary work led to the First World Missionary Conference, which was held in Edinburgh, Scotland, in 1910. John Raleigh Mott, an American Methodist who had been active in the Young Men's Christian Association (YMCA), served as chairman.

As a result of the conference, many denominations began cooperating in their missions' programs. A committee established to encourage such cooperation became the International Missionary Council in 1921.

Missionary cooperation inspired interest in other kinds of cooperation among Christian groups. Some Christian leaders wanted their groups to meet with others to discuss where they agreed and differed regarding Christian beliefs. Mostly through the determination of Charles

Henry Brent, a bishop of the Protestant Episcopal Church of the United States, this desire led to the forming of the Faith and Order movement. Faith and Order discussions were based on the idea that though church organizations might operate separately, their members still had unity in Christ.

Though some Christians criticized the Faith and Order movement, it became a formal organization during a conference in Lausanne, Switzerland, in 1927. One of its main goals was to bring churches together to study theology.

The Life and Work movement, which aimed to unite churches to address social, economic, and political problems, was led by Nathan Söderblom, Lutheran archbishop of Uppsala, Sweden. The first Universal Conference on Life and Work met at Stockholm in 1925, with Söderblom as chair. Conference delegates considered some of the problems of modern society and appointed a continuing committee to study these problems and to help churches respond to them.

By 1937, leaders of the Faith and Order and the Life and Work movements realized that they should work together. Both groups scheduled conferences for 1937: Faith and Order at Edinburgh and Life and Work at Oxford, England. The two conferences agreed that they should combine into one "world council of churches."

A conference at Utrecht, the Netherlands, in 1938 settled questions of which churches could become members of the new council and what authority it would have. A provisional committee was set up, and a first assembly of the World Council of Churches was planned for 1941. Because World War II broke out in 1939, however, the assembly was not held until 1948. During the war years, the provisional committee helped organize Christian relief and refugee services.

Formed at Amsterdam in August, 1948, the World Council of Churches (WCC) unified the

Faith and Order movement with the Life and Work movement. It also established a relationship with the International Missionary Council, and these two organizations were unified in 1961.

The council declared that it had several purposes: ecumenism, or carrying on the work of the uniting movements; helping churches to work together; promoting cooperation in theological study; stimulating awareness of missions and ecumenism in all the churches; supporting the churches in missions and evangelism; maintaining ties with national and regional councils and other ecumenical organizations; and calling world conferences on specific subjects.

## The WCC Is Organized

The WCC's first membership statement declared that "the World Council of Churches is a fellowship of churches which accept the Lord Jesus Christ as God and Saviour." Some members believed that this statement was too ambiguous, however, and it was modified in 1961 with more specific language: "The World Council of Churches is a fellowship of churches which confess the Lord Jesus Christ as God and Saviour according to the Scriptures and therefore seek to fulfill together their common calling to the glory of the one God, Father, Son and Holy Spirit." By making references to the Scriptures (the Bible) and to God as "Father, Son and Holy Spirit" (the Trinity), the council had carefully recognized orthodox Christian teachings.

At the first assembly in 1948, 145 churches accepted membership in the council. Many major Protestant and Orthodox denominations were included or have since become affiliated with the council, though some fast-growing evangelical denominations have not. The Roman Catholic Church was not included in the council, but the WCC has kept good communication with the Vatican. After the Second Vatican Council in 1962, this relationship grew closer. Together the Catholics and the WCC established a Joint Working Group.

The WCC Assembly meets about every five years, gathering official representatives of member churches. The council does not have the authority to impose any laws or rules on its member churches, though it does make recommendations and puts out policy statements.

The council has six presidents, who make up a central committee along with up to 120 members elected by the assembly. The committee acts on the WCC's behalf between assembly meetings. A general secretary, an assistant general secretary, and five associate general secretaries run the organization from its offices in Geneva, Switzerland.

## Consequences

With the formation of the WCC, Christians around the world began to work with instead of against one another. Through the WCC, younger churches in Asia, Africa, and Latin America have gained a voice alongside the traditional denominations of Europe and North America. Even during the years of the Cold War, the council helped maintain relationships between the churches of the East and those of the West.

The WCC has also been a means for the international Christian church to express concerns about issues of social justice. In South Africa, for example, the South African Council of Churches, a national organization that is part of the WCC, was widely respected for its strong stand against apartheid.

# Truman Is Elected U.S. President

*Harry S. Truman, who had become president after Franklin D. Roosevelt's death, fought an uphill battle to keep his office in the elections of 1948.*

**What:** National politics
**When:** November 2, 1948
**Where:** United States
**Who:**

HARRY S. TRUMAN (1884-1972), president of the United States from 1945 to 1948, and Democratic presidential candidate in 1948

ALBEN W. BARKLEY (1877-1956), U.S. senator from Kentucky from 1927 to 1949, and Truman's running mate in 1948

THOMAS E. DEWEY (1902-1971), governor of New York from 1943 to 1955, and Republican presidential candidate in 1948

EARL WARREN (1891-1974), governor of California from 1943 to 1953, and Dewey's running mate in 1948

## Truman Loses Ground

When Harry S. Truman took the presidential oath of office in April, 1945, after the death of Franklin D. Roosevelt, a great majority of Americans rallied around him. He was an ordinary man who seemed to be trying to do the best he could, and his simple, straightforward manner won the sympathy of the American people. That summer he became involved in the great decisions that were part of ending World War II, and he took on a tougher attitude toward the Soviet Union.

On September 6, only four days after Japan's formal surrender, Truman sent Congress a program for social and economic reforms in the spirit of Roosevelt's New Deal. He called for the expansion of social security, a higher minimum wage, laws to ensure that all American workers could find jobs, a permanent Fair Employment Practices Commission to protect African American workers, and new public housing to replace city slums.

By this time, however, Truman's initial burst of popularity was petering out. Congress was dominated by Republicans and conservative Democrats who were not ready to enact liberal reforms; they gave Truman the cut and amended Employment Act of 1946, but not much else.

In the meantime, Truman struggled to help the country's economy make the transition from war to peace. All the troubles that followed the war—inflation, strikes, shortages, problems with the Soviet Union, sharp debates over the use of atomic weapons—prompted criticism of the president and of his party. "To err is Truman," people joked.

In the fall of 1946, Republicans adopted the campaign slogan "Had Enough?" The question caught the mood of the country. The Republicans made great gains in the mid-term elections, winning both houses of Congress for the first time since 1928. The size of the victory was the real surprise: They had a 245-188 majority in the House and 51-45 in the Senate. Not surprisingly, the new Eightieth Congress was even less interested in Truman's domestic programs than the previous Congress had been.

## Fighting Back

As the 1948 election approached, Republicans seemed confident of another victory. When their national convention met, they nominated New York governor Thomas E. Dewey. Earl Warren, the popular governor of California, received the vice-presidential nomination.

The Democratic Party seemed to be disintegrating. In late 1947, Henry A. Wallace, who had recently been fired from the Truman cabinet, announced that he would run on a third-party ticket. The Northern liberals who supported him

**1085**

were unhappy that social reforms were coming so slowly, and they wanted more cooperation with the Soviet Union.

Other Democrats hoped to find a glamorous candidate to replace Truman. The first choice of the "dump Truman" movement was General Dwight D. Eisenhower, whose party affiliation was not yet known. He was flattered, but not interested in running.

At Philadelphia, a gloomy Democratic Convention accepted Truman for lack of any alternative; Senator Alben W. Barkley was named as his running mate. When a liberal group managed to add a statement on civil rights to the party platform, Southern conservatives protested by leaving the party. Meeting at Birmingham, Alabama, they formed the States' Rights Democratic Party (Dixiecrats), with Strom Thurmond of South Carolina as their candidate.

Confident that the prize would be his, Dewey campaigned as though he had already won. The polls showed him far ahead, even though ordinary Republicans were not excited about him. His speeches were well written, and he delivered them with dignity. Though he accused the Truman administration of corruption, he accepted the New Deal reforms and even promised to carry them out more efficiently.

In contrast, Truman entered the campaign in a fighting mood. Ignoring Dewey, he lashed out at the "no-account, do-nothing" Eightieth Congress and accused Republicans of representing selfish special-interest groups. Hurling himself into exhausting campaign tours that covered 31,700 miles and included 356 speeches, he became a symbol of stubborn courage, a scrappy underdog making an uphill fight. Truman's spontaneous style of speaking delighted his audi-

Library of Congress

*Harry S. Truman exultantly holds up a newspaper that prematurely announced Dewey as the winner of the presidential election.*

ences. "Give 'em hell, Harry!" voices called out.

To almost everyone's surprise, Truman defeated Dewey by 24,105,812 to 21,970,065 votes. In the electoral college he won 303 votes to 189 for Dewey and 39 for Thurmond. Wallace and Thurmond each won about 1,150,000 votes, far fewer than had been predicted. The Democrats also regained control of both houses of Congress. Truman had pulled off one of the most remarkable upsets in the history of presidential politics.

## Consequences

Once he was president in his own right—by election—Truman began again to fight for his domestic program, which he called the Fair Deal. After the seven years of his presidency, however, the results were slim: The National Housing Act of 1949 had been enacted, the minimum hourly wage had been increased to seventy-five cents, a social security law had increased benefits and given coverage to more people, and more liberal immigration laws had been passed. Congress turned down Truman's requests for civil rights legislation, national health insurance, a new system of subsidies to help farmers, and federal aid to education.

Through executive action, Truman did manage to attack racial segregation within the government and the armed services. He kept the values of the New Deal alive in American politics and had been able to extend some New Deal programs. Some of his concerns, such as national health insurance, continued to be debated for many decades after he had left the presidency.

*Donald Holley*

# Land Invents Instant Photography

*Edwin Herbert Land and his coworkers invented a camera that could produce a finished photograph as soon as its film was exposed.*

**What:** Photography
**When:** November 26, 1948
**Where:** Boston, Massachusetts
**Who:**
EDWIN HERBERT LAND (1909-1991), an American physicist and chemist
HOWARD G. ROGERS (1915-     ), a senior researcher at Polaroid and Land's collaborator
WILLIAM J. MCCUNE (1915-     ), an engineer and head of the Polaroid team
ANSEL ADAMS (1902-1984), an American photographer and Land's technical consultant

## The Daughter of Invention

Because he was a chemist and physicist interested primarily in research relating to light and vision, and to the materials that affect them, it was inevitable that Edwin Herbert Land should be drawn into the field of photography. Land founded the Polaroid Corporation in 1929. During the summer of 1943, while Land and his wife were vacationing in Santa Fe, New Mexico, with their three-year-old daughter, Land stopped to take a picture of the child. After the picture was taken, his daughter asked to see it. When she was told she could not see the picture immediately, she asked how long it would be. Within an hour after his daughter's question, Land had conceived a preliminary plan for designing the camera, the film, and the physical chemistry of what would become the instant camera. Such a device would, he hoped, produce a picture immediately after exposure.

Within six months, Land had solved most of the essential problems of the instant photography system. He and a small group of associates at Polaroid secretly worked on the project. Howard G. Rogers was Land's collaborator in the laboratory. Land conferred the responsibility for the engineering and mechanical phase of the project on William J. McCune, who led the team that eventually designed the original camera and the machinery that produced both the camera and Land's new film.

The first Polaroid Land camera, the Model 95, produced photographs measuring 8.25 by 10.8 centimeters; there were eight pictures to a roll. Rather than being black and white, the original Polaroid prints were sepia-toned (producing a warm, reddish-brown color). The reasons for the sepia coloration were chemical rather than aesthetic; as soon as Land's researchers could devise a workable formula for sharp black-and-white prints (about ten months after the camera was introduced commercially), they replaced the sepia film.

## A Sophisticated Chemical Reaction

Although the mechanical process involved in the first demonstration camera was relatively simple, this process was merely the means by which a highly sophisticated chemical reaction—the diffusion transfer process—was produced.

In the basic diffusion transfer process, when an exposed negative image is developed, the undeveloped portion corresponds to the opposite aspect of the image, the positive. Almost all self-processing instant photography materials operate in three phases—negative development, diffusion transfer, and positive development. These occur simultaneously, so that positive image formation begins instantly. With black-and-white materials, the positive was originally completed in about sixty seconds; with color materials (introduced later), the process took somewhat longer.

**1088**

The basic phenomenon of silver in solution diffusing from one emulsion to another was first observed in the 1850's, but no practical use of this action was made until 1939. The photographic use of diffusion transfer for producing normal-contrast, continuous-tone images was investigated actively from the early 1940's by Land and his associates. The instant camera using this method was demonstrated in 1947 and marketed in 1948.

The fundamentals of photographic diffusion transfer are simplest in a black-and-white peel-apart film. The negative sheet is exposed in the camera in the normal way. It is then pulled out of the camera, or film pack holder, by a paper tab. Next, it passes through a set of rollers, which press it face-to-face with a sheet of receiving material included in the film pack. Simultaneously, the rollers rupture a pod of reagent chemicals that are spread evenly by the rollers between the two layers. The reagent contains a strong alkali and a silver halide solvent, both of which diffuse into the negative emulsion. There the alkali activates the developing agent, which immediately reduces the exposed halides to a negative image. At the same time, the solvent dissolves the unexposed halides. The silver in the dissolved halides forms the positive image.

## Consequences

The Polaroid Land camera had a tremendous impact on the photographic industry as well as on the amateur and professional photographer. Ansel Adams, who was known for his monumental, ultrasharp black-and-white panoramas of the American West, suggested to Land ways in which the tonal value of Polaroid film could be enhanced, as well as new applications for Polaroid photographic technology.

Soon after it was introduced, Polaroid photography became part of the American way of life and changed the face of amateur photography. By the 1950's, Americans had become accustomed to the world of recorded visual information through films, magazines, and newspapers; they also had become enthusiastic picture-takers as a result of the growing trend for simpler and more convenient cameras. By allowing these photographers not only to record their perceptions but also to see the results almost immediately, Polaroid brought people closer to the creative process. However, the 1990's would bring to instant photography a new challenge: electronic digital cameras. The advent of inexpensive systems of recording still and moving pictures on reusable media threatened to make Polaroid's instant-photography technology obsolete.

*Genevieve Slomski*

# First Multistage Rocket Is Launched

The first successful liquid-fuel rocket with more than one stage was launched, heralding a new era in rocket science.

**What:** Space and aviation
**When:** February 24, 1949
**Where:** White Sands Proving Ground, New Mexico
**Who:**
H. N. Toftoy, the U.S. military official who first conceived Project Bumper
Frank J. Malina, the Jet Propulsion Laboratory scientist who was responsible for the design of the second stage of the rocket

## HERMES

The "Ordnance Guided Missile and Rocket Programs—HERMES," initiated as World War II was drawing to a close, investigated nearly every aspect of rocketry. The purpose of HERMES was to develop long-range guided missiles for use against ground targets and high-altitude aircraft. As part of the project, a group of scientists was assigned to study German missiles used in World War II and to collect important pieces for shipment to the United States. Their job was complicated by the fact that almost all the captured rocket components had been deliberately damaged by the retreating German army. After the scientists had assembled and test-launched eight German V-2 rockets, the remaining identified hardware was divided between England and the United States. As a result, three hundred train carloads of V-2 parts were unloaded at Las Cruces, New Mexico, at the end of July, 1945.

The second critical phase of HERMES was the acquisition of top German rocket scientists. At the end of the war, four hundred German scientists surrendered to the United States; of those, one hundred were selected to be sent to the United States to perform scientific research. Twenty were assigned to the White Sands Proving Ground, later called the White Sands Missile Range, located between Alamogordo and Las Cruces, New Mexico. These scientists deciphered drawings and specifications, and the General Electric Company built V-2 rockets according to their directions. Before the General Electric contract expired on June 30, 1951, sixty-seven rockets had been launched.

## Project Bumper

The idea for the multistage rocket project, which came to be known as "Project Bumper," was first conceived by H. N. Toftoy, a colonel in the Office of the Chief of Ordnance. Toftoy suggested that a V-2 could be combined with a WAC Corporal, a small American-made rocket, to produce a long-range multistage missile. In October, 1946, Project Bumper was instituted to investigate two-stage rockets and to set a new high-altitude record. The WAC Corporal, which had a range of only about 40 kilometers, was 4.88 meters long. The WAC Corporal was fitted as deeply as possible into the V-2 booster rocket, with enough space in the instrument compartment for the necessary guidance equipment. Guide rails and expulsion cylinders (powered by compressed air) for the WAC Corporal's mid-air launch were also included. At the end of the V-2 rocket's burn, a compressed-air valve would open, causing the fin of the second stage rocket to slide out of slots in the nose cone of the V-2.

The forward portion of the V-2 was modified for Project Bumper, with General Electric given overall responsibility for the V-2 modifications. The Jet Propulsion Laboratory's Frank J. Malina was responsible for theoretical investigations, the design of the second stage, and the basic design of the separation system. The Douglas Aircraft company was responsible for the fabrication of the second stage and the detail design and fabrication of special V-2 parts.

**1090**

On February 24, 1949, a rocket known as "Bumper Five" reached a height of 410 kilometers, becoming the first true space vehicle. Thirty seconds after takeoff, the V-2 reached a speed of 5,904 kilometers per hour, just prior to separation. The WAC Corporal then attained a speed of 8,446 kilometers per hour, setting both velocity and altitude records. Instrumentation in the nose cone measured temperatures and transmitted technical data pertaining to conditions during the flight. The flight also marked the first time that radio equipment had been operated at such extreme altitudes. The flight was a complete success, although the crashed remains of the rocket were not found for more than a year.

## Consequences

Project Bumper was significant in the history of space for several reasons. First of all, Bumper's program of origin, the HERMES project, was in-valuable for the immense amount of research and development it achieved with relatively little monetary investment, particularly after the war, when cutbacks in research funding were common. Also, the projects encouraged organizational growth, both in government and the private sector, to accommodate the tremendous amounts of data collected. The heritage of research data provided a strong base for future developments in space technology, especially the data concerning the many problems associated with high-altitude rocket separations, attachments, and ignitions. With its integration of two separate stages and the technology required to launch a second stage in mid-flight, Project Bumper thus left a rich heritage of technology and a vital base of information that helped to make space travel possible within little more than a decade.

*Ellen F. Mitchum*

# Twelve Nations Sign North Atlantic Treaty

> *In a break with a tradition of isolationism, the Truman administration made the United States a member of the North Atlantic Treaty Organization, with Canada and ten European nations.*

**What:** International relations
**When:** April 4, 1949
**Where:** Western Europe and Washington, D.C.
**Who:**
HARRY S. TRUMAN (1884-1972), president of the United States from 1945 to 1953
DEAN G. ACHESON (1893-1971), U.S. secretary of state from 1949 to 1953

## The Question of Alliance

For nearly two centuries, American leaders had heeded George Washington and Thomas Jefferson's advice to keep their country out of "entangling alliances"—peacetime military agreements with European powers. After World War II, however, President Harry S. Truman and his cabinet began to rethink this policy.

Trying to keep the Soviet Union from extending its control of Eastern Europe, the United States had given military aid to Greece and Turkey. Yet that aid could only be temporary, and Greece and Turkey were not strong enough to continue standing alone against the Soviet threat. Meanwhile, the Marshall Plan had brought economic aid to Europe after the war, but many American leaders believed that Europe would not really begin to thrive until the Europeans considered themselves safe from the threat of the Red Army. Military security was necessary for economic recovery.

In 1948, a Communist coup in Czechoslovakia, the Soviet blockade of Berlin, and other Soviet moves convinced the Truman administration that the United States would have to stay involved in Europe for a long time. Though European leaders asked the United States to join them in a defensive alliance, however, the administration was not yet sure that the American people would accept that kind of commitment.

In June, 1948, the Senate did approve the Vandenberg Resolution, which supported American participation in regional agreements for "self-help and mutual aid." It was not clear, however, whether this resolution was really meant to encourage mutual-security pacts or to limit the president's power to form such agreements.

Meanwhile, in March, 1948, five European nations—Great Britain, France, Belgium, the Netherlands, and Luxembourg—had shown the way by signing the Brussels Pact, a fifty-year economic, cultural, and defensive alliance. Under this pact, each of the five countries was obligated to help any of the others who was attacked by an aggressor. The Brussels Pact nations had invited the United States to participate, but Americans were not yet ready.

After Truman won the presidential election of 1948, however, he began taking the possibility of alliance more seriously. In his inaugural address of January, 1949, he promised that the United States would contribute to the defense of friendly nations. Meanwhile, Washington began negotiations with European states to create a system of cooperation against the Soviet threat to Western Europe.

The Soviet Union and some Americans criticized these negotiations, arguing that the United States was hampering the United Nations by forming a bloc of states for aggressive purposes. American officials answered, however, by pointing out that article 51 of the U.N. Charter allowed for regional defense pacts, and that the proposed alliance was to be defensive.

## NATO Is Formed

Secretary of State Dean G. Acheson believed

that the United States should rely on military force and shows of power to hold back the Soviet Union; he considered the United Nations an idealistic organization that could not do much against the aggressive Communists. Confident that an alliance with Western Europe was necessary, he helped bring the European negotiations to a successful conclusion.

On April 4, 1949, representatives of twelve nations—Canada, the United States, the United Kingdom, France, Italy, Belgium, the Netherlands, Luxembourg, Norway, Denmark, Iceland, and Portugal—signed the North Atlantic Treaty in Washington, D.C. In the treaty they promised to support the United Nations, to cooperate in maintaining stability in the North Atlantic region, and to work together for mutual defense and preservation of peace and security.

Although the treaty bound its members to settle international disputes by peaceful means, it also stated that "the Parties agree that an armed attack against one or more of them in Europe or North America shall be considered an attack against them all." Any such attack would be met, if necessary, by armed force.

The treaty also set up the Council of the North Atlantic Treaty Organization (NATO); each of the member states would be represented on this council. It was to create other NATO organizations for specific purposes—especially a defense committee to make recommendations about how the member nations could best defend themselves.

Neither the United States nor any other member was absolutely committed to go to war, but the treaty was a powerful commitment to help members threatened by aggression. The treaty was to be in effect for at least twenty years.

## Consequences

When the Senate began hearings on the North Atlantic Treaty, there were bitter debates about whether it was wise for the United States to become involved with Europe in this type of pact.

*President Harry S. Truman signs the North Atlantic Treaty as representatives from other nations watch.*

Hulton Archive

These arguments did not really threaten the treaty itself, however, and on July 21, 1949, the Senate ratified it by a vote of 82-13.

Though in the treaty NATO members stated their willingness to make a real military commitment to common security, as things turned out all of them put strict limitations on the number of troops they would contribute to the NATO forces. Thus a large NATO force was never created, but the organization continued to exist for a number of decades, bringing together North American and European leaders to discuss matters of defense and security.

Greece and Turkey joined in 1952, and West Germany was admitted soon after the beginning of the Korean War. For political reasons, Yugoslavia and Spain, which were both ruled by dictators, were excluded; yet various NATO members signed individual treaties with these countries, so that for practical purposes they did become part of NATO.

*Theodore A. Wilson*

# BINAC Computer Uses Stored Programs

The BINAC demonstrated the usefulness of several engineering innovations, the most important of which is the stored-program concept, that have become standard features of many modern computers.

**What:** Computer science
**When:** August, 1949
**Where:** Philadelphia, Pennsylvania
**Who:**

JOHN PRESPER ECKERT (1919-1995), an American electrical engineer

JOHN W. MAUCHLY (1907-1980), an American physicist

JOHN VON NEUMANN (1903-1957), a Hungarian American mathematician

ALAN MATHISON TURING (1912-1954), an English mathematician

## Computer Evolution

In the 1820's, there was a need for error-free mathematical and astronomical tables for use in navigation, unreliable versions of which were being produced by human "computers." The problem moved English mathematician and inventor Charles Babbage to design and partially construct some of the earliest prototypes of modern computers, with substantial but inadequate funding from the British government. In the 1880's, the search by the U.S. Bureau of the Census for a more efficient method of compiling the 1890 census led American inventor Herman Hollerith to devise a punched-card calculator, a machine that reduced by several years the time required to process the data.

The emergence of modern electronic computers began during World War II (1939-1945), when there was an urgent need in the American military for reliable and quickly produced mathematical tables that could be used to aim various types of artillery. The calculation of very complex tables had progressed somewhat since Babbage's day, and the human computers were being assisted by mechanical calculators. Still, the growing demand for increased accuracy and efficiency was pushing the limits of these machines. Finally, in 1946, following three years of intense work at the University of Pennsylvania's Moore School of Engineering, John Presper Eckert and John W. Mauchly presented their solution to the problems in the form of the Electronic Numerical Integrator and Computer (ENIAC), the world's first electronic general-purpose digital computer.

The ENIAC, built under a contract with the Army's Ballistic Research Laboratory, became a great success for Eckert and Mauchly, but even before it was completed, they were setting their sights on loftier targets. The primary drawback of the ENIAC was the great difficulty involved in programming it. Whenever the operators needed to instruct the machine to shift from one type of calculation to another, they had to reset a vast array of dials and switches, unplug and replug numerous cables, and make various other adjustments to the multiple pieces of hardware involved. Such a mode of operation was deemed acceptable for the ENIAC because, in computing firing tables, it would need reprogramming only occasionally. Yet if instructions could be stored in a machine's memory, along with the data, such a machine would be able to handle a wide range of calculations with ease and efficiency.

## The Turing Concept

The idea of a stored-program computer had first appeared in a paper published by English mathematician Alan Mathison Turing in 1937. In this paper, Turing described a hypothetical machine of quite simple design that could be used to solve a wide range of logical and mathematical problems. One significant aspect of this imaginary Turing machine was that the tape that would run through it would contain both information to be processed and instructions on how to process it. The tape would thus be a type of memory device, storing both the data and the program as sets of symbols that the machine could "read" and understand. Turing never attempted to construct this machine, and it was not until 1946 that he developed a design for an electronic stored-program computer, a prototype of which was built in 1950.

In the meantime, John von Neumann, a Hungarian American mathematician acquainted with Turing's ideas, joined Eckert and Mauchly in 1944 and contributed to the design of ENIAC's successor, the Electronic Discrete Variable Automatic Computer (EDVAC), another project financed by the Army. The EDVAC was the first computer designed to incorporate the concept of the stored program.

In March of 1946, Eckert and Mauchly, frustrated by a controversy over patent rights for the ENIAC, resigned from the Moore School. Several months later, they formed the Philadelphia-based Electronic Control Company on the strength of a contract from the National Bureau of Standards and the Census Bureau to build a much grander computer, the Universal Automatic Computer (UNIVAC). They thus abandoned the EDVAC project, which was finally completed by the Moore School in 1952, but they incorporated the main features of the EDVAC into the design of the UNIVAC.

Building the UNIVAC, however, proved to be much more involved and expensive than anticipated, and the funds provided by the original contract were inadequate. Eckert and Mauchly, therefore, took on several other smaller projects in an effort to raise funds. On October 9, 1947, they signed a contract with the Northrop Corporation of Hawthorne, California, to produce a relatively small computer to be used in the guidance system of a top-secret missile called the "Snark," which Northrop was building for the Air Force. This computer, the Binary Automatic Computer (BINAC), turned out to be Eckert and Mauchly's first commercial sale and the first stored-program computer completed in the United States.

The BINAC was designed to be at least a preliminary version of a compact, airborne com-

*John W. Mauchly, at the central control of BINAC.*

puter. It had two main processing units. These contained a total of fourteen hundred vacuum tubes, a drastic reduction from the eighteen thousand used in the ENIAC. There were also two memory units, as well as two power supplies, an input converter unit, and an input console, which used either a typewriter keyboard or an encoded magnetic tape (the first time such tape was used for computer input). Because of its dual processing, memory, and power units, the BINAC was actually two computers, each of which would continually check its results against those of the other in an effort to identify errors.

The BINAC became operational in August, 1949. Public demonstrations of the computer were held in Philadelphia from August 18 through August 20.

### Consequences

The design embodied in the BINAC is the real source of its significance. It demonstrated successfully the benefits of the dual processor design for minimizing errors, a feature adopted in many subsequent computers. It showed the suitability of magnetic tape as an input-output medium. Its most important new feature was its ability to store programs in its relatively spacious memory, the principle that Eckert, Mauchly, and von Neumann had originally designed into the EDVAC. In this respect, the BINAC was a direct descendant of the EDVAC.

In addition, the stored-program principle gave electronic computers new powers, quickness, and automatic control that, as they have continued to grow, have contributed immensely to the aura of intelligence often associated with their operation.

The BINAC successfully demonstrated some of these impressive new powers in August of 1949 to eager observers from a number of major American corporations. It helped to convince many influential leaders of the commercial segment of society of the promise of electronic computers. In doing so, the BINAC helped to ensure the further evolution of computers.

*Gordon L. Miller*

# Soviet Union Explodes Its First Nuclear Bomb

*Soviet success in building an atomic bomb marked the beginning of a long, expensive American-Soviet arms race.*

**What:** Technology; Military
**When:** August 29, 1949
**Where:** Soviet Central Asia
**Who:**
HARRY S. TRUMAN (1884-1972), president of the United States from 1945 to 1953
IGOR VASILEVICH KURCHATOV (1903-1960) and
SERGEI I. VAVILOV (1891-1951), Soviet nuclear scientists
BORIS LVOVICH VANNIKOV (1897-1962), administrative director of the Soviet nuclear arms project after World War II

## Soviet Advances

In the years before World War II, nuclear science was mainly viewed as simply the newest branch of physics, and information flowed freely between Western and Soviet scientists. Scientists in the West soon realized that the Soviets had made real progress in nuclear research. Frédéric Joliot, a French scientist who had helped to found nuclear science, visited the Soviet Union in 1932 and found the Soviets' knowledge of the field to be good.

In 1934, Sergei Vavilov made discoveries that led to Nobel prizes for other Soviet researchers in 1958. The Soviet government in 1938 set up a nuclear physics laboratory at Leningrad (St. Petersburg), under the direction of Igor Vasilevich Kurchatov, the Soviet Union's most outstanding nuclear scientist. Three of the cyclotrons needed for nuclear research were being built when the Germans invaded the Soviet Union in 1941.

Under the pressure of war, Soviet scientists put aside their nuclear research and concentrated on meeting military needs. Even they were not sure that a nuclear bomb could be built, and Soviet dictator Joseph Stalin was in no mood to fund research that might lead nowhere. Defeating Germany had first priority, not research in physics. The very fact that the Soviet press published articles about nuclear physics showed that the Soviets were not particularly interested in using their atomic research to build bombs.

It was not until late 1945 to early 1946 that the Soviet government created a program for developing an atomic bomb. Little is known of the details of the program, except that it was directed by General Boris Lvovich Vannikov; Western experts are sure that Kurchatov was quite involved, probably with the help of V. S. Fursov and Lev Landau.

## The Scientists' Warnings

Meanwhile, in the United States, scientists had begun warning that the Soviet Union was capable of building a nuclear weapon. They argued that the United States should not build an atomic weapon, for that would set in motion an endless arms race. American political leaders, however, found it convenient to ignore these warnings. Commonly the Soviets were described as having only primitive science—an idea that was reinforced when the Soviets made ridiculous claims about being the first to invent certain devices. In any case, Americans were confident that the "free enterprise system" would always produce superior technology.

This ignorant attitude, however, was not as disturbing to scientists as the general American indifference toward nuclear weapons. Some physicists had already warned against the danger of a nuclear holocaust that could destroy humanity. In the spring of 1945, the physicist Leo Szilard argued that because the United States no longer needed to fear a German atomic bomb, it was unnecessary to explode the American weapon. He and other scientists, including Albert Einstein, cautioned that once the United States detonated a nuclear bomb, a nuclear arms race would be almost inevitable.

Though the scientists proposed that nuclear weapons be internationally controlled, with inspections by both American and Soviet officials, the U.S. government assured them that the United States would have a monopoly on nuclear weapons for decades to come. The secretary of state informed Szilard that there was no uranium in Russia anyway.

On September 23, 1949, President Harry S. Truman had to eat his administration's words: He informed the public that the Soviets had detonated an atomic bomb on August 29. Because the scientists' concerns had not been taken seriously, the American people reacted with surprise. There was a new reason to fear the "Reds."

## Consequences

Until 1949, the cover of the *Bulletin of Nuclear Scientists* showed a clock, which represented humanity's existence on earth as a twenty-four-hour day; the clock's hands showed eight minutes before midnight. After 1949, however, the time shown on the clock was changed to three minutes before midnight. Now that everyone knew that the Soviet Union possessed nuclear weapons, the scientists began a new campaign for arms control—but to no avail.

Concerned by the Soviet nuclear test, the American government hurried along work on a thermonuclear weapon. Surely this fearful bomb would guarantee American security. An unexpectedly powerful thermonuclear device was exploded successfully in 1952, and American leaders became confident that soon the Soviets would lag far behind in weapons development.

On November 23, 1955, however, the Soviet Union dropped a thermonuclear device from an airplane, and the United States was put in the role of a follower, duplicating the Soviet achievement on May 21, 1956. The scientists had been right: In the years that followed, the American and Soviet governments raced to build and stockpile enough nuclear weapons to blow up the world several times over.

# Germany Is Split into Two Countries

*The division of Germany became a symbol of the Cold War, dividing the world into Eastern and Western blocs.*

**What:** International relations
**When:** September 21, and October 7, 1949
**Where:** Bonn and Berlin
**Who:**
KONRAD ADENAUER (1876-1967), chancellor of the Federal Republic of Germany and chairman of the Christian Democratic Union from 1949 to 1963
THEODOR HEUSS (1884-1963), president of the Federal Republic of Germany and chairman of the Free Democratic Party from 1949 to 1959
WALTER ULBRICHT (1893-1973), head of the ruling Socialist Unity Party of the German Democratic Republic

## East-West Crisis

At the close of World War II, the Allied Powers did not make formal arrangements for Germany's long-term future. For the time being, the territories that had belonged to Germany before the war were divided into eight zones. Great Britain, France, the United States, and the Soviet Union each took responsibility for one major zone; Berlin, the capital city, was also split into four zones, each of which was governed by one of the Big Four. An Allied Control Council was given responsibility over matters having to do with all of occupied Germany.

At the Potsdam Conference in the summer of 1945, the victorious Allies stated their intentions for Germany: It would be disarmed and demilitarized; all traces of Nazism would be erased, and a new democratic political system would be built; industrial cartels and monopolies would be destroyed; reparations (war damages) would be taken by the Allies.

Yet the United States and Great Britain disagreed with the Soviet Union about the question of reparations. Concerned about the growth of Soviet power in Eastern and Central Europe, the Western Allies wanted to see Germany rebuilt to serve as a barrier to Soviet expansion. They were not interested in demanding large reparations. The Soviet Union, which had lost huge numbers of its troops and suffered other damages during the war, wanted the benefit of Germany's resources.

In the end, it was agreed that each occupying nation could take reparations from its own zone in Germany. Since the western zones (occupied by the Western Allies) had more industry, the Soviet Union would receive from these zones a certain percentage of industrial machinery. Yet this agreement was quite vague, and the Big Four continued to quarrel over how it was to be carried out.

The Allied Control Council was unable to continue its work after the Soviet commandant walked out on March 20, 1948. Within a few weeks, the Soviets blockaded Berlin, and the Western Allies began to fly huge shipments of food, medicine, and fuel into the western zones of the city. The Berlin Airlift continued through September, 1949, even though the Soviets lifted the blockade in May.

## Two Nations Are Formed

As early as December, 1946, the United States and Great Britain had agreed that their German occupation zones would become an economic unity. The area, known as Bizonia, was to begin supporting itself in three years, and German leaders helped organize a Bizonian administration.

By 1947 there was an Economic Council whose members were chosen by the new provincial legislatures within Bizonia. The Economic

**1100**

Council had limited lawmaking authority, and soon an executive committee and a German high court were added. Because of the growing division between the Soviet Union and the three Western Allies, France joined its zone to Bizonia so that a central German government could be established.

In September, 1946, U.S. secretary of state James F. Byrnes had declared that the United States would grant the German people the right to manage their own affairs, as soon as they could do so democratically. At London in February, 1948, the Council of Foreign Ministers, which included representatives of Belgium, the Netherlands, Luxembourg, and the three Western Allies, took up this matter again and agreed that the three western zones of Germany would be fused into one. Because the Soviets left the Allied Control Council the next month, their zone was not included.

The military governors of these zones gave the heads of the German provinces the authority to call a constituent assembly, which was to write a democratic, federal constitution. Reluctant to see the division of Germany as final, delegates to the German assembly called themselves "Parliamentary Council," and the new constitution came to be called "Basic Law." Konrad Adenauer, chairman of the Christian Democratic Union, was elected to preside over the council.

On May 23, 1949, the Basic Law was formally adopted. The Allies approved it and began to make arrangements for the new German government to take over. After the first postwar elections, the new Parliament convened in Bonn on September 7, 1949. A federal convention elected Theodor Heuss, chairman of the Free Democratic Party, as federal president; Heuss then nominated Konrad Adenauer for federal chancellor. Elected by Parliament with a one-vote margin, Adenauer formed a coalition government. On

September 21, 1949, the Federal Republic of Germany came into being with a formal ceremony.

The Soviet Union strongly protested the establishment of the West German state. Yet in the Soviet zone social and economic changes were being made—changes aimed at eliminating capitalism and building a socialist society. The Soviets had allowed the Germans to create administrative organizations, which were controlled by the Socialist Unity Party; this party was dominated by Communists and led by Walter Ulbricht.

Since these organizations already existed, the Soviet Union was able to respond quickly to the creation of the Federal Republic of Germany. A constitution was drafted for an East German state. On October 7, 1949, the "German People's Council" met in Berlin and voted unanimously to name itself the People's Chamber of the German Democratic Republic. The People's Chamber adopted the new constitution and thus gave official birth to the German Democratic Republic.

## Consequences

After the creation of the two German states, relations between East and West became even colder. There were severe travel restrictions, and the Soviets built the Berlin Wall to keep East Germans from fleeing to the West. Over the years, the wall became a legendary marker, a reminder of the deep divisions between the Communist East and the free West. Stories were circulated of people who had managed to cross the wall—and of those who were shot trying to cross.

As the decades passed, many Germans still hoped that one day their country would be reunited. This dream finally became a reality on September 12, 1990, when the two Germanies became one by signing a treaty along with representatives of the Big Four Allies—the United States, Great Britain, France, and the Soviet Union. By this time concrete chunks of the Berlin Wall were being sold as souvenirs in the West, and the Cold War had come to an end.

*Frederick B. Chary*

# People's Republic of China Is Formed

*Defeating the Nationalist forces of Chiang Kai-shek, Mao Zedong and his Communist Party established a new government in China, the People's Republic.*

**What:** National politics
**When:** October 1, 1949
**Where:** Beijing, China
**Who:**
MAO ZEDONG (1893-1976), chairman of the Chinese Communist Party from 1949 to 1976
CHIANG KAI-SHEK (1887-1975), leader of the Nationalist Party
ZHOU ENLAI (CHOU EN-LAI; 1898-1976), a founder and first vice chairman of the Chinese Communist Party

## Civil War, World War

For more than a century before 1949, China had little peace at home and received little respect abroad. The Manchu Dynasty, which ruled from 1644 to 1911, was weak and allowed foreign powers to have considerable control over Chinese affairs. First the British (1839-1842), then the French, Germans, and Russians (1844-1860), moved into China, seeking spheres of interest and trade preferences.

In its first war with Japan (1894-1985), China lost Taiwan and the Pescadores Islands. The Boxer Rebellion of 1898-1900 was a protest against foreign interference in China, but it ended when the European powers and Japan were able to occupy Beijing.

After the Manchu dynasty was overthrown in 1911 and a republic was established, there were two decades of civil war. The Chinese Communist Party was formed in 1921 and immediately began competing with other groups for leadership. In 1923, however, the Communists began cooperating with the Kuomintang (Nationalist Party), then led by Sun Yat-sen, who had led the overthrow of the Manchus. Sun died in 1925 and was succeeded by Chiang Kai-shek, who became concerned that the Communists had gained too much power within the Kuomintang.

In 1927, Chiang began to send raids against the Communists and their supporters. Those Communists who could escape went into hiding throughout the countryside, where they continued to wage guerrilla warfare. Beginning in 1931, Chiang attacked Mao's base of operations in Southeast China.

Barely surviving these fierce assaults, Mao was eventually forced to take his one hundred thousand men by foot on a difficult six-thousand-mile trek north into the hills of Shaanxi (Shensi). This fighting retreat, known as the "Long March," began on October 16, 1934, and lasted more than a year. Many lives were lost, and all the marchers suffered greatly. For Chinese Communists, the Long March became one of the heroic landmarks of their history.

The survivors of the Long March became the core of both the Communist Party and the Red Army. Leaders who emerged from this ordeal included not only Mao but also Zhu De (Chu Teh), Zhou Enlai, Lin Biao (Lin Piao), and Peng Dehuai (P'eng Teh-hauai).

Meanwhile, Japan had invaded Manchuria in 1931 and turned it into one huge military base. The Japanese also pushed into Inner Mongolia and North China. Full-scale war broke out between China and Japan in 1937, and the Communists and Nationalists realized that they must fight together to defeat this enemy. At the same time, the Communists used the war years as a chance to build up their own forces.

## Cease-Fire and More War

When Japan was defeated in World War II, it had to leave China. The Western Allies hoped that the Communists and Nationalists would finally be able to resolve their differences and build a new, peaceful China. Chiang and Mao met several times to discuss the problems. The United States was willing to supply financial aid, which China badly needed, but only on the condition that there be no civil war.

To help the Chinese work toward unification and democracy, U.S. president Harry S. Truman sent General George C. Marshall as a special envoy. On January 10, 1946, the Guomindang (Kuomintang) and the Communist Party did sign a cease-fire. Soon a Political Consultative Conference, including other political parties besides the Nationalists and Communists, met in Chungking to form a coalition government. Yet the cease-fire was soon violated; the Communists and Nationalists did not trust each other, and each believed the other could be defeated. Marshall's mission had failed.

For a time, it seemed that Chiang would probably beat the Communists. Armed with heavy military equipment that was supplied by the United States, his troops spread across China and Manchuria and managed to push the Communists out of several strongholds, including their capital, Yan'an (Yenan).

Yet the Communists were well disciplined and determined, and they had won the loyalty of many Chinese peasants with their promises of land reform, democracy, and civil rights. Also, the Communists were successful in giving Chiang the image of a corrupt power-seeker who took orders from the West. The Chinese people wanted no more outside interference in their country.

By 1948, the Communists began defeating the Nationalists in open battles. Yan'an was recovered on April 22, 1948, and in the fall, Mukden in southern Manchuria surrendered to Mao's forces. Soon the Communists had captured Shanghai, Nanjing (Nanking), and Hanyang (Hankow); Peiping (later Beijing) was surrendered without a fight. Discouraged, many of Chiang's troops deserted and joined the Communists, taking with them their American equipment.

In August, 1949, much of China was in Communist hands. The U.S. State Department issued a statement that it accepted no responsibility for the Nationalist defeat. Chiang's forces had failed because of their own mistakes, the United States said, not because they lacked American help. Unable to stop the Communists, and no longer able

to count on American support, Chiang and the Guomindang fled to the island of Taiwan. There they established the Republic of China.

Meanwhile, on May 1, 1948, the Communists had invited all democratic groups and persons who opposed the Guomindang to a conference, which was held in Manchuria on November 25, 1948. There a Preparatory Committee was selected to convene a People's Congress and make plans for a democratic coalition government.

The committee, made up of 134 delegates representing twenty-three political organizations, met first in June and again in September, 1949. From this group, twenty-one persons were chosen to make up a Standing Committee; Mao Zedong was elected chair, and Zhou Enlai became vice chair. The committee decided to call its new organization the Chinese People's Political Consultative Conference (CPPCC).

*Mao Zedong during his revolutionary years.*

The CPPCC met for the first time on September 21, 1949, with 588 voting delegates. They approved a provisional constitution and set standards for foreign, political, military, economic, and educational policy. Before it adjourned on September 30, the CPPCC also adopted a national anthem, selected a national flag, and changed the name of its new capital from Peiping to Beijing.

The next day, October 1, Mao stood on a platform at Tiananmen Square (Gate of Heavenly Peace) in Beijing and officially proclaimed the establishment of the People's Republic of China.

## Consequences

The effects of Communist rule in China have been enormous. The civil war ended, and soon Mao's government began bringing about reforms that ended inflation, foreign domination, and illiteracy. It was a brutal and repressive government; those who spoke up against it were imprisoned and sometimes killed. Yet Mao had insisted that his nation must never again be insulted and that it be treated with full respect by other governments. In this he succeeded.

# Li Isolates Human Growth Hormone

*Choh Hao Li purified and characterized human growth hormone, ushering in a new era of advances in clinical medicine.*

**What:** Biology
**When:** 1950's
**Where:** Berkeley, California
**Who:**
CHOH HAO LI (1913-1987), a Chinese
    American biochemist
HAROLD PAPKOFF (1925-    ), an
    American endocrinologist
HERBERT MCLEAN EVANS (1882-1971), an
    experimental biologist

## Peptide Differences

The pituitary is a small gland lying underneath the lower surface of the brain. Controlled by the nearby brain region known as the hypothalamus, the pituitary gland secretes into the bloodstream many hormonal substances that regulate basic biological functions. Growth hormone, which is a peptide, or small protein, is one of these pituitary hormones; it is responsible for proper growth and development in young animals, including humans, and for other complex processes in adults.

In a series of classic experiments in the 1920's and 1930's, Herbert McLean Evans had demonstrated the growth-promoting potential of crude pituitary extracts from dogs. By the early 1940's, Li and Evans of the University of California, Berkeley, had succeeded in isolating and partially characterizing growth hormone from bovine (beef) pituitaries. This work was particularly noteworthy because of the exceedingly small amounts of the hormone present in the diminutive pituitary and because of the creative use of classical methods and emerging technologies, such as electrophoresis (a method in which molecules are separated according to differences in electrical charge), to purify the hormone effectively.

Because insufficient levels of growth hormone in human infants lead to retarded growth, availability of growth hormone supplements is important in clinical medicine. Despite the basic importance of Li and Evans's work with animal growth hormones, it had no direct impact on human medicine; animal growth hormones, except for those of the higher primates such as monkeys, do not stimulate growth in humans. It was thus important to isolate the human growth hormone. Human growth hormone differs from other growth hormones in the number and the sequence of the amino acids that compose it. These physical differences not only lead to differences in biological activity but also cause difficulties in working with the various hormones, as the technology to distinguish among the molecular differences is rudimentary.

## Charting the Way

Up until the mid-1950's, none of the previously isolated animal hormones had been thoroughly characterized, meaning that their amino acid compositions and sequences were unknown. Li's work, however, had established some physical properties of the bovine growth hormone. With this information in hand, Li and a colleague, Harold Papkoff, began to isolate and characterize the human growth hormone to test the hypothesis that primate hormones, including human ones, differed in chemical composition from other growth hormones, thus explaining the differences in biological activities.

Both human and monkey pituitary glands are rare items, so gifts of these tissues to Li were critical to the work's success. Human tissue was provided from researchers in Stockholm, Sweden, while monkey pituitaries came from the United States pharmaceutical firm Eli Lilly.

Li and Papkoff used previously tested techniques as well as new tools to effect the purifica-

**1106**

tion of growth hormone from these scarce tissues. Using the purified material, Li and Papkoff performed physical studies showing that the human and monkey growth hormones were significantly different in amino acid composition and molecular size from the growth hormones of other animals. In 1956, Li and Papkoff summarized their work in an article published in the American journal *Science*.

The final piece of standard chemical characterization of growth hormone awaited development of efficient technology for analyzing the exact sequence of amino acids in the hormone. Li and his coworkers published the sequence data in 1966. By 1970, Li had succeeded in producing human growth hormone using the sequence information and new techniques of laboratory synthesis. The synthesis brought to fruitful conclusion more than forty years of revolutionary thinking and experimentation in peptide chemistry. Li's work represents the power of scientific work that effectively utilizes the cutting edge of scientific design and method. Although many new approaches have arisen, these early studies were responsible for charting the way.

## Consequences

The most obvious application of the human growth hormone work has been in clinical medicine, where availability of pure, highly characterized molecules has produced dramatic positive changes for growth-hormone-deficient people, usually children. The pituitary is a complex organ containing many other peptide hormones and other substances. Other peptide hormones, present in even small quantities, may activate bodily processes that are undesirable or even dangerous. Pure preparations are thus essential in minimizing untoward effects. Availability of pure human growth hormone is also critical because growth hormones from other species do not effectively stimulate growth in humans.

The unraveling of the structural mysteries of the growth hormone has led to a greater understanding of the pituitary gland and its relationship to the brain as well as an improved knowledge of the dynamics of hormone function.

*Choh Hao Li.*

Mrs. C. H. Li

Better understanding of the pituitary gland has also been instrumental in exploring not only growth-related disorders but also many other physiological conditions.

By the early 1980's, the commercial application of human growth hormones had begun, as corporations started to produce genetically engineered human growth hormone, vastly increasing the supply. Therein may lie Li's most lasting contribution: Before the genetic engineering process could succeed, small amounts of the pure hormone had to be available. Without the pure hormone and the accompanying knowledge of how to isolate the eventual protein product, the modern technology would not work. Li's work is thus likely to have lasting practical consequences.

*Keith Krom Parker*

**1107**

# McCarthy Conducts Red-Baiting Campaign

> *Having stated publicly that the U.S. State Department was full of Communist Party members, Senator Joseph McCarthy began a crusade to root out spies in the federal government.*

**What:** National politics

**When:** 1950

**Where:** Wheeling, West Virginia, and Washington, D.C.

**Who:**

JOSEPH MCCARTHY (1908-1957), senator from Wisconsin from 1947 to 1957

HARRY S. TRUMAN (1884-1972), president of the United States from 1945 to 1953

ALGER HISS (1904-1996) an accused Communist spy and former U.S. government official

OWEN LATTIMORE (1900-1989), a professor and expert on Far Eastern affairs

MILLARD TYDINGS (1890-1961), senator from Maryland from 1927 to 1951

## Rising Fears

On February 9, 1950, Senator Joseph R. McCarthy of Wisconsin stood up to address the Ohio County Women's Republican Club in Wheeling, West Virginia. According to one reporter, he said, "While I cannot take the time to name all of the men in the State Department who have been named as members of the Communist Party and members of a spy ring, I have here in my hand a list of 205 that were known to the secretary of state as being members of the Communist Party and who nevertheless are still working and shaping the policy of the State Department." Later McCarthy changed his figure to 81 and then to 57, but he continued to insist that known Communists and spies were employed in the State Department.

This was a sensitive and emotional issue, for it was a known fact that just before, during, and after World War II there had been Communists and spies in the government, the labor movement, some intellectual circles, and other organizations. Communist cells had met in Washington, D.C., during the 1930's. The most important of these cells, headed by Harold Ware and later by Nathan Witt and John Abt, included government officials such as Alger Hiss and Julian Wadleigh.

After the Soviet purge trials and executions of 1936-1938, and especially after the Soviets signed a pact with Nazi Germany in 1939, the Communist Party in the United States no longer seemed so believable as an antifascist movement, and it lost many members. Yet the Soviet Union became the United States' ally during World War II, and Communists continued to enter the government.

As the war came to a close in 1945, American leaders realized that there had been espionage. In March, 1945, federal officers raided the offices of *Amerasia*, a Communist-sponsored magazine on Far Eastern affairs, and found many secret diplomatic and military documents. Meanwhile, a Canadian royal commission revealed that several Soviet spy rings had operated in Canada during the war, and that a leading scientist had sent atomic secrets to the Soviet Union.

President Harry S. Truman, on March 22, 1947, began a four-year investigation into the loyalty of all federal employees. As new cases of disloyalty became public, the American people grew more fearful.

The most important of these cases was that of Alger Hiss. During the hearing of the Hiss case, between 1948 and 1950, there were other disturbing events. In September, 1949, President Truman announced that the Soviets had exploded an atomic bomb. Up to this time, Americans had believed that the design of atomic weapons was their country's carefully guarded secret. That same year, Chiang Kai-shek and his Nationalist government had to flee to Taiwan from the Chinese mainland.

To many Americans, these events seemed related. Surely Communists in the U.S. government had sold out the Nationalists to the Chinese Communists and had also given atomic secrets to the Soviets. These suspicions seemed to be justified when early in 1950 the British arrested an atomic scientist, Klaus Fuchs, for having sent secret information to the Soviets from 1943 to 1947.

Americans were uneasy, filled with anxieties and fears about the Cold War, and troubled because World War II had obviously not been the "war to end all wars." They wanted answers to explain the situation, and the Communist "conspiracy" in Washington seemed to be the most logical explanation.

## McCarthy's Accusations

McCarthy had been looking for a political issue to help in his reelection campaign, and now he had found one. Since he was not able to name 205, 81, or even 57 Communists in the State Department, he changed his attack to focus on

Owen Lattimore, a professor who was an expert on Far Eastern affairs; McCarthy called him the "top espionage agent" in America.

Some Republicans believed that their fellow party member had come upon an important issue, one that might help the Republican Party in the upcoming national elections. Yet other Republicans, including Senator Margaret Chase Smith of Maine, disagreed. Smith declared that the Senate was being used for "selfish political gain at the sacrifice of individual reputations and national unity."

Soon a committee headed by Millard Tydings, a Democratic senator from Maryland, began investigating McCarthy's accusations. Scolding McCarthy for making charges based on distorted evidence, or no evidence at all, the Tydings Committee concluded that his anticommunist campaign had been "a fraud and a hoax . . . perhaps the most nefarious campaign of half-truths and untruth in the history of the Republic."

## Consequences

After the Tydings report was issued, a number of people rushed to McCarthy's defense. Far from being silenced, McCarthy was able to continue his crusade for another four years. As head of a Senate subcommittee on "Un-American Affairs," he accused many prominent Americans of disloyalty and of giving aid to Communism. The consequences for those accused were sometimes quite serious; for example, a number of Hollywood actors suspected of Communist sympathies were "blacklisted" and could not find work for years.

In the spring of 1954, however, McCarthy caused his own downfall. During nationally televised hearings, he ruthlessly attacked members of the U.S. Army; those who watched, more than twenty million people, were disgusted. On December 22, 1954, the Senate formally criticized McCarthy's behavior for bringing the Senate "into dishonor and disrepute" and damaging its dignity.

*William M. Tuttle*

*Joseph McCarthy.*

Library of Congress

# Abbott Labs Introduces Artificial Sweetener Cyclamate

> Abbott Laboratories launched the artificial sweetener Sucaryl, which contained cyclamate; Du Pont marketed a similar product, Cyclan.

**What:** Food science
**When:** 1950
**Where:** Abbott Laboratories, Illinois
**Who:**
MICHAEL SVEDA (1912-          ), an American chemist

## A Foolhardy Experiment

The first synthetic sugar substitute, saccharin, was developed in 1879. It became commercially available in 1907 but was banned for safety reasons in 1912. Sugar shortages during World War I (1914-1918) resulted in its reintroduction. Two other artificial sweeteners, Dulcin and P-4000, were introduced later but were banned in 1950 for causing cancer in laboratory animals.

In 1937, Michael Sveda was a young chemist working on his Ph.D. at the University of Illinois. A flood in the Ohio valley had ruined the local pipe-tobacco crop, and Sveda, a smoker, had been forced to purchase cigarettes. One day while in the laboratory, Sveda happened to brush some loose tobacco from his lips and noticed that his fingers tasted sweet. Having a curious, if rather foolhardy, nature, Sveda tasted the chemicals on his bench to find which one was responsible for the taste. The culprit was the forerunner of cyclohexylsulfamate, the material that came to be known as "cyclamate." Later, on reviewing his career, Sveda explained the serendipitous discovery with the comment: "God looks after . . . fools, children, and chemists."

Sveda joined E. I. du Pont de Nemours and Company in 1939 and assigned the patent for cyclamate to his employer. In June of 1950, after a decade of testing on animals and humans, Abbott Laboratories announced that it was launching Sveda's artificial sweetener under the trade name Sucaryl. Du Pont followed with its sweetener product, Cyclan. A *Time* magazine article in 1950 announced the new product and noted that Abbott had warned that because the product was a sodium salt, individuals with kidney problems should consult their doctors before adding it to their food.

Cyclamate had no calories, but it was thirty to forty times sweeter than sugar. Unlike saccharin, cyclamate left no unpleasant aftertaste. The additive was also found to improve the flavor of some foods, such as meat, and was used extensively to preserve various foods. By 1969, about 250 food products contained cyclamates, including cakes, puddings, canned fruit, ice cream, salad dressings, and its most important use, carbonated beverages.

It was originally thought that cyclamates were harmless to the human body. In 1959, the chemical was added to the GRAS (generally recognized as safe) list. Materials on this list, such as sugar, salt, pepper, and vinegar, did not have to be rigorously tested before being added to food. In 1964, however, a report cited evidence that cyclamates and saccharin, taken together, were a health hazard. Its publication alarmed the scientific community. Numerous investigations followed.

## Shooting Themselves in the Foot

Initially, the claims against cyclamate had been that it caused diarrhea or prevented drugs from doing their work in the body. By 1969, these claims had begun to include the threat of cancer. Ironically, the evidence that sealed the fate of the artificial sweetener was provided by Abbott itself.

A private Long Island company had been hired by Abbott to conduct an extensive toxicity study to determine the effects of long-term expo-

**1110**

sure to the cyclamate-saccharin mixtures often found in commercial products. The team of scientists fed rats daily doses of the mixture to study the effect on reproduction, unborn fetuses, and fertility. In each case, the rats were declared to be normal. When the rats were killed at the end of the study, however, those that had been exposed to the higher doses showed evidence of bladder tumors. Abbott shared the report with investigators from the National Cancer Institute and then with the U.S. Food and Drug Administration (FDA).

The doses required to produce the tumors were equivalent to an individual drinking 350 bottles of diet cola a day. That was more than one hundred times greater than that consumed even by those people who consumed a high amount of cyclamate. A six-person panel of scientists met to review the data and urged the ban of all cyclamates from foodstuffs. In October, 1969, amid enormous media coverage, the federal government announced that cyclamates were to be withdrawn from the market by the beginning of 1970.

In the years following the ban, the controversy continued. Doubt was cast on the results of the independent study linking sweetener use to tumors in rats, because the study was designed not to evaluate cancer risks but to explain the effects of cyclamate use over many years. Bladder parasites, known as "nematodes," found in the rats may have affected the outcome of the tests. In addition, an impurity found in some of the saccharin used in the study may have led to the problems observed. Extensive investigations such as the three-year project conducted at the National Cancer Research Center in Heidelberg, Germany, found no basis for the widespread ban.

In 1972, however, rats fed high doses of saccharin alone were found to have developed bladder tumors. At that time, the sweetener was removed from the GRAS list. An outright ban was averted by the mandatory use of labels alerting consumers that certain products contained saccharin.

## Consequences

The introduction of cyclamate heralded the start of a new industry. For individuals who had to restrict their sugar intake for health reasons, or for those who wished to lose weight, there was now an alternative to giving up sweet food.

The Pepsi-Cola company put a new diet drink formulation on the market almost as soon as the ban was instituted. In fact, it ran advertisements the day after the ban was announced showing the Diet Pepsi product boldly proclaiming "Sugar added—No Cyclamates."

Sveda, the discoverer of cyclamates, was not impressed with the FDA's decision on the sweetener and its handling of subsequent investigations. He accused the FDA of "a massive cover-up of elemental blunders" and claimed that the original ban was based on sugar politics and bad science.

For the manufacturers of cyclamate, meanwhile, the problem lay with the wording of the Delaney amendment, the legislation that regulates new food additives. The amendment states that the manufacturer must prove that its product is safe, rather than the FDA having to prove that it is unsafe. The onus was on Abbott Laboratories to deflect concerns about the safety of the product, and it remained unable to do so.

*Susan J. Mole*

# Oort Postulates Origin of Comets

*Jan Hendrik Oort advanced the theory that comets originate in a cloud of comets located a light-year from the sun.*

**What:** Astronomy
**When:** 1950
**Where:** Leiden, the Netherlands
**Who:**
JAN HENDRIK OORT (1900-1992), a Dutch astronomer
FRED WHIPPLE (1906-      ), an American astronomer

## What Are Comets?

A comet is an object consisting of a dense nucleus of frozen gases and dust. It orbits the sun and develops a luminous halo and tail as it nears the sun. Throughout history, the appearance of a comet has been considered a portent of disaster. When the great comet of 1066 C.E. made its appearance, William the Conqueror, claimant to the throne of England, shrewdly informed his troops that the comet signified the defeat of the English defenders; it also implied that he would become king of England. In the 1500's, scientists began to study comets. More than one hundred years later, the English astronomer Edmond Halley, a friend of Sir Isaac Newton, the English physicist and mathematician, noticed that comets appearing in 1456, 1531, 1607, and 1682 had similar orbits around the sun and came close to the sun every seventy-six years. He concluded that these were not several comets, but rather one comet with a period of revolution around the sun of seventy-six years. In 1705, he predicted that this comet would appear again in 1758. It arrived as scheduled and was christened Halley's comet in his honor.

As technology improved, additional information about comets was gathered: the shapes of their orbits, the directions in which they orbit the sun, why they have two tails, and why the tails always point away from the sun. Fred Whipple first described comets as dirty snowballs in 1949. They are composed of ices of ammonia, methane, and other compounds. Mixed in with ices are pieces of rocky material. Each time a comet comes close to the sun, the heat from the sun melts some of the icy material. This forms an atmosphere around the comet body, known as a "coma," that may be tens of kilometers in diameter. The solar wind, which is composed of gases expelled by the sun at high speeds, pushes these gases away from the coma to form one of the comet's tails. Light pressure from the sun forms the other tail by pushing dust matter away from the coma. This explains why the tails always point away from the sun.

With the melting and vaporizing of its ices, the comet loses some of its mass each time it passes by the sun. At a loss of one-millionth of its mass per orbit, Halley's comet would disappear in 76 million years. There are also objects that orbit the sun that may be burned-out comets. This is possible, since once the ices melt, most of the rocky material remains.

The length of a comet's orbital period (the time it takes to orbit the sun) is used to classify it. Long-period comets have a period greater than two hundred years and short-period comets less than two hundred years. Short-period comets were once long-period comets, but the gravitational attraction of the outer planets, such as Jupiter, has altered their orbits. Long-period comets have greatly elongated elliptical orbits that take them far beyond the orbit of Pluto. Even the orbit of Halley's comet, with a period of only seventy-six years, reaches beyond the orbit of Neptune.

## The Oort Cloud

Jan Hendrik Oort first described the source of comets in 1950. He reasoned that comets actually originate in a cloud of comets located a light-year from the sun. The cloud should be a spherical shell and billions of kilometers thick. The number of comets in this group could be in the trillions. This represents a total mass similar to that of the sun. The individual orbits should be elliptical but not as elongated as those of comets that closely approach the sun. The orbital period for a comet at a distance of a light-year is 15 million years. Its temperature should be close to absolute zero, the lowest possible temperature, since very little energy is received from the sun.

To verify this theory, the known facts about comets can be checked. A comet in Oort's cloud is moving around the sun at a velocity of several meters per second. (By comparison, an object such as Earth orbits the sun at tens of kilometers per second.) If a comet comes close to another comet, the two will gravitationally interact and will change orbital direction and speed. If a comet slows down, it will move closer to the sun, and its orbit will become elongated. More gravitational encounters with the outer planets will change its orbit even more, and it will move closer still. Now it would be recognized as a long-period comet. Additional encounters with Jupiter will alter the orbit to that of a short-period comet.

In the inner solar system, comets burn out over time and need to be replaced by new comets from the Oort cloud. Only several comets per year need to start their journey toward the inner solar system to keep it supplied with comets.

## Consequences

The idea of Oort's cloud being the source of the solar system's comets fits well with the accepted model of the origin of the solar system. In this model, a huge cloud of dust and gas condensed to form the sun and planets. An intermediate stage was the formation of "planetesimals," smaller lumps of matter that came together to form planets. Some of these planetesimals were thrown away from the sun by gravitational interaction with other planetesimals. They escaped totally or became part of Oort's cloud.

One implication of Oort's cloud is its possible involvement in the periodic mass extinctions that have occurred on Earth. Sixty-five million years ago, the dinosaurs and many other creatures became extinct. The impact of a 10-kilometer-diameter asteroid or comet may have been the cause. Other mass extinctions have occurred at roughly 26-million-year intervals. An unknown planet with a 26-million-year period of revolution may pass through the Oort cloud and cause many comets to start their journeys toward the inner solar system. On the way there, some could collide with Earth and cause mass extinctions by drastically changing the climate of Earth.

*Stephen J. Shulik*

# Morel Multiplies Plants In Vitro, Revolutionizing Agriculture

> *Morel successfully cultured virus-free monocot plant tissue on an artificial medium and was the first to apply such tissue culture techniques to the propagation of plants.*

**What:** Agriculture
**When:** 1950-1964
**Where:** Versailles, France
**Who:**
GEORGES MICHEL MOREL (1916-1973), a French physiologist
PHILIP CLEAVER WHITE (1913-    ), an American chemist

## Plant Tissue Grows "In Glass"

In the mid-1800's, biologists began pondering whether a cell isolated from a multicellular organism could live separately if it were provided with the proper environment. In 1902, with this question in mind, the German plant physiologist Gottlieb Haberlandt attempted to culture (grow) isolated plant cells under sterile conditions on an artificial growth medium. Although his cultured cells never underwent cell division under these "in vitro" (in glass) conditions, Haberlandt is credited with originating the concept of cell culture.

Subsequently, scientists attempted to culture plant tissues and organs rather than individual cells and tried to determine the medium components necessary for the growth of plant tissue in vitro. In 1934, Philip White grew the first organ culture, using tomato roots. The discovery of plant hormones, which are compounds that regulate growth and development, was crucial to the successful culture of plant tissues; in 1939, Roger Gautheret, P. Nobécourt, and White independently reported the successful culture of plant callus tissue. "Callus" is an irregular mass of dividing cells that often results from the wounding of plant tissue. Plant scientists were fascinated by

the perpetual growth of such tissue in culture and spent years establishing optimal growth conditions and exploring the nutritional and hormonal requirements of plant tissue.

## Plants by the Millions

A lull in botanical research occurred during World War II, but immediately afterward there was a resurgence of interest in applying tissue culture techniques to plant research. Georges Morel, a plant physiologist at the National Institute for Agronomic Research in France, was one of many scientists during this time who had become interested in the formation of tumors in plants as well as in studying various pathogens such as fungi and viruses that cause plant disease.

To further these studies, Morel adapted existing techniques in order to grow tissue from a wider variety of plant types in culture, and he continued to try to identify factors that affected the normal growth and development of plants. Morel was successful in culturing tissue from ferns and was the first to culture monocot plants. Monocots have certain features that distinguish them from the other classes of seed-bearing plants, especially with respect to seed structure. More important, the monocots include the economically important species of grasses (the major plants of range and pasture) and cereals.

For these cultures, Morel utilized a small piece of the growing tip of a plant shoot (the shoot apex) as the starting tissue material. This tissue was placed in a glass tube, supplied with a medium containing specific nutrients, vitamins, and plant hormones, and allowed to grow in the light. Under these conditions, the apex tissue grew roots and buds and eventually developed into a complete plant. Morel was able to generate whole plants from pieces of the shoot apex that were only 100 to 250 micrometers in length.

Morel also investigated the growth of parasites such as fungi and viruses in dual culture with host-plant tissue. Using results from these studies and culture techniques that he had mastered, Morel and his colleague Claude Martin regenerated virus-free plants from tissue that had been taken from virally infected plants. Tissues from certain tropical species, dahlias, and potato plants were used for the original experiments, but after Morel adapted the methods for the generation of virus-free orchids, plants that had previously been difficult to propagate by any means, the true significance of his work was recognized.

Morel was the first to recognize the potential of the in vitro culture methods for the mass propagation of plants. He estimated that several million plants could be obtained in one year from a single small piece of shoot-apex tissue. Plants generated in this manner were clonal (genetically identical organisms prepared from a single plant). With other methods of plant propagation, there is often a great variation in the traits of the plants produced, but as a result of Morel's ideas, breeders could select for some desirable trait in a particular plant and then produce multiple clonal plants, all of which expressed the desired trait. The methodology also allowed for the production of virus-free plant material, which minimized both the spread of potential pathogens during shipping and losses caused by disease.

## Consequences

Variations on Morel's methods are used to propagate plants used for human food consumption; plants that are sources of fiber, oil, and livestock feed; forest trees; and plants used in landscaping and in the floral industry. In vitro stocks are preserved under deep-freeze conditions, and disease-free plants can be proliferated quickly at any time of the year after shipping or storage.

The in vitro multiplication of plants has been especially useful for species such as coconut and certain palms that cannot be propagated by

*Morel's technique is especially useful for palm trees and other species that cannot be propagated by common methods such as sowing seeds or grafting.*

**1115**

other methods, such as by sowing seeds or grafting, and has also become important in the preservation and propagation of rare plant species that might otherwise have become extinct. Many of these plants are sources of pharmaceuticals, oils, fragrances, and other valuable products.

The capability of regenerating plants from tissue culture has also been crucial in basic scientific research. Plant cells grown in culture can be studied more easily than can intact plants, and scientists have gained an in-depth understanding of plant physiology and biochemistry by using this method. This information and the methods of Morel and others have made possible the genetic engineering and propagation of crop plants that are resistant to disease or disastrous environmental conditions such as drought and freezing. In vitro techniques have truly revolutionized agriculture.

*Diane White Husic*

# United States Enters Korean War

*Determined to stop the spread of Communism, President Harry S. Truman sent American troops to fight North Korean forces that had invaded South Korea.*

**What:** War
**When:** June, 1950
**Where:** Korea
**Who:**

KIM IL SUNG (1912-1994), premier of the North Korean People's Republic from 1948

SYNGMAN RHEE (1875-1965), president of the Republic of South Korea from 1948 to 1960

HARRY S. TRUMAN (1884-1972), president of the United States from 1945 to 1953

DOUGLAS MACARTHUR (1880-1964), U.S. commander in chief of the Far East and supreme commander of the United Nations forces

DWIGHT D. EISENHOWER (1890-1969), president of the United States from 1953 to 1961

## Pushing Back Invasion

At the end of World War II, Korea was divided into two zones of occupation. The "Hermit Kingdom," which had been under Japanese control for many years, was occupied by Soviet and American forces, and the 38th parallel was set as the temporary dividing line between zones.

In their zone, which lay north of this parallel, the Soviets organized a Communist government, which in 1948 was named the Democratic People's Republic of Korea, with an old-time Communist, Kim Il Sung, as its first premier. Meanwhile, in the South, various groups struggled for power until the party of the "father of Korean nationalism," Syngman Rhee, won an election supervised by United Nations observers. On August 15, 1948, Rhee became president of the Republic of Korea.

Both Korean governments were determined to reunify Korea according to their own plans. The North Koreans sent guerrilla forces to invade the south, and South Korean forces raided the North in return, so that the divided country continued in a state of crisis.

Nevertheless, the United States withdrew its troops from the South in June, 1949, leaving behind only a small group of technical advisers. South Korea, whose army was small, ill-trained, and badly equipped, faced an enemy whose army of 135,000 members used modern Soviet weapons. North Korea also had between 150 and 200 combat airplanes.

Although South Korean leaders and some Americans feared that North Korea might attack across the 38th parallel at any time, U.S. secretary of state Dean Acheson announced on January 12, 1950, that Korea was not within the "defensive perimeter" of U.S. "vital interests" in the Far East. It seemed that the United States did not intend to take action to protect the South Koreans.

The expected attack came on June 25, 1950. North Korean armed forces—armored units and mechanized divisions supported by artillery—struck without warning across the demarcation line. The South Koreans were not able to fight back effectively, and within thirty-six hours North Korean tanks were moving into the outer suburbs of Seoul, South Korea's capital.

In spite of Acheson's January statement, the United States reacted quickly and with great determination. At the urging of U.S. leaders, the United Nations Security Council met in special session on the day of the attack and unanimously passed a resolution calling for North Korea to withdraw. (The Soviet Union was boycotting the Security Council at this time.) The North Koreans ignored this message, and the Security Council met again on June 27 and passed another reso-

## KOREAN WAR, 1950-1953

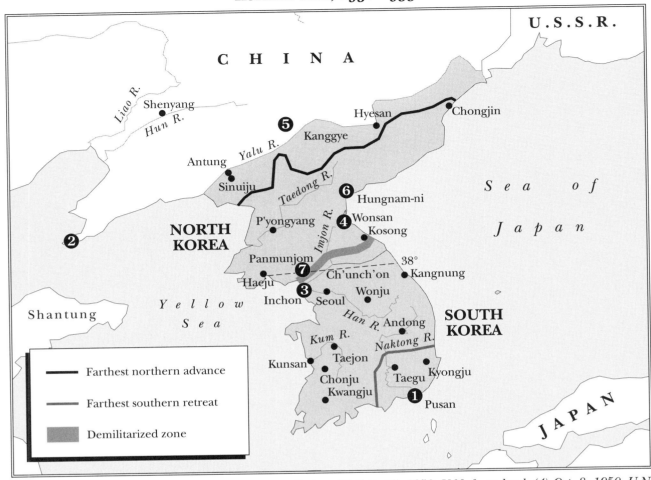

*(1) Main U.N. base. (2) Russian-Chinese naval installation. (3) Sept. 15, 1950, U.N. forces land. (4) Oct. 8, 1950, U.N. forces land. (5) Nov. 26, 1950, Chinese attack. (6) Dec. 9, 1950, U.N. forces evacuate. (7) July 27, 1953, armistice signed.*

lution recommending that U.N. members help the Republic of Korea "repel the armed attack."

American president Harry S. Truman decided to enforce the U.N. resolutions. On June 27, he authorized U.S. air and naval forces to become involved in Korea's struggle, along with U.S. ground forces that had been stationed in Japan.

By the end of June, however, more than half the Republic of Korea (ROK) Army had been destroyed, and American units were fighting on a southward retreat. In early August, American and ROK forces were able to create a defensive line around the important port of Pusan.

As forces poured into South Korea from the United States and fifteen other nations, General Douglas MacArthur, commander in chief of the Far East and supreme commander of the

U.N. forces, made a bold plan. He would use the troops not in a direct attack from the Pusan line but in a landing at Inchon, a port just a few miles from Seoul. This risky operation, begun on September 15, 1950, was very successful. Threatened with encirclement, the North Korean army was forced to retreat back across the 38th parallel.

### The War Drags On

Now came the most important decision of the Korean War. Should the defeated North Koreans be followed across the 38th parallel? With many people in the United States calling for a complete victory, the Truman administration allowed General MacArthur to continue the fight.

The first crossings came on October 1. U.N. and ROK forces hurried northward, and by late November they were nearing the Yalu River boundary between North Korea and the People's Republic of China.

The Communist leaders of China had warned that they would not allow North Korea to be invaded, and they were not bluffing. By late October, thousands of Chinese soldiers had crossed the Yalu. A month later, they struck at MacArthur's armies, and by early December, the U.N. troops were again in headlong retreat.

A new defensive line was organized south of the 38th parallel, and through the remaining months of winter and early spring, the lines moved from south of Seoul to north of the parallel. By July, 1951, neither side could budge the other. The conflict settled down to trench warfare—bloody battles that were never really won. This situation lasted for two cruel years. During this time more than a million Americans served in Korea.

Meanwhile, at Panmunjom, South Korea, the two sides negotiated to try to bring about a cease-fire and an armistice. There were many heated arguments and accusations, and the talks seemed to be getting nowhere.

Within the United States, growing frustration over the Korean War influenced the rise of McCarthyism and the election of Dwight D. Eisenhower to the presidency. Having promised to go to Korea if elected, Eisenhower kept his word, but his visit did not succeed in moving the peace talks along.

Finally, however, an armistice agreement was signed at Panmunjom on July 27, 1953. There was to be a cease-fire, with both armies withdrawing two kilometers from the existing battle line, which ran from coast to coast from just below the 38th parallel in the west to thirty miles north of it in the east. A U.N. Supervisory Commission would carry out the armistice terms, and a political conference was to settle all remaining questions.

## Consequences

In the months that followed, the United Nations returned more than 70,000 North Korean and Communist prisoners, but received in return only 3,597 Americans, 7,848 South Koreans, and 1,315 prisoners of other nationalities. The political conference was never held, and North and South Korea continued to view each other as enemies.

The Korean War had lasted three years and one month. It cost the United States about 140,000 casualties including 22,500 dead; the financial cost to Americans was $22 billion.

No one had really won this war, but U.N. forces had been able to keep the Communists from conquering South Korea, and the United States had shown that it was willing to fight to prevent Communism from spreading further. In American military policy, the Korean War also marked a change from war for total victory to limited war with no demand for total victory.

*Theodore A. Wilson*

# Truman and MacArthur Clash over Korea Policy

> *The conflict between President Harry S. Truman and General Douglas MacArthur over how to handle the Korean War became a constitutional issue: Would a military leader be able to override the decisions of the U.S. president?*

**What:** International relations; National politics

**When:** July 8, 1950-April 11, 1951

**Where:** Korea and Washington, D.C.

**Who:**

HARRY S. TRUMAN (1884-1972), president of the United States from 1945 to 1953

DOUGLAS MACARTHUR (1880-1964), U.S. commander in chief in the Far East and supreme commander of United Nations forces in Korea

## MacArthur's Successes

In June, 1950, General Douglas MacArthur was serving as commanding general, U.S. Army, Far East; commander in chief, Far East; and supreme commander for the Allied Powers. Since September, 1945, he had governed Japan, carrying out many reforms intended to make the former Axis Power into a modern democratic nation.

With North Korean armies overrunning South Korea, MacArthur was appointed Supreme Commander of United Nations forces in Korea on July 8, 1950. Operating out of his Tokyo headquarters and using staff officers who were intensely loyal to him, MacArthur soon began to plan a bold counterstrike that would put the U.N. forces on the way to a complete victory against North Korea.

MacArthur planned to land forces at Inchon, a port on South Korea's west coast a few miles from the capital, Seoul. There were strong objections from U.S. Marine Corps and Navy officers in Korea, but MacArthur brushed them aside. An ambitious and confident man, he was accustomed to keeping control. He believed that Korea provided an opportunity for the United States to regain prestige in the world and to stop Communism in Asia once and for all.

Suggesting that the Communist government of China was ready to collapse and that its military forces were pitifully weak, MacArthur advocated sending troops into China if necessary. He also thought that it would be wise to encourage Chinese Nationalist leader Chiang Kai-shek and to bring some of Chiang's army from Taiwan to fight in Korea. All these ideas went against the Truman administration's goal of keeping the Korean War from spreading.

MacArthur's outspokenness regarding Taiwan caused his first dispute with the White House. The president was so furious that he began to think of firing the general. Instead, Truman ordered MacArthur to take back his statement, and MacArthur did so.

Yet Truman had to be careful in his dealings with MacArthur, because the general was very popular and had supporters in Congress. The remarkable success of the Inchon landings added to the general's reputation. Soon the U.N. forces were able to push the North Koreans back across the 38th parallel, into their own territory. This had been Truman's original goal—simply restoring the boundary line between North and South Korea.

With this goal reached, however, Truman agreed with MacArthur that U.N. and South Korean forces should press ahead into North Korea and occupy the entire country. The U.N. Security Council had already recommended that, unless Soviet or Chinese forces became involved, all North Korea should be occupied.

**1120**

American troops crossed the 38th parallel on October 7 and moved quickly northward. By mid-November, the leading units were nearing the Yalu River, which divided North Korea from China. Discounting rumors that the Chinese Communists were about to enter the war, MacArthur was confident of complete victory.

## General Versus President

In mid-October, Truman had flown to Wake Island to discuss the situation with MacArthur. There was no reason to fear Chinese intervention, MacArthur assured the president. Yet for once he was wrong.

On November 26, Chinese forces crossed the Yalu River and attacked MacArthur's troops. There was a difficult retreat, and by Christmas, 1950, U.N. forces were once again fighting below the 38th parallel.

At this point, Truman and MacArthur took opposite sides. Frightened by China's action, the Truman administration wished to limit the war. MacArthur wanted to go on the offensive against

Chinese troops and supplies in Manchuria—and eventually in China itself. The president refused and decided to allow only the Korean halves of the Yalu bridges to be bombed. This compromise infuriated MacArthur.

In January, 1951, MacArthur recommended a naval blockade of China, air attacks to destroy Chinese military and industrial bases, and using Nationalist Chinese forces in Korea. The president again restrained him, explaining that the Soviet Union posed a worldwide threat and that the war in Korea was simply a war of "containment."

Finally, the general tried to bypass the president in order to get Congress and the American people to support his plans. On several occasions, MacArthur spoke publicly of his disagreements with Truman, angering and embarrassing the president.

The break came in late March, 1951. When MacArthur learned that Truman planned to issue a peace offer, he made public a "military appraisal"—a document that was actually an ulti-

*President Harry S. Truman (left) and General Douglas MacArthur on Wake Island in 1950.*

Hulton Archive

**1121**

matum to the Chinese and to Truman. It made negotiating a settlement in Korea much more difficult and forced Truman to realize that he must dismiss MacArthur. "By this act," Truman stated, "MacArthur left me no choice—I could no longer tolerate his insubordination." Now the president simply waited for the right moment to fire the general.

On April 5, Representative Joseph Martin read on the floor of the House a letter from Mac-Arthur, in which the general again criticized the president's policy of limited war and called for total victory in Asia. A series of meetings began in the White House on the following day. On April 11, President Truman sent a telegram to MacArthur in Tokyo and at the same time informed the press that the general was being relieved of his commands because he did not give "wholehearted support" to the president's policies.

## Consequences

Always self-confident, General MacArthur returned to the United States as a triumphant hero. He addressed a joint session of Congress, and across the nation there was a great surge of support for him. For his part, however, President Truman was sure that he had done the right thing. He had turned back a major challenge to the American principle of civilian control of the military, and he had stood by his doctrine of "containment"—fighting limited wars to keep Communism from expanding.

*Theodore A. Wilson*

# United Nations Forces Land at Inchon

> *With North Korean forces pushing far into South Korea, General Douglas MacArthur made a bold plan for landing United Nations troops at the western port of Inchon.*

**What:** War
**When:** September 15, 1950
**Where:** Inchon, South Korea
**Who:**
DOUGLAS MACARTHUR (1880-1964), U.S. commander in chief in the Far East and supreme commander of United Nations forces in Korea
OLIVER P. SMITH (1893-1977), commander of the U.S. First Marine Division in Korea

## Arguing the Plan

In June, 1950, all of South Korea was suddenly under siege by Communist forces that had swept in from North Korea. With a poorly trained and badly equipped army, the Republic of South Korea seemed about to collapse.

The United Nations Security Council reacted by condemning the invasion and recommending that U.N. members come to South Korea's aid. On instructions from United States president Harry S. Truman, General Douglas MacArthur left his headquarters in Japan and flew to Seoul to survey the crisis.

Seoul, the South Korean capital, had been overrun by the North Koreans, and Syngman Rhee's government had been forced to flee farther south. About forty-four thousand Republic of Korea troops had been lost, along with many weapons. It was evident that the Communists intended to keep moving south toward the important port of Pusan.

A veteran commander of the American victory over Japan in World War II, MacArthur had faced difficult situations before. In the early months of 1942, he had been driven out of the Philippines by an aggressive Japanese offensive. Now, eight years later, he faced a similar dilemma, for he did not have readily available troops to allow him to act quickly. Certainly there were U.S. troops, ships, and aircraft in the Far East, but bringing them to South Korea would take time. Meanwhile, the North Koreans were advancing quickly.

MacArthur realized, however, that the North Koreans' rapid advance had made it hard for them to bring enough supplies in for their leading units. If the thread of their supply line could be broken, the invasion might be thrown into confusion, and they might be forced to withdraw.

During his week's survey of the battlefront, MacArthur chose the port of Inchon for his surprise attack. Lying close to Seoul, Inchon was a risky choice. Veteran marine and navy officers, as well as army chief of staff Joseph L. Collins and Admiral Forrest Sherman of Naval Operations, immediately questioned the possibility of an Inchon strike. The terrain in the Inchon area was rugged, and tides in the Inchon area were the second deepest in the world, so that any landing there would be difficult. And no one knew how the Chinese forces in Manchuria would react.

At a meeting of the Joint Chiefs of Staff in Tokyo on August 23, General MacArthur argued, however, that the Inchon landing was a good idea simply because it would be unexpected. On August 29, he received the go-ahead from Washington.

## The Landing

On September 15, 1950, the First Marine Division under Major General Oliver P. Smith successfully seized the tiny Wolmi-do Island (also known as Moontip Island) inside Inchon harbor. Naval guns and marine airpower had prepared the way for the assault with a series of bombardments. There were no American casualties.

Meanwhile, to keep Communist forces distracted from the Inchon site, there was bombard-

Hulton Archive

*U.S. assault craft move toward the beach at Inchon.*

ment in other areas: in Chinnampo to the north (by a British task force) and on Korea's east coast, near Samchok (by the USS *Missouri*).

As MacArthur had hoped, the port of Inchon was taken quickly, with almost no resistance. Once on shore, the marine units captured the important Kimpo Airfield, while backup support from the Seventh Division (which came ashore at Inchon as a second wave) made it possible to seize the airstrip at Suwon, south of Seoul. Most important of all, however, was that the U.N. forces were able to cut off the Seoul-Pusan roadway.

Within two weeks of the Inchon landings, most of the North Korean invasion force had been dealt with. Entire units had surrendered, and the roads were cluttered with hastily abandoned vehicles and weapons. In the panicked retreat northward, some 130,000 North Korean troops were taken prisoner. Almost all of South Korea was restored to the Syngman Rhee government, which had moved back into Seoul.

## Consequences

Naturally, the U.N. expedition at Inchon did not bring an end to the conflict between North and South Korea; there were two more years of seesaw battles both north and south of the 38th parallel, the dividing line between North and South Korea. The People's Republic of China sent troops to help North Korea, and President Truman barely kept General MacArthur from turning the conflict into a much bigger war. Peace talks at Panmunjom seemed to go on endlessly.

Still, the Inchon landing was a watershed of the Korean War—partly because it was such a daring move, but mainly because it had a great influence over the rest of the war. Without the success of this surprising plan of attack, it is very possible that Korea would have come under complete Communist domination.

# China Occupies Tibet

---

*Claiming that Tibet had always been a part of China, the Chinese government invaded Tibet and began destroying its religion and culture.*

---

**What:** Military conflict; Human rights; Political aggression
**When:** October 7, 1950
**Where:** Tibet
**Who:**
TENZIN GYATSO (1935-      ), the fourteenth Dalai Lama, Tibet's spiritual leader and head of state
MAO ZEDONG (1898-1976), founder of the People's Republic of China
JAWAHARLAL NEHRU (1889-1964), prime minister of India from 1947 to 1964

## China Moves In

Tibet did not really exist as a country before the seventh century C.E., when it was unified by a powerful tribal leader named Songtsen Gompo. He took wives from the royal families of Nepal and China, and was responsible for bringing Buddhism into Tibet.

A treaty between Tibet and China in 821-822 C.E. laid out the boundary between the two countries, and each promised that it would respect the other's sovereignty. Between the tenth and thirteenth centuries, power in Tibet became concentrated in the Buddhist monasteries, and in 1244, Basba became the first Lama king.

In the eighteenth and nineteenth centuries, the Chinese Manchus gained some influence in Tibet. Yet China had trade and diplomatic contacts with Europe and the rest of Asia during this time, while Tibet preferred to remain isolated.

In 1904, a British expedition made its way into Tibet and forced the Tibetans to accept trade agreements with the outside world. Soon the Manchus began using force against the Tibetans, but in 1912 the Tibetans drove out the Manchu forces. In 1913, the thirteenth Dalai Lama declared Tibet independent.

The thirteenth Dalai Lama died in 1933. According to Tibetan tradition, after the Dalai Lama's death he is reincarnated within a few months. Traditional rules were used to choose a two-year-old successor in 1937. He was lifted onto the golden throne at the age of five.

Meanwhile, China's civil war was coming to an end. The Communists, led by Mao Zedong, defeated the Nationalists, and the People's Republic of China was established on October 1, 1949. On January 1, 1950, the new government announced that one of the main tasks of the People's Liberation Army (PLA) would be to "liberate" Tibet.

On October 7, 1950, the PLA invaded Tibet. The Tibetan army resisted but was soon overwhelmed. The Chinese government insisted that it was trying to free the people of Tibet from "imperialist oppression," but no one could quite figure out who the "imperialists" were. Only a few European traders were in Tibet at the time of the invasion.

## Atrocities and Exile

The Dalai Lama finished his training at the age of fifteen and was named head of Tibet's government on November 17, 1950. On May 23, 1951, the People's Republic of China announced that a Seventeen-Point Agreement between the two countries had been signed in Beijing. China claimed sovereignty over Tibet but promised that it would not change Tibet's political system or take power from the Dalai Lama. China also pledged to allow religious freedom in Tibet and to leave the monasteries alone.

Soon the Dalai Lama himself went to Beijing and served as a vice president of the Steering Committee of the People's Republic of China. He found that many parts of Marxism fit with Buddhist philosophy, and he got along well with Chairman Mao. Yet eventually he realized that the Chinese leaders were determined to destroy Tibet's religion and culture.

**1125**

In 1956, revolt broke out in eastern Tibet, an area that was not ruled directly by the Dalai Lama. As the unrest spread into the central regions, Mao announced that Tibet was not yet ready for Communist reforms. The Chinese tried to control the unrest, but it became even stronger in 1959. The Dalai Lama was forced to flee into northern India, and in the first year of his exile about twenty thousand Tibetans followed him.

India's prime minister, Jawaharlal Nehru, was later criticized for his response to the Tibetan crisis. His government did offer land for refugee settlements, but the Dalai Lama was kept from engaging in political activities and was not allowed to bring his people's case before the world. In fairness, however, it should be said that India gave not only land but also education and health care to the Tibetan refugees, who eventually numbered nearly 100,000.

In Tibet, the Communists named a new head of state, the Panchen Lama, who had been a rival to the Dalai Lama. In 1965, China made Tibet an "autonomous region" under the direct rule of the Beijing government.

## Consequences

According to the Legal Inquiry Committee on Tibet, which reported to the International Commission of Jurists in Geneva, Switzerland, in 1960, the Chinese violated many of the Tibetan people's human rights. Execution, torture, and rape were used by the PLA; children were taken from their parents, and many shrines and monasteries were destroyed. The Tibetans were forced to work longer hours for less food, and agricultural "reforms" led to a famine that brought death to hundreds of thousands of people.

During the Cultural Revolution in China in the late 1960's, the Chinese Red Guard led a new attack on the religious life of Tibet. About thirty-seven hundred monasteries had been operating in Tibet before the invasion, and only seven hundred survived the Chinese attacks.

The Chinese brought secular public education to Tibet, provided clinics and hospitals, and built wells, canals, and roads. Yet the Tibetan people remained determined to regain their independence. More than eighty-seven thousand died in the revolt between March, 1959, and September, 1960, and even more were killed in an uprising in 1969. Another rebellion in 1989 left about one hundred dead in Lhasa, Tibet's capital. It is estimated that 1.2 million Tibetans have died since the 1950 invasion.

After the Cultural Revolution, Chinese leaders admitted that "mistakes" had been made in Tibet. Policy toward Tibet became more open, and the Dalai Lama's government was allowed to send a fact-finding mission to Tibet in 1979.

The plight of the Tibetan people eventually gained sympathy around the world. The Dalai Lama, still living in northern India, received the Nobel Peace Prize in 1989 for his attempts to gain independence without violence.

*L. B. Shriver*

# President Truman Escapes Assassination Attempt

*Two Puerto Rican nationalists attempted to storm Blair House, temporary replacement for the White House, where President Harry S. Truman was living. One assailant and one police officer were killed, and others were injured.*

**What:** Assassination; National politics
**When:** November 1, 1950
**Where:** Blair House, Washington, D.C.
**Who:**
Oscar Collazo (1914-1994), Puerto Rican nationalist assailant
Griselio Torresola (1925-1950), Puerto Rican nationalist assailant
Harry S. Truman (1884-1972), president of the United States

## The Roots of Nationalism

The attempt to assassinate President Harry S. Truman on November 1, 1950, was the result of simmering anger among Puerto Rican nationalists demanding Puerto Rico's independence from the United States. Independence was advocated by the Puerto Rican Nationalist Party, led by Pedro Albizu Campos. Campos had become violently hostile to the United States after experiencing discrimination in the U.S. Army in World War I. His party, however, failed to attract votes in its electoral campaigns and, as a result, did not run candidates after its poor showing in 1932. Instead, Campos declared himself head of a new independent Puerto Rican government, supported by a fascistlike black-shirted liberation army. The "army" proceeded to commit a series of violent acts, including bombings and assassinations, which failed to endear the party to the populace. Most of the island, in any case, did not favor independence.

The two would-be assassins were members of the Puerto Rican Nationalist Party living in New York City. Thirty-six-year-old Oscar Collazo had been converted to the Nationalist Party's cause while visiting the island after hearing a speech by Campos. Twenty-five-year-old Griselio Torresola was born into a family steeped in the tradition of revolutionary violence and had been a devoted follower of Campos all his life. Both men's ardor for their cause was renewed in 1943 when Campos arrived in New York City after his release from Atlanta Federal Penitentiary, where he had served a sentence for political crimes in Puerto Rico.

The immediate cause of the assassination plot was the failure of yet another revolutionary *coup d'état* in Puerto Rico. On October 28, 1950, an attempt to assassinate the island's governor in San Juan and take over the government fell through. In the failed coup, Torresola's brother killed a police officer, and his sister was wounded. Three days later, Collazo and Torresola bade farewell to their families and bought one-way train tickets to Washington, D.C. Arriving at a hotel near the station, the two neatly dressed men were mistaken by the desk clerk for divinity students.

## The Assassination Attempt

In November, 1950, President Truman was living in Blair House, across the street from the White House while it was gutted and rebuilt. Collazo and Torresola did not plan their attack on the president well. Had they consulted Washington, D.C., newspapers, they would have known that the president would be leaving his residence at 2:50 P.M. for an engagement, relieving them of the necessity of shooting their way into his quarters.

At 2:20 P.M. on an unseasonably hot autumn afternoon, the two conspirators unfolded their ill-founded plan. Approaching Blair House from opposite directions, they proceeded to open fire

*Police killed Griselio Torresola as he approached from the west (A), and Oscar Collazo was wounded as he approached the entrance from the east and opened fire on a police officer (B). After the shooting was over, President Harry S. Truman peered out the window to see what had happened (C).*

on guards. Collazo, who knew little about firearms apart from what he had gleaned from Torresola's hurried instructions in their hotel room, attempted to shoot one guard at point-blank range, but his gun misfired with a loud click. As he pounded the gun in desperation, it went off, hitting the guard in the leg. Collazo then ran to the Blair House entrance but was felled by one of several shots by Secret Service Agent Floyd M. Bowring. Collazo lay bleeding on the ground but survived his wound to stand trial, resulting in a death sentence.

Meanwhile, trying to reach the rear entrance, Torresola came upon Washington Metropolitan police officer Private Leslie Coffelt and shot him at close range. Though mortally wounded, Coffelt managed to draw his revolver and put a bullet into Torresola's head, killing him in-

stantly. The entire incident, in which nearly thirty shots were fired, took less than three minutes. Truman was never in any immediate danger. Had Torresola managed to gain entrance to the building, however, only two guards would have stood between him and the president, who was upstairs taking a nap. Accounts stating that the president went to the window during the shooting are contradicted by Agent Bowring, who later insisted that the president appeared at the window only afterward.

**Consequences**

On July 24, 1952, a week before Collazo's scheduled execution President Truman commuted his sentence to life imprisonment. Twenty-nine years later, an unrepentant Collazo was granted a pardon for his crimes by President

Jimmy Carter and returned to Puerto Rico, where he lived quietly until his death in 1994.

More important than the fate of one individual, however, was the growing isolation of American presidents from the people that resulted from such threats to their lives. Before the assassination attempt, Truman enjoyed the relative freedom of being able to walk in public with only a few Secret Service guards. Afterward, he was always driven from his residence to his While House office and was less frequently permitted to walk outside the security of the While House grounds, mingling freely with the public. When he did venture out, he was surrounded by a much enlarged swarm of bodyguards. Truman and later presidents found themselves virtual prisoners of security arrangements.

## Consequences

Apart from the end of relatively lackadaisical presidential security, the assassination attempt gave renewed force to the trend toward violence among the extremist wing of Puerto Rican nationalists. It was a trend that never entirely died out. In 1954. just four years after the Blair House incident, four more violent Puerto Rican nationalist extremists opened fire from the visitors' gallery of the U.S. House of Representatives in the Capitol, wounding seven members.

Later, in the 1970's and 1980's, more than a dozen Puerto Rican nationalists were convicted of committing a variety of crimes, including bank robbery, weapons violations, and seditious conspiracy, which resulted in the deaths of six people. At the outset of the twenty-first century, these nationalists remained in jail, and the status of Puerto Rico, despite successive free elections allowing the island to choose among independence, statehood, and continued status as a commonwealth associated with the United States, remained unresolved.

*Charles F. Bahmueller*

# United States Explodes Its First Hydrogen Bomb

*Though some scientists had argued against developing a hydrogen bomb, for political reasons the research was continued, and the first American hydrogen bomb was exploded on March 1, 1954.*

**What:** Technology; Military capability
**When:** Early 1951
**Where:** Elugelab Island; Siberia; and the Bikini atoll
**Who:**
HARRY S. TRUMAN (1884-1972), president of the United States from 1945 to 1953
J. ROBERT OPPENHEIMER (1904-1967), a theoretical physicist, and director of the Los Alamos Laboratories
EDWARD TELLER (1908-    ), an atomic physicist who became the "father" of the hydrogen bomb
ERNEST O. LAWRENCE (1901-1958), an atomic physicist who won a Nobel Prize for the invention of the cyclotron
LEWIS L. STRAUSS (1896-1974), a member of the Atomic Energy Commission

## Fission Versus Fusion

In the summer of 1942, J. Robert Oppenheimer, a theoretical physicist, invited several other leading American physicists to Berkeley, California, to decide how to develop atomic weapons for the war against Germany. At this meeting, Edward Teller, a Hungarian-born physicist, suggested that a fusion, or hydrogen, bomb be developed.

The idea for the fusion bomb came from the type of thermonuclear reactions thought to exist in the interior of stars: two ions of deuterium (or "heavy" hydrogen) fused to make one atom of helium, releasing about four million volts of energy. Yet most of the scientists at the Berkeley meeting had been more interested in a fission bomb, using the enormous energies set free when heavy atoms are split in a nuclear reaction. Although the fusion, or hydrogen, bomb had the potential to be more powerful, developing it would involve greater problems. Some scientists feared that a hydrogen bomb might ignite the oceans in one vast chain reaction—winning the war by demolishing the earth.

So it was that the scientists at Berkeley decided to begin work on the fission bomb and to put off considering the problem of fusion. Oppenheimer became head of the Los Alamos Laboratories, where the first working atomic bomb was designed and built. Teller also went to work at Los Alamos, but never accepted the decision to concentrate on the fission bomb.

The other Los Alamos scientists worked hard and successfully on fission; in July, 1945, the first atomic bomb was exploded in Alamogordo, New Mexico. By this time, the Germans had surrendered and the Japanese were trying to find a way to negotiate peace without losing their dignity. Nevertheless, U.S. president Harry S. Truman decided to drop the bomb; some have thought, perhaps unfairly, that his main concern was to impress the Soviet Union.

The explosions were impressive: The cities of Hiroshima and Nagasaki, Japan, were destroyed, and 100,000 people were killed.

Some American scientists had already urged that nuclear weapons should not be used, and after the war they continued lobbying against them. Teller was by then almost alone in his desire to create the hydrogen bomb, or "Super"; he was opposed by Albert Einstein and the Emergency Committee of Atomic Scientists.

In August, 1949, however, American leaders learned that the Soviet Union had exploded its first atomic weapon, "Joe 1." Many military and government officials were shocked to realize that the United States was no longer the only nuclear power.

With his coworkers Ernest O. Lawrence (inventor of the cyclotron) and Luis Alvarez, Teller began lobbying hard for the "Super" and gained many influential friends. Supporters of the "Super" included Senator Brien McMahon, chair of the Joint Congressional Committee on Atomic Energy; General Hoyt Vandenberg, Air Force chief of staff; Omar Bradley, chair of the Joint Chiefs of Staff; and Lewis L. Strauss, the most conservative and powerful member of the Atomic Energy Commission.

The General Advisory Committee of the Atomic Energy Commission, made up of nine leading scientists under Oppenheimer's chairmanship, all advised against production of the "Super." By a vote of four to one, the Atomic Energy Commission agreed with this decision.

Yet the scientists' advice was soon overruled. On January 31, 1950, a committee of three met at the White House: Louis A. Johnson, secretary of defense; Dean Acheson, secretary of state; and David Lilienthal, chair of the Atomic Energy Commission. They voted two to one in favor of the "Super" (Lilienthal was opposed). That afternoon, President Truman announced his decision to go ahead with the rapid production of the hydrogen bomb.

## Making It Work

In 1950, it was not yet clear whether the hydrogen bomb would work. Heavy hydrogen would fuse when ignited by the heat of an atomic blast, and a mixture of deuterium and tritium (two forms of heavy hydrogen) would ignite more quickly than deuterium alone. The heavy hydrogen had to be kept in liquid form and required refrigeration.

The first model of the "Super" was a fission bomb coated with layers of deuterium and tritium and enclosed in a huge refrigeration unit, weighing sixty-five tons. Since this thermonuclear icebox was not a practical weapon, it was

*Explosion of the first hydrogen bomb on the Eniwetok Atoll's Elugelab Island.*

AP/Wide World Photos

called a thermonuclear "device" rather than a "bomb"; some scientists called it "the Superfluous." This device was exploded on the island of Elugelab in the South Pacific on November 1, 1952; it sank the entire island and carved a mile-long crater in the ocean floor.

The American scientists did not know that the Soviet Union had already exploded a similar thermonuclear device in 1951. This fact, known only to President Truman and a few of his top advisers, was kept a secret, even from Teller.

By 1951, however, Teller and Stanislaw Ulam, a mathematician, had already developed a new idea for a workable bomb that would be made with lithium deuteride and would need no refrigeration. Neutrons produced by the first fission explosion would turn lithium into tritium; the result would be a practical "dry" bomb. The scientists worked night and day to refine the design of this bomb.

On August 8, 1953, however, Soviet premier Georgi Malenkov announced that the Americans no longer had a monopoly on the hydrogen bomb. Several days later, radiation traces showed that the Soviet Union had exploded a "dry" lithium hydride bomb.

On March 1, 1954, the United States responded with a new design: the three-stage fission-fusion-fission bomb, coated with uranium isotopes for maximum destructive power. This bomb, dubbed "Shoot 1," was exploded in the Bikini atoll with the power of about twenty million tons of TNT. The uranium coating caused radioactive fallout to spray over a wide area. A crew of Japanese fishermen on the *Lucky Dragon*, 120 miles from the explosion, became sick and had to be hospitalized; one of them died.

## Consequences

Senator Joseph McCarthy had been involving many people in his obsession with rooting out Communist sympathizers from American government. Many Americans were outraged that the Soviet Union had exploded a hydrogen bomb before the United States managed to do so. Looking for a scapegoat, security officials accused Oppenheimer of having opposed the development of the hydrogen bomb.

A three-man Personnel Security Board judged Oppenheimer's case: Thomas A. Morgan, an industrialist, and Gordon Grey, a newspaper and broadcasting executive, voted against him, while Ward Evans, a professor of chemistry, voted in his favor. In 1954, Oppenheimer was denied security clearance and lost his influence over nuclear science in the United States. Teller, Lawrence, and Strauss became the new scientific advisers for the Cold War.

*Elizabeth Fee*

# Lipmann Discovers Acetyl Coenzyme A

*Fritz Albert Lipmann discovered the molecule acetyl coenzyme A, the link between the two most important energy cycles in living cells.*

**What:** Biology
**When:** 1951
**Where:** Boston, Massachusetts
**Who:**

FRITZ ALBERT LIPMANN (1899-1986), a German American biochemist who won the 1953 Nobel Prize in Physiology or Medicine

SIR HANS ADOLF KREBS (1900-1981), a German biochemist who won the 1953 Nobel Prize in Physiology or Medicine

OTTO HEINRICH WARBURG (1883-1970), a German biochemist who won the 1931 Nobel Prize in Physiology or Medicine

OTTO MEYERHOF (1884-1951), a German biochemist and physiologist

## The Life Cycle

All life on Earth is sustained by the constant recycling of matter and energy through the earth's environment. Plants and phytoplankton absorb light from the sun to produce their own food (sugar) from sunlight, water, and carbon dioxide. Organisms that cannot make their own food must scavenge for food; these include animals, protozoa, and fungi. These organisms must obtain food either by eating other organisms or by decomposing dead organisms into their component molecules, which are returned to the soil to be recycled into new plants. The energy cycle starts all over again, the entire purpose being to provide energy for two very important life chemical reaction pathways: glycolysis and the citric acid cycle. (A pathway is the sequence of intracellular chemical reactions by which a substance is made to yield life-sustaining energy.)

During the 1920's and 1930's, German scientists Gustav Embden, Otto Meyerhof, and Otto Heinrich Warburg solved the complete enzymatic pathway for the breakdown of glucose, a series of chemical reactions called "glycolysis." The purpose of glycolysis in living cells is the production of adenosine triphosphate (ATP), a high-energy molecule, from adenosine diphosphate (ADP), a low-energy molecule.

Pyruvic acid, the end product of glycolysis, can be converted into one of three possible molecules: ethyl alcohol, lactic acid, or acetyl coenzyme A. Acetyl coenzyme A was discovered by one of Meyerhof's students, Fritz Albert Lipmann. Lipmann, a native of Königsberg, Germany, began his biochemical research in Meyerhof's laboratory at the University of Heidelberg in 1927. He played an important supporting role in Meyerhof's discovery of glycolysis.

In 1941, Lipmann became director of the Biochemistry Research Department at Massachusetts General Hospital in Boston. He continued to study the way that living organisms, particularly animals, generate energy. Animal ATP production was much higher than could be explained simply by glycolysis. Seeking other enzymatic pathways for ATP production, Lipmann and other biochemists attempted to derive an enzymatic pathway to explain ATP production in the presence of oxygen.

Among the key scientists in this research effort were Lipmann, Sir Hans Adolf Krebs, Warburg, Albert Szent-Györgyi, Franz Knoop, and Carl Martius. Krebs had discovered several of the intermediate molecules (for example, citric acid, succinate, fumarate) for aerobic metabolism in the early 1930's. He had discovered these molecules because of their high reactivity with oxygen in muscles. Szent-Györgyi had obtained similar results. After these and other discoveries, Krebs pieced together the various intermediates of aerobic ATP production. By 1937, he had deciphered the enzymatic pathway known as the "citric acid cycle," also called the "Krebs cycle" or the "tricarboxylic acid (TCA) cycle."

## The Intermediary Molecule

In 1945, Lipmann discovered a molecule that accelerated aerobic respiration without being part of Krebs's famous citric acid cycle. By 1947, he had isolated the molecule and determined its structure. The molecule, which he named "coenzyme A," is a large molecule. Coenzyme A is important because it is highly reactive, particularly with acetic acid.

In 1951, Lipmann and his colleagues demonstrated that coenzyme A can chemically react with pyruvic acid, the end product of anaerobic glycolysis, to produce acetyl coenzyme A (coenzyme A plus acetic acid). Acetyl coenzyme A dumps the acetate into the Krebs cycle, thereby leaving coenzyme A free to react with another pyruvic acid. Therefore, coenzyme A is the intermediate between the two most important energy-generating reaction pathways in the cells of living organisms: glycolysis and the Krebs cycle.

Once acetate enters the Krebs cycle, it combines with oxaloacetate to produce citric acid. Citric acid in turn undergoes a series of intermediate conversions, which end with the production of more oxaloacetate. Oxaloacetate picks up an acetate from coenzyme A to produce citric acid and thus the Krebs cycle begins over again. Each time through the cycle, more life-sustaining ATP molecules are generated.

## Consequences

Lipmann's discovery of coenzyme A, his determination of its structure, and his deciphering of its role as acetyl coenzyme A in cellular respiration represent a great breakthrough in twentieth century biochemistry. The discovery of coenzyme A linked the anaerobic and aerobic phases of cellular respiration, helped to determine the amount of energy production for each glucose molecule an organism consumes, and helped to link cellular respiration to other metabolic reaction pathways (for example, carbohydrate and protein metabolism). Lipmann and Krebs shared the 1953 Nobel Prize in Physiology or Medicine for their contributions to the understanding of energy metabolism in living organisms.

Cellular respiration has been the core of modern biochemistry since the field was established by Krebs, Lipmann, Meyerhof, and others. The central focus of every introductory biochemistry course is cellular respiration. Lipmann's contribution is of great value because it helped to resolve how cells stay alive.

*David Wason Hollar, Jr.*

# UNIVAC I Computer Is First to Use Magnetic Tape

*John Presper Eckert and John W. Mauchly became the first to exploit the commercial applications of computers with the introduction of the UNIVAC I.*

**What:** Computer science
**When:** 1951
**Where:** Philadelphia, Pennsylvania
**Who:**
JOHN PRESPER ECKERT (1919-1995), an American electrical engineer
JOHN W. MAUCHLY (1907-1980), an American physicist
JOHN VON NEUMANN (1903-1957), a Hungarian American mathematician
HOWARD H. AIKEN (1900-1973), an American physicist
GEORGE STIBITZ (1904-1995), a scientist at Bell Labs

## The Origins of Computing

On March 31, 1951, the U.S. Census Bureau accepted delivery of the first Universal Automatic Computer (UNIVAC). This powerful electronic computer, far surpassing anything then available in technological features and capability, ushered in the first computer generation and pioneered the commercialization of what had previously been the domain of academia and the interest of the military. The fanfare that surrounded this historic occasion, however, masked the turbulence of the previous five years for the young upstart Eckert-Mauchly Computer Corporation (EMCC), which by this time was a wholly owned subsidiary of Remington Rand Corporation.

John Presper Eckert and John W. Mauchly met in the summer of 1941 at the University of Pennsylvania. A short time later, Mauchly, then a physics professor at Ursinus College, joined the Moore School of Engineering at the University of Pennsylvania and embarked on a crusade to con-vince others of the feasibility of creating electronic digital computers. Up to this time, the only computers available were called "differential analyzers," which were used to solve complex mathematical equations known as "differential equations." These slow machines were good only for solving a relatively narrow range of mathematical problems.

Eckert and Mauchly landed a contract that eventually resulted in the development and construction of the world's first operational general-purpose electronic computer, the Electronic Numerical Integrator and Computer (ENIAC). This computer, used eventually by the Army for the calculation of ballistics tables, was deficient in many obvious areas, but this was caused by economic rather than engineering constraints. One major deficiency was the lack of automatic program control; the ENIAC did not have stored program memory. This was addressed in the development of the Electronic Discrete Variable Automatic Computer (EDVAC), the successor to the ENIAC.

## Fighting the Establishment

A symbiotic relationship had developed between Eckert and Mauchly that worked to their advantage on technical matters. They worked well with each other, and this contributed to their success in spite of external obstacles. They both were interested in the commercial applications of computers and envisioned uses for these machines far beyond the narrow applications required by the military.

This interest brought them into conflict with the administration at the Moore School of Engineering as well as with the noted mathematician John von Neumann, who "joined" the ENIAC/EDVAC development team in 1945. Von Neu-

**1135**

mann made significant contributions and added credibility to the Moore School group, which often had to fight against the conservative scientific establishment characterized by Howard H. Aiken at Harvard University and George Stibitz at Bell Labs. Philosophical differences between von Neumann and Eckert and Mauchly, as well as patent issue disputes with the Moore School administration, eventually caused the resignation of Eckert and Mauchly on March 31, 1946.

Eckert and Mauchly, along with some of their engineering colleagues at the University of Pennsylvania, formed the Electronic Control Company and proceeded to interest potential customers (including the Census Bureau) in an "EDVAC-type" machine. On May 24, 1947, the EDVAC-type machine became the UNIVAC. This new computer would overcome the shortcomings of the ENIAC and the EDVAC (which was eventually completed by the Moore School in 1951). It would be a stored-program computer and would allow input to and output from the computer via magnetic tape. The prior method of input/output used punched paper cards that were extremely slow compared to the speed at which data in the computer could be processed.

A series of poor business decisions and other unfortunate circumstances forced the newly renamed Eckert-Mauchly Computer Corporation to look for a buyer. They found one in Remington Rand in 1950. Remington Rand built tabulating equipment and was a competitor of International Business Machines Corporation (IBM). IBM was approached about buying EMCC, but the negotiations fell apart. EMCC became a division of Remington Rand and had access to the resources necessary to finish the UNIVAC.

## Consequences

Eckert and Mauchly made a significant contribution to the advent of the computer age with the introduction of the UNIVAC I. The words "computer" and "UNIVAC" entered the popular vocabulary as synonyms. The efforts of these two visionaries were rewarded quickly as contracts started to pour in, taking IBM by surprise and propelling the inventors into the national spotlight.

This spotlight shone brightest, perhaps, on the eve of the national presidential election of 1952, which pitted war hero General Dwight D. Eisenhower against statesman Adlai Stevenson. At the suggestion of Remington Rand, CBS was invited to use UNIVAC to predict the outcome of the election. Millions of television viewers watched as CBS anchor Walter Cronkite "asked" UNIVAC for its predictions. A program had been written to analyze the results of thousands of voting districts in the elections of 1944 and 1948. Based on only 7 percent of the votes coming in, UNIVAC had Eisenhower winning by a landslide, in contrast with all the prior human forecasts of a close election.

Surprised by this answer and not willing to suffer the embarrassment of being wrong, the programmers quickly directed the program to provide an answer that was closer to the perceived situation. The outcome of the election, however, matched UNIVAC's original answer. This prompted CBS commentator Edward R. Murrow's famous quote, "The trouble with machines is people."

The development of the UNIVAC I produced many technical innovations. Primary among these is the use of magnetic tape for input and output. All machines that preceded the UNIVAC (with one exception) used either paper tape or cards for input and cards for output. These methods were very slow and created a bottleneck of information. The great advantage of magnetic tape was the ability to store the equivalent of thousands of cards of data on one 30-centimeter reel of tape. Another advantage was its speed.

*Paul G. Nyce*

# Watson and Crick Develop Double-Helix Model of DNA

> *By showing precisely the double-helix configuration of DNA, James D. Watson and Francis Crick essentially discovered the chemical nature of the gene.*

**What:** Biology; Genetics
**When:** 1951-1953
**Where:** Cambridge and London, England
**Who:**

JAMES D. WATSON (1928-    ), an American molecular biologist who shared, with Crick and Wilkins, the 1962 Nobel Prize in Physiology or Medicine

FRANCIS CRICK (1916-    ), an English physicist

MAURICE H. F. WILKINS (1916-    ), an English biophysicist

ROSALIND FRANKLIN (1920-1958), an English physical chemist and X-ray crystallographer

LINUS PAULING (1901-1994), an American theoretical physical chemist

JERRY DONOHUE (1920-1985), an American crystallographer

## An Odd Team

In 1944, the American bacteriologist Oswald T. Avery and his colleagues published a paper demonstrating that deoxyribonucleic acid (DNA) is the carrier of genetic information. Largely through the work of the English organic chemist Alexander Todd, DNA's construction from nucleotide building blocks was understood in terms of how all its atoms are linked together. Each nucleotide consists of one of four bases (adenine, thymine, guanine, or cytosine).

While he was working out the detailed bonding of the nucleic acids, Erwin Chargaff, a biochemist at the College of Physicians and Surgeons in New York, was investigating the differences in the base compositions of DNA from various plants and animals. By 1950, his careful analyses had revealed that the amounts of various bases in different DNA varied widely, but his data also yielded the significant result that the ratios of bases adenine to thymine and guanine to cytosine were always close to one.

Linus Pauling, along with many other American chemists, was slow to accept DNA as the genetic material. For a long time it seemed to him that the molecule was too simple to handle the formidable task of transferring huge amounts of information from one generation of living things to the next. Proteins, with their large variety of amino acids, seemed much better adapted to this information-carrying role.

Nevertheless, while he was a visiting professor at the University of Oxford in 1948, Pauling discovered one of the basic structures of proteins: the alpha-helix. This three-dimensional structure, which was based on the planarity of the group of atoms holding the amino acids together (the peptide bond), was secured in its twisting turns by hydrogen bonds (links whereby a hydrogen atom serves as a bridge between certain neighboring atoms).

At the Cavendish Laboratory at the University of Cambridge, in the early 1950's, James D. Watson, a young postdoctoral fellow, and Francis Crick, a graduate student, were trying to work out the structure of DNA. They were deeply impressed by Pauling's work. Crick believed that Pauling's method would allow biologists to build accurate three-dimensional models of the complex molecules in living things, particularly DNA.

At first, Watson and Crick seemed an odd team for this task. Neither had much knowledge of chemistry. Crick's previous training was in

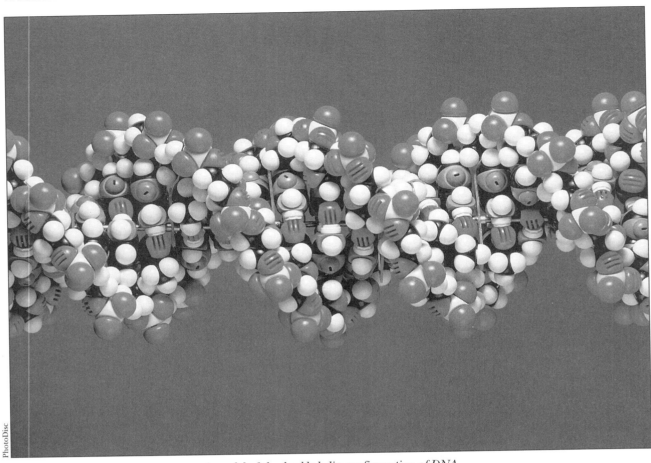

PhotoDisc

*A model of the double-helix configuration of DNA.*

physics and work on mines during World War II (1939-1945). Watson's education in ornithology and zoology did not prepare him for dealing with the daunting complexities of the DNA molecule.

## X Rays and Cardboard Models

At King's College of the University of London, Maurice H. F. Wilkins and Rosalind Franklin were engaged in an experimental approach to the same problem: They were trying to determine DNA's structure through improved X-ray diffraction photographs of a carefully prepared DNA sample. They were able to show that DNA exists in two forms, which they called the "B form" and the "A form."

Chargaff visited Cambridge in 1952. After talking with Watson and Crick about DNA, he was shocked to learn that neither of them had precise knowledge about the chemical differences among the four bases. He explained to them his findings about the base ratios, but he

did not see them as skilled chemists who could solve the problem of DNA's three-dimensional structure. Although Crick had not been aware of Chargaff's discovery, he had been thinking about piling complementary bases on top of one another in a DNA model, so he was fascinated by Chargaff's information. By pairing adenine with thymine, and guanine with cytosine, a rational explanation for the ratios emerged.

Wilkins showed Watson the X-ray picture of the B form, which clearly revealed the presence of a helix. On his return to Cambridge, Watson began building models of DNA. Like Pauling, he began arranging the phosphoric-acid groups in the center.

Fortunately, Jerry Donohue, an American crystallographer and protégé of Pauling, was sharing an office with Watson. He saw from Watson's cardboard cutouts that Watson did not understand the proper chemical structures of the four bases. When Watson—following Donohue's advice—put the bases into their correct forms, he was able to pair adenine and thymine as well as guanine and cytosine by means of hydrogen bonds whose locations were natural, not forced. Watson immediately sensed that something was right about these pairings, since both pairs (adenine-thymine, guanine-cytosine) had nearly the same size and shape and could be neatly stacked, like a pile of plates, in the interior of a double helix, while the regular sugar-phosphate backbone at the molecule's exterior could account for its acidity and its interactions with water. It turned out that this pairing of bases is the pivotal feature of DNA's structure and the reason for its complementary nature.

In March of 1953, Watson and Crick constructed a detailed model of their double helix to show that all the atoms were in sensible locations. They published their model in *Nature* on April 25, 1953. Their brief paper, which quickly achieved classic status, succinctly described their discovery and noted its implications, especially as a mechanism for transferring genetic material.

## Consequences

Scholars have compared the discovery of the double helix with Charles Darwin's discovery of natural selection and Gregor Mendel's discovery of the laws of heredity. The importance of the double helix also can be measured by the prolif-

eration of significant discoveries in molecular biology that followed it. Pauling stated that the double helix was "the most important discovery in the field of biology that has been made in the last hundred years." He saw it as a culmination of molecular biology, since no problem is more fundamental than the mechanism of heredity. The double helix did not disappoint scientists eager to understand this mechanism. In the years after the model was unveiled, it proved surpassingly suitable for explaining molecular details about how cells replicate.

The most obvious feature of the model was its natural explanation of how DNA could make an exact copy of itself. In this process of replication, DNA's two complementary strands "unzip" and separate, and each half takes on a new "partner." Thus, both halves of the separated double helix act as the templates or molds on which complementary strands are then synthesized. When the process is completed, two identical double helices appear where one previously existed. This explanation of how, at the molecular level, genes can duplicate themselves exactly would eventually lead to our current understanding of many things, from genetic diseases to genetic "engineering": the ability to clone and genetically manipulate living organisms.

*Robert J. Paradowski*

# First Breeder Reactor Begins Operation

*Engineers and physicists at the Idaho National Engineering Laboratory successfully produced electricity from nuclear fission and in the process generated new fuel.*

**What:** Energy
**When:** December 20, 1951
**Where:** Arco, Idaho
**Who:**
WALTER HENRY ZINN (1906-    ), the first director of the Argonne National Laboratory

## Producing Electricity with More Fuel

The discovery of nuclear fission involved both the discovery that the nucleus of a uranium atom would split into two lighter elements when struck by a neutron and the observation that additional neutrons, along with a significant amount of energy, were released at the same time. These neutrons might strike other atoms and cause them to fission (split) also. That, in turn, would release more energy and more neutrons, triggering a chain reaction as the process continued to repeat itself, yielding a continuing supply of heat.

Besides the possibility that an explosive weapon could be constructed, early speculation about nuclear fission included its use in the generation of electricity. The occurrence of World War II (1939-1945) meant that the explosive weapon would be developed first.

Both the weapons technology and the basic physics for the electrical reactor had their beginnings in Chicago with the world's first nuclear chain reaction. The first self-sustaining nuclear chain reaction occurred in a laboratory at the University of Chicago on December 2, 1942.

It also became apparent at that time that there was more than one way to build a bomb. At this point, two paths were taken: One was to build an atomic bomb with enough fissionable uranium in it to explode when detonated, and another was to generate fissionable plutonium and build a bomb. Energy was released in both methods, but the second method also produced another fissionable substance.

The observation that plutonium and energy could be produced together meant that it would be possible to design electric power systems that would produce fissionable plutonium in quantities as large as, or larger than, the amount of fissionable material consumed. This is the breeder concept, the idea that while using up fissionable uranium 235, another fissionable element can be made. The full development of this concept for electric power was delayed until the end of World War II.

## Electricity from Atomic Energy

On August 1, 1946, the Atomic Energy Commission (AEC) was established to control the development and explore the peaceful uses of nuclear energy. The Argonne National Laboratory was assigned the major responsibilities for pioneering breeder reactor technologies. Walter Henry Zinn was the laboratory's first director. He led a team that planned a modest facility (Experimental Breeder Reactor I, or EBR-I) for testing the validity of the breeding principle.

Planning for this had begun in late 1944 and grew as a natural extension of the physics that developed the plutonium atomic bomb. The conceptual design details for a breeder-electric reactor were reasonably complete by late 1945. On March 1, 1949, the AEC announced the selection of a site in Idaho for the National Reactor Station (later to be named the Idaho National Engineering Laboratory, or INEL). Construction at the INEL site in Arco, Idaho, began in October, 1949. Critical mass was reached in August, 1951. ("Critical mass" is the amount and concentration of fissionable material required to produce a self-sustaining chain reaction.)

The system was brought to full operating power, 1.1 megawatts of thermal power, on De-

**1140**

cember 19, 1951. The next day, December 20, at 11:00 A.M., steam was directed to a turbine generator. At 1:23 P.M., the generator was connected to the electrical grid at the site, and "electricity flowed from atomic energy," in the words of Zinn's console log of that day. Approximately 200 kilowatts of electric power were generated most of the time that the reactor was run. This was enough to satisfy the needs of the EBR-I facilities. The reactor was shut down in 1964 after five years of use primarily as a test facility. It had also produced the first pure plutonium.

With the first fuel loading, a conversion ratio of 1.01 was achieved, meaning that more new fuel was generated than was consumed by about 1 percent. When later fuel loadings were made with plutonium, the conversion ratios were more favorable, reaching as high as 1.27. EBR-I was the first reactor to generate its own fuel and the first power reactor to use plutonium for fuel.

The use of EBR-I also included pioneering work on fuel recovery and reprocessing. During its five-year lifetime, EBR-I operated with four different fuel loadings, each designed to establish specific benchmarks of breeder technology. This reactor was seen as the first in a series of increasingly larger reactors in a program designed to develop breeder technology. The reactor was replaced by EBR-II, which had been proposed in 1953 and was constructed from 1955 to 1964.

EBR-II was capable of producing 20 megawatts of electrical power. It was approximately fifty times more powerful than EBR-I but still small compared to light-water commercial reactors of 600 to 1,100 megawatts in use toward the end of the twentieth century.

## Consequences

The potential for peaceful uses of nuclear fission were dramatized with the start-up of EBR-I in 1951: It was the first in the world to produce electricity, while also being the pioneer in a breeder reactor program. The breeder program was not the only reactor program being developed, however, and it eventually gave way to the light-water reactor design for use in the United States. Still, if energy resources fall into short supply, it is likely that the technologies first developed with EBR-I will find new importance. In France and Japan, commercial reactors make use of breeder reactor technology; these reactors require extensive fuel reprocessing.

Following the completion of tests with plutonium loading in 1964, EBR-I was shut down and placed in standby status. In 1966, it was declared a national historical landmark under the stewardship of the U.S. Department of the Interior. The facility was opened to the public in June, 1975.

*Donald H. Williams*

# Bell Labs Introduces Electronically Amplified Hearing Aids

*The application of electronic amplifiers to hearing aid technology made possible the miniaturization of hearing aids and permitted some hearing-impaired persons to participate more fully in mainstream society.*

**What:** Communications
**When:** 1952
**Where:** United States
**Who:**
BELL LABS, the research and development arm of the American Telephone and Telegraph Company

## Trapped in Silence

Until the middle of the twentieth century, people who experienced hearing loss had little hope of being able to hear sounds without the use of large, awkward, heavy appliances. For many years, the only hearing aids available were devices known as ear trumpets. The ear trumpet tried to compensate for hearing loss by increasing the number of sound waves funneled into the ear canal. A wide, bell-like mouth similar to the bell of a musical trumpet narrowed to a tube that the user placed in his or her ear. Ear trumpets helped a little, but they could not truly increase the volume of the sounds heard.

Beginning in the nineteenth century, inventors tried to develop electrical devices that would serve as hearing aids. The telephone was actually a by-product of Alexander Graham Bell's efforts to make a hearing aid. Following the invention of the telephone, electrical engineers designed hearing aids that employed telephone technology, but those hearing aids were only a slight improvement over the old ear trumpets. They required large, heavy battery packs and used a carbon microphone similar to the receiver in a telephone. More sensitive than purely physical devices such as the ear trumpet, they could transmit a wider range of sounds but could not amplify them as effectively as electronic hearing aids now do.

## Transistors Make Miniaturization Possible

Two types of hearing aids exist: body worn and head worn. Body-worn hearing aids permit the widest range of sounds to be heard, but because of the devices' larger size, many hearing-impaired persons do not like to wear them. Head-worn hearing aids, especially those worn completely in the ear, are much less conspicuous. In addition to in-ear aids, the category of head-worn hearing aids includes both hearing aids mounted in eyeglass frames and those worn behind the ear.

All hearing aids, whether head worn or body worn, consist of four parts: a microphone to pick up sounds, an amplifier, a receiver, and a power source. The microphone gathers sound waves and converts them to electrical signals; the amplifier boosts, or increases, those signals; and the receiver then converts the signals back into sound waves. In effect, the hearing aid is a miniature radio. After the receiver converts the signals back to sound waves, those waves are directed into the ear canal through an earpiece or ear mold. The ear mold generally is made of plastic and is custom fitted from an impression taken from the prospective user's ear.

Effective head-worn hearing aids could not be built until the electronic circuit was developed in the early 1950's. The same invention—the transistor—that led to small portable radios and tape players allowed engineers to create miniaturized, inconspicuous hearing aids. Depending on the degree of amplification required, the amplifier in a hearing aid contains three or more transistors. Transistors first replaced vacuum tubes in devices such as radios and phonographs, and then engineers realized that they could be used in devices for the hearing impaired.

The research at Bell Labs that led to the invention of the transistor rose out of military research during World War II. The vacuum tubes used in, for example, radar installations to amplify the strength of electronic signals were big, were fragile because they were made of blown glass, and gave off high levels of heat when they were used. Transistors, however, made it possible to build solid-state, integrated circuits. These are made from crystals of metals such as germanium or arsenic alloys and therefore are much less fragile than glass. They are also extremely small (in fact, some integrated circuits are barely visible to the naked eye) and give off no heat during use.

The number of transistors in a hearing aid varies depending upon the amount of amplification required. The first transistor is the most important for the listener in terms of the quality of sound heard. If the frequency response is set too high—that is, if the device is too sensitive—the listener will be bothered by distracting background noise. Theoretically, there is no limit on the amount of amplification that a hearing aid can be designed to provide, but there are practical limits. The higher the amplification, the more power is required to operate the hearing aid. This is why body-worn hearing aids can convey a wider range of sounds than head-worn devices can. It is the power source—not the electronic components—that is the limiting factor. A body-worn hearing aid includes a larger battery pack than can be used with a head-worn device. Indeed, despite advances in battery technology, the power requirements of a head-worn hearing aid are such that a 1.4-volt battery that could power a wristwatch for several years will last only a few days in a hearing aid.

## Consequences

The invention of the electronic hearing aid made it possible for many hearing-impaired persons to participate in a hearing world. Prior to the invention of the hearing aid, hearing-impaired children often were unable to participate in routine school activities or function effectively in mainstream society. Instead of being able to live at home with their families and enjoy the same experiences that were available to other children their age, often they were forced to attend special schools operated by the state or by charities.

Hearing-impaired people were singled out as being different and were limited in their choice of occupations. Although not every hearing-impaired person can be helped to hear with a hearing aid—particularly in cases of total hearing loss—the electronic hearing aid has ended restrictions for many hearing-impaired people. Hearing-impaired children are now included in public school classes, and hearing-impaired adults can now pursue occupations from which they were once excluded.

At the beginning of the twenty-first century, many deaf and hearing-impaired persons chose to live without the help of hearing aids. They believe that they are not disabled but simply different, and they point out that their "disability" often allows them to appreciate and participate in life in unique and positive ways. For them, the use of hearing aids is a choice, not a necessity. For those who choose, hearing aids make it possible to participate in the hearing world.

*Nancy Farm Mannikko*

# Wilkins Introduces Reserpine to Treat High Blood Pressure

> *Wilkins studied reserpine's unique hypertension-decreasing effects, providing clinical medicine with a versatile and effective tool.*

**What:** Medicine
**When:** 1952
**Where:** Boston, Massachusetts
**Who:**
ROBERT WALLACE WILKINS (1906-     ), an American physician and clinical researcher
WALTER E. JUDSON (1916-     ), an American clinical researcher

## Treating Hypertension

Excessively elevated blood pressure, clinically known as "hypertension," has long been recognized as a pervasive and serious human malady. In a few cases, hypertension is recognized as an effect brought about by particular pathologies (diseases or disorders). Often, however, hypertension occurs as the result of unknown causes. Despite the uncertainty about its origins, unattended hypertension leads to potentially dramatic health problems, including increased risk of kidney disease, heart disease, and stroke.

Recognizing the need to treat hypertension in a relatively straightforward and effective way, Robert Wallace Wilkins, a clinical researcher at Boston University's School of Medicine and the head of Massachusetts Memorial Hospital's Hypertension Clinic, began to experiment with reserpine in the early 1950's. Initially, the samples that were made available to Wilkins were crude and unpurified. Eventually, however, a purified version was used.

Reserpine has a long and fascinating history of use—both clinically and in folk medicine—in India. The source of reserpine is the root of the shrub *Rauwolfia serpentina*, first mentioned in Western medical literature in the 1500's but virtually unknown, or at least unaccepted, outside India until the mid-twentieth century. Crude preparations of the shrub had been used for a variety of ailments in India for centuries prior to its use in the West.

Wilkins's work with the drug did not begin on an encouraging note, because reserpine does not act rapidly—a fact that had been noted in Indian medical literature. The standard observation in Western pharmacotherapy, however, was that most drugs work rapidly; if a week has elapsed without positive effects being shown by a drug, the conventional Western wisdom is that it is unlikely to work at all. Additionally, physicians and patients alike tend to look for rapid improvement or at least positive indications. Reserpine is deceptive in this temporal context, and Wilkins and his coworkers were nearly deceived. In working with crude preparations of *Rauwolfia serpentina*, they were becoming very pessimistic, when a patient who had been treated for many consecutive days began to show symptomatic relief. Nevertheless, only after months of treatment did Wilkins become a believer in the drug's beneficial effects.

## The Action of Reserpine

When preparations of pure reserpine became available in 1952, the drug did not at first appear to be the active ingredient in the crude preparations. When patients' heart rate and blood pressure began to drop after weeks of treatment, however, the investigators saw that reserpine was indeed responsible for the improvements.

Once reserpine's activity began, Wilkins observed a number of important and unique consequences. Both the crude preparations and pure reserpine significantly reduced the two most meaningful measures of blood pressure. These two measures are systolic blood pressure and dia-

**1144**

stolic blood pressure. Systolic pressure represents the peak of pressure produced in the arteries following a contraction of the heart. Diastolic pressure is the low point that occurs when the heart is resting. To lower the mean blood pressure in the system significantly, both of these pressures must be reduced. The administration of low doses of reserpine produced an average drop in pressure of about 15 percent, a figure that was considered less than dramatic but still highly significant. The complex phenomenon of blood pressure is determined by a multitude of factors, including the resistance of the arteries, the force of contraction of the heart, and the heartbeat rate. In addition to lowering the blood pressure, reserpine reduced the heartbeat rate by about 15 percent, providing an important auxiliary action.

In the early 1950's, various therapeutic drugs were used to treat hypertension. Wilkins recognized that reserpine's major contribution would be as a drug that could be used in combination with drugs that were already in use. His studies established that reserpine, combined with at least one of the drugs already in use, produced an additive effect in lowering blood pressure. Indeed, at times, the drug combinations produced a "synergistic effect," which means that the combination of drugs created an effect that was more effective than the sum of the effects of the drugs when they were administered alone. Wilkins also discovered that reserpine was most effective when administered in low dosages. Increasing the dosage did not increase the drug's effect significantly, but it did increase the likelihood of unwanted side effects. This fact meant that reserpine was indeed most effective when administered in low dosages along with other drugs.

Wilkins believed that reserpine's most unique effects were not those found directly in the cardiovascular system but those produced indirectly by the brain. Hypertension is often accompanied by neurotic anxiety, which is both a consequence of the justifiable fears of future negative health changes brought on by prolonged hypertension and contributory to the hypertension itself. Wilkins's patients invariably felt better mentally, were less anxious, and were sedated, but in an unusual way. Reserpine made patients drowsy but did not generally cause sleep, and if sleep did occur, patients could be awakened easily. Such effects are now recognized as characteristic of tranquilizing drugs, or antipsychotics. In effect, Wilkins had discovered a new and important category of drugs: tranquilizers.

*Robert Wallace Wilkins.*

## Consequences

Reserpine holds a vital position in the historical development of antihypertensive drugs for two reasons. First, it was the first drug that was discovered to block activity in areas of the nervous system that use norepinephrine or its close relative dopamine as transmitter substances. Second, it was the first hypertension drug to be widely accepted and used. Its unusual combination of characteristics made it effective in most patients.

Since the 1950's, medical science has rigorously examined cardiovascular functioning and diseases such as hypertension. Many new factors, such as diet and stress, have been recognized as factors in hypertension. Controlling diet and lifestyle help tremendously in treating hypertension, but if the nervous system could not be partially controlled, many cases of hypertension would continue to be problematic. Reserpine has made that control possible.

*Keith Krom Parker*

# Aserinsky Discovers REM Sleep

*Eugene Aserinsky's discovery of rapid eye movements (REMs) in normal human sleep provided the first objective method of studying neural function and behavioral patterns associated with dreaming.*

**What:** Medicine
**When:** 1952
**Where:** Chicago, Illinois
**Who:**
EUGENE ASERINSKY (1921-    ), a graduate student of physiology at the University of Chicago
NATHANIEL KLEITMAN (1895-1999), a professor of physiology at the University of Chicago
WILLIAM DEMENT (1928-    ), an American physiologist

## Laboratory Dreaming

As early as 1867, German psychiatrist Wilhelm Griesinger speculated on the occurrence of eye movements during dreams. These eye movements, he believed, occurred both during the transition from wakefulness to sleep and during dreaming. From these observations, he concluded that sleep was not a passive but rather an active state. It was another eighty-five years before Eugene Aserinsky discovered that sleep is not a homogeneous process but is organized in rhythmic cycles of different stages.

In 1952, Aserinsky, a graduate student working on his dissertation in the physiology laboratory of Nathaniel Kleitman at the University of Chicago, achieved a breakthrough in modern sleep research. Aserinsky turned his focus to the study of attention in children, using his young son, Armond, as one of his subjects. While making clinical observations of his young subject's efforts to pay attention, he noticed that eye closure was associated with attention lapse, and thus decided to record these eyelid movements using the electrooculogram (EOG).

Aserinsky and Kleitman observed that a series of bursts of rapid eye movements (REMs) oc-curred about four to six times during the night. The first such REM period took place about an hour after the onset of sleep and lasted from five to ten minutes. Succeeding REM periods occurred at intervals of about ninety minutes each and lasted progressively longer; the final period occupied approximately thirty minutes.

Suspecting a correlation of eye movements with dreaming, Aserinsky and Kleitman awakened subjects during REM periods and asked them whether they had been dreaming. In a large majority of such awakenings, the subjects acknowledged that they had been dreaming and proceeded to relate their dreams. When subjects were awakened while their eyes were motionless, they could rarely remember a dream. Therefore, Aserinsky and Kleitman concluded that rapid eye movements were an objective signal of dreaming. Although investigators still had to rely upon the dreamer's verbal report to ascertain the content of the dream, the process of dreaming was now opened up to objective study under laboratory conditions.

## The Sleep of Cats and Children

In order to obtain a more complete picture of the mental state of his subjects, Aserinsky also recorded brain-wave activity with an electroencephalograph (EEG). Using both the EEG and the EOG enabled Aserinsky to register brain-wave activity during sleep from the moment it began, regardless of the time of day. This combination proved fortuitous because, unlike adults, children often enter the REM phase immediately at sleep onset, and such sleep-onset REM periods are especially likely to occur during daytime naps.

When Aserinsky's subjects lost attentional focus and fell asleep, their EEGs showed an activation pattern, and their EOGs showed rapid eye movements. Kleitman quickly deduced that this

**1146**

brain-activated sleep state, with its rapid eye movements, might be associated with dreaming. The two investigators immediately applied the combined EEG and EOG measures to the sleep of adult humans and were able to observe the periodic alternations of REM and non-REM sleep throughout the night. In addition, when the investigators awakened their subjects during REM sleep, these subjects related accounts of dreams.

In 1953, Aserinsky and Kleitman reported their findings in the journal *Science* in an article titled "Regularly Occurring Periods of Eye Motility and Concomitant Phenomena During Sleep." As is the case with many breakthrough articles, this one was relatively brief (barely two pages). Yet it included the observation that during REM sleep, other physiological functions vary according to the state of the brain: Respiratory frequency and heart rate fluctuate, and their rhythm becomes irregular.

Physiologist William Dement later confirmed Aserinsky and Kleitman's hypothesis. Following Aserinsky and Kleitman's groundbreaking 1953 article and their recognition of REM as the physiological basis of dreaming, Dement established that an identical phase of sleep occurs in cats; he published his results in the 1958 *EEG Journal*.

## Consequences

Studies that have attempted to show a relationship between the subject matter of dreams and the physiological changes that occur during REM periods have not established any close correlation between the two phenomena. Although early investigations indicated that the pattern of eye movements is correlated with the directions in which the dreamer is looking in the dream, subsequent evidence raised doubts concerning this hypothesis.

More conclusive evidence exists to support the theory that dreaming can sometimes occur during non-REM periods. This possibility suggests that dreaming may be more or less continuous during sleep but that conditions for the recall of dreams are most favorable following REM awakenings. In any case, the prevailing view is that REMs are not an objective sign of all dreaming but that they do indicate when a dream is most likely to be recalled.

Aserinsky's discovery of a stage of sleep during which most dreaming seems to occur led to experiments to investigate what would happen if a sleeping person were deprived of REM sleep. These studies concluded that there is an overwhelming demand for REM sleep. Because REM sleep usually accompanies dreaming, it was also concluded from these studies that there is a strong need to dream. Later studies with prolonged deprivation of REM sleep, however, did not confirm the degree of behavioral changes noted earlier. Thus, it can be concluded that there is definitely a need for REM sleep, but the question of whether there is also a need to dream is still open to debate.

By using an EEG to monitor sleep during the night and by awakening subjects during REM periods, it has been conclusively established that everyone normally dreams every night. Even a person who has never remembered a dream in his or her life will typically do so if awakened during a REM period.

*Genevieve Slomski*

# Müller Invents Field Ion Microscope

*Erwin Wilhelm Müller achieved atomic resolution with the field ion microscope, an improved version of his field emission microscope.*

**What:** Physics
**When:** 1952-1956
**Where:** University Park, Pennsylvania
**Who:**
ERWIN WILHELM MÜLLER (1911-1977), a physicist, engineer, and research professor
J. ROBERT OPPENHEIMER (1904-1967), an American physicist

## To See Beneath the Surface

In the early twentieth century, developments in physics, especially quantum mechanics, paved the way for the application of new theoretical and experimental knowledge to the problem of viewing the atomic structure of metal surfaces. Of primary importance were American physicist George Gamow's 1928 theoretical explanation of the field emission of electrons by quantum mechanical means and J. Robert Oppenheimer's 1928 prediction of the quantum mechanical ionization of hydrogen in a strong electric field.

In 1936, Erwin Wilhelm Müller developed his field emission microscope, the first in a series of instruments that would exploit these developments. It was to be the first instrument to view atomic structures—although not the individual atoms themselves—directly. Müller's subsequent field ion microscope utilized the same basic concepts used in the field emission microscope yet proved to be a much more powerful and versatile instrument. By 1956, Müller's invention allowed him to view the crystal lattice structure of metals in atomic detail; it actually showed the constituent atoms.

The field emission and field ion microscopes make it possible to view the atomic surface structures of metals on fluorescent screens. The field ion microscope is the direct descendant of the field emission microscope. In the case of the field emission microscope, the images are projected by electrons emitted directly from the tip of a metal needle, which constitutes the specimen under investigation. These electrons produce an image of the atomic lattice structure of the needle's surface. The needle serves as the electron-donating electrode in a vacuum tube, also known as the "cathode." A fluorescent screen that serves as the electron-receiving electrode, or "anode," is placed opposite the needle. When sufficient electrical voltage is applied across the cathode and anode, the needle tip emits electrons, which strike the screen. The image produced on the screen is a projection of the electron source—the needle surface's atomic lattice structure.

Müller studied the effect of needle shape on the performance of the microscope throughout much of 1937. When the needles had been properly shaped, Müller was able to realize magnifications of up to 1 million times. This magnification allowed Müller to view what he called "maps" of the atomic crystal structure of metals, since the needles were so small that they were often composed of only one simple crystal of the material. While the magnification may have been great, however, the resolution of the instrument was severely limited by the physics of emitted electrons, which caused the images Müller obtained to be blurred.

## Improving the View

In 1943, while working in Berlin, Müller realized that the resolution of the field emission microscope was limited by two factors. The electron velocity, a particle property, was extremely high and uncontrollably random, causing the

**1148**

micrographic images to be blurred. In addition, the electrons had an unsatisfactorily high wavelength. When Müller combined these two factors, he was able to determine that the field emission microscope could never depict single atoms; it was a physical impossibility for it to distinguish one atom from another.

By 1951, this limitation led him to develop the technology behind the field ion microscope. In 1952, Müller moved to the United States and founded the Pennsylvania State University Field Emission Laboratory. He perfected the field ion microscope between 1952 and 1956.

The field ion microscope utilized positive ions instead of electrons to create the atomic surface images on the fluorescent screen. When an easily ionized gas—at first hydrogen, but usually helium, neon, or argon— was introduced into the evacuated tube, the emitted electrons ionized the gas atoms, creating a stream of positively charged particles, much as Oppenheimer had predicted in 1928. Müller's use of positive ions circumvented one of the resolution problems inherent in the use of imaging electrons. Like the electrons, however, the positive ions traversed the tube with unpredictably random velocities. Müller eliminated this problem by cryogenically cooling the needle tip with a supercooled liquefied gas such as nitrogen or hydrogen.

By 1956, Müller had perfected the means of supplying imaging positive ions by filling the vacuum tube with an extremely small quantity of an inert gas such as helium, neon, or argon. By using such a gas, Müller was assured that no chemical reaction would occur between the needle tip and the gas; any such reaction would alter the surface atomic structure of the needle and thus alter the resulting microscopic image. The imaging ions allowed the field ion microscope to image the emitter surface to a resolution of between two and three angstroms, making it ten

*Erwin Wilhelm Müller holds his field ion microscope in front of a model of tungsten crystal atoms.*

times more accurate than its close relative, the field emission microscope.

## Consequences

The immediate impact of the field ion microscope was its influence on the study of metallic surfaces. It is a well-known fact of materials science that the physical properties of metals are influenced by the imperfections in their constituent lattice structures. It was not possible to view the atomic structure of the lattice, and thus the finest detail of any imperfection, until the field ion microscope was developed. The field ion microscope is the only instrument powerful enough to view the structural flaws of metal specimens in atomic detail.

**1149**

Although the instrument may be extremely powerful, the extremely large electrical fields required in the imaging process preclude the instrument's application to all but the heartiest of metallic specimens. The field strength of 500 million volts per centimeter exerts an average stress on metal specimens in the range of almost 1 ton per square millimeter. Metals such as iron and platinum can withstand this strain because of the shape of the needles into which they are formed. Yet this limitation of the instrument makes it extremely difficult to examine biological materials, which cannot withstand the amount of stress that metals can. A practical by-product in the study of field ionization—field evaporation—eventually permitted scientists to view large biological molecules.

Field evaporation also allowed surface scientists to view the atomic structures of biological molecules. By embedding molecules such as phthalocyanine within the metal needle, scientists have been able to view the atomic structures of large biological molecules by field evaporating much of the surrounding metal until the biological material remains at the needle's surface.

*William J. McKinney*

# Bevis Shows How Amniocentesis Can Reveal Fetal Problems

*After Douglas Bevis described amniocentesis, the removal of amniotic fluid from a pregnant woman, it became a tool for diagnosing fetal maturity, health, and genetic defects.*

**What:** Medicine
**When:** February 23, 1952
**Where:** Manchester, England
**Who:**
DOUGLAS BEVIS, an English physician
AUBREY MILUNSKY, (1936-     ), an
American pediatrician

## How Babies Grow

For thousands of years, the inability to see or touch a fetus in the uterus was a staggering problem in obstetric care and in the diagnosis of the future mental and physical health of human offspring. A beginning to the solution of this problem occurred on February 23, 1952, when *The Lancet* published a study called "The Antenatal Prediction of a Hemolytic Disease of the Newborn." This study, carried out by physician Douglas Bevis, described the use of amniocentesis to assess the risk factors found in the fetuses of Rh-negative women impregnated by Rh-positive men. The article is viewed by many as a landmark in medicine that led to the wide use of amniocentesis as a tool for diagnosing fetal maturity, fetal health, and fetal genetic defects.

At the beginning of a human pregnancy (conception) an egg and a sperm unite to produce the fertilized egg that will become a new human being. After conception, the fertilized egg passes from the oviduct into the uterus, while dividing and becoming an organized cluster of cells capable of carrying out different tasks in the nine-month-long series of events leading up to birth.

About a week after conception, the cluster of cells, now a "vesicle" (a fluid-filled sac containing the new human cells), attaches to the uterine lining, penetrates it, and becomes intimately inter-twined with uterine tissues. In time, the merger between the vesicle and the uterus results in formation of a placenta that connects the mother and the embryo, and an amniotic sac filled with the amniotic fluid in which the embryo floats.

Eight weeks after conception, the embryo (now a fetus) is about 2.5 centimeters long and possesses all the anatomic elements it will have when it is born. At this time, about two and one-half months after her last menstruation, the expectant mother typically visits a physician and finds out she is pregnant. Also at this time, expecting mothers often begin to worry about possible birth defects in the babies they carry. Diabetic mothers and mothers older than thirty-five years have higher than usual chances of delivering babies who have birth defects.

Many other factors inferred from the medical history an expecting mother provides to her physician can indicate the possible appearance of birth defects. In some cases, knowledge of possible physical problems in a fetus may allow their treatment in the uterus and save the newborn from problems that could persist throughout life or lead to death in early childhood. Information is obtained through the examination of the amniotic fluid in which the fetus is suspended throughout pregnancy. The process of obtaining this fluid is called "amniocentesis."

## Diagnosing Diseases Before Birth

Amniocentesis is carried out in several steps. First, the placenta and the fetus are located by the use of ultrasound techniques. Next, the expecting mother may be given a local anesthetic; a long needle is then inserted carefully into the amniotic sac. As soon as amniotic fluid is seen, a small sample (about four teaspoons) is drawn into a hypodermic syringe and the syringe is re-

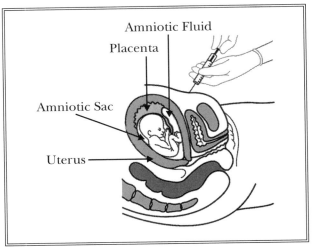

*The physician extracts amniotic fluid from the womb and examines it to determine the health of the fetus.*

moved. Amniocentesis is nearly painless, and most patients feel only a little abdominal pressure during the procedure.

The amniotic fluid of early pregnancy resembles blood serum. As pregnancy continues, its content of substances from fetal urine and other fetal secretions increases. The fluid also contains fetal cells from skin and from the gastrointestinal, reproductive, and respiratory tracts. Therefore, it is of great diagnostic use. Immediately after the fluid is removed from the fetus, the fetal cells are separated out. Then, the cells are used for genetic analysis and the amniotic fluid is examined by means of various biochemical techniques.

One important use of the amniotic fluid from amniocentesis is the determination of its lecithin and sphingomyelin content. Lecithins and sphingomyelins are two types of body lipids (fatty molecules) that are useful diagnostic tools. Lecithins are important because they are essential components of the so-called pulmonary surfactant of mature lungs. The pulmonary surfactant acts at lung surfaces to prevent the collapse of the lung air sacs (alveoli) when a person exhales.

Subnormal lecithin production in a fetus indicates that it most likely will exhibit respiratory distress syndrome or a disease called "hyaline membrane disease" after birth. Both diseases can be fatal, so it is valuable to determine whether fetal lecithin levels are adequate for appropriate lung function in the newborn baby. This is particularly important in fetuses being carried by diabetic mothers, who frequently produce newborns with such problems. Often, when the risk of respiratory distress syndrome is identified through amniocentesis, the fetus in question is injected with hormones that help it produce mature lungs. This effect is then confirmed by the repeated use of amniocentesis. Many other problems can also be identified by the use of amniocentesis and corrected before the baby is born.

**Consequences**

In the years that have followed Bevis's original observation, many improvements in the methodology of amniocentesis and in the techniques used in gathering and analyzing the genetic and biochemical information obtained have led to good results. Hundreds of debilitating hereditary diseases can be diagnosed—and some ameliorated—by the examination of amniotic fluid and fetal cells isolated by amniocentesis. For many parents who have had a child afflicted by some hereditary disease, the use of the technique has become a major consideration in family planning. Furthermore, many physicians recommend strongly that all mothers over the age of thirty-four be tested by amniocentesis to assist in the diagnosis of Down syndrome, a congenital but nonhereditary form of mental deficiency.

There remains the question of whether such solutions are morally appropriate, but parents—and society—now have a choice resulting from the techniques that have developed since Bevis's 1952 observation. It is also hoped that these techniques will lead to means for correcting and preventing diseases and preclude the need for considering the therapeutic termination of any pregnancy.

*Sanford S. Singer*

# Salk Develops First Polio Vaccine

> *Jonas Edward Salk developed the first vaccine that could prevent polio, resulting in the virtual eradication of polio epidemics.*

**What:** Medicine
**When:** July 2, 1952
**Where:** Pittsburgh, Pennsylvania
**Who:**

JONAS EDWARD SALK (1914-1995), an American physician, immunologist, and virologist

THOMAS FRANCIS, JR. (1900-1969), an American microbiologist

## Cause for Celebration

Poliomyelitis (polio) is an infectious disease that can adversely affect the central nervous system, causing paralysis and great muscle wasting due to the destruction of motor neurons (nerve cells) in the spinal cord. Epidemiologists believe that polio has existed since ancient times, and evidence of its presence in Egypt, circa 1400 B.C.E., has been presented. Fortunately, the Salk vaccine and the later vaccine developed by the American virologist Albert Bruce Sabin can prevent the disease. Consequently, except in underdeveloped nations, polio is rare. Moreover, although once a person develops polio, there is still no cure for it, a large number of polio cases end without paralysis or any observable effect. Polio is often called "infantile paralysis." This results from the fact that it is seen most often in children. It is caused by a virus and begins with body aches, a stiff neck, and other symptoms that are very similar to those of a severe case of influenza. In some cases, within two weeks after its onset, the course of polio begins to lead to muscle wasting and paralysis.

On April 12, 1955, the world was thrilled with the announcement that Jonas Edward Salk's poliomyelitis vaccine could prevent the disease. It was reported that schools were closed in celebration of this event. Salk, the son of a New York City garment worker, has since become one of the most well-known and publicly venerated medical scientists in the world.

Vaccination is a method of disease prevention by immunization, whereby a small amount of virus is injected into the body to prevent a viral disease. The process depends on the production of antibodies (body proteins that are specifically coded to prevent the disease spread by the virus) in response to the vaccination. Vaccines are made of weakened or killed virus preparations.

## Electrifying Results

The Salk vaccine was produced in two steps. First, polio viruses were grown in monkey kidney tissue cultures. These polio viruses were then killed by treatment with the right amount of formaldehyde to produce an effective vaccine. The killed-virus polio vaccine was found to be safe and to cause the production of antibodies against the disease, a sign that it should prevent polio.

In early 1952, Salk tested a prototype vaccine against Type I polio virus on children who were afflicted with the disease and were thus deemed safe from reinfection. This test showed that the vaccination greatly elevated the concentration of polio antibodies in these children. On July 2, 1952, encouraged by these results, Salk vaccinated forty-three children who had never had polio with vaccines against each of the three virus types (Type I, Type II, and Type III). All inoculated children produced high levels of polio antibodies, and none of them developed the disease. Consequently, the vaccine appeared to be both safe in humans and likely to become an effective public health tool.

In 1953, Salk reported these findings in the *Journal of the American Medical Association*. In

**1153**

April, 1954, nationwide testing of the Salk vaccine began, via the mass vaccination of American schoolchildren. The results of the trial were electrifying. The vaccine was safe, and it greatly reduced the incidence of the disease. In fact, it was estimated that Salk's vaccine gave the schoolchildren 60 to 90 percent protection against polio.

Salk was instantly praised. Then, several cases of polio occurred as a consequence of the vaccine. Its use was immediately suspended by the U.S. surgeon general, pending a complete examination. Soon, it was evident that all the cases of vaccine-derived polio were attributable to faulty batches of vaccine made by one pharmaceutical company. Salk and his associates were in no way responsible for the problem. Appropriate steps were taken to ensure that such an error would not be repeated, and the Salk vaccine was again released for use by the public.

## Consequences

The first reports on the polio epidemic in the United States had occurred on June 27, 1916, when one hundred residents of Brooklyn, New York, were afflicted. Soon, the disease had spread. By August, twenty-seven thousand people had developed polio. Nearly seven thousand afflicted people died, and many survivors of the epidemic were permanently paralyzed to varying extents. In New York City alone, nine thousand people developed polio and two thousand died. Chaos reigned as large numbers of terrified people attempted to leave and were turned back by police. Smaller polio epidemics occurred throughout the nation in the years that followed (for example, the Catawba County, North Carolina, epidemic of 1944). A particularly horrible aspect of polio was the fact that more than 70 percent of polio victims were small children. Adults

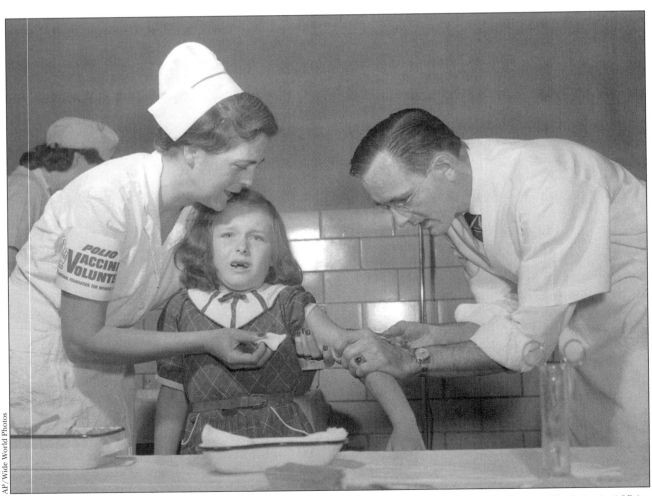

*Dr. Richard Mulvaney inoculates seven-year-old Mimi Meade with the new Salk polio vaccine in McClean, Virginia, in 1954.*

caught it too; the most famous of these adult polio victims was U.S. president Franklin D. Roosevelt. There was no cure for the disease. The best available treatment was physical therapy.

By August, 1955, more than four million polio vaccines had been given. The Salk vaccine appeared to work very well. There were only half as many reported cases of polio in 1956 as there had been in 1955. It appeared that polio was being conquered. By 1957, the number of cases reported nationwide had fallen below six thousand. Thus, in two years, its incidence had dropped by about 80 percent.

This was very exciting, and soon other countries clamored for the vaccine. By 1959, ninety other countries had been supplied with the Salk vaccine. Worldwide, the disease was being eradicated. The introduction of an oral polio vaccine by Albert Bruce Sabin supported this progress.

Salk received many honors, including honorary degrees from American and foreign universities, the Lasker Award, a Congressional Medal for Distinguished Civilian Service, and membership in the French Legion of Honor, yet he received neither the Nobel Prize nor membership in the American National Academy of Sciences. It is believed by many that this neglect was a result of the personal antagonism of some of the members of the scientific community who strongly disagreed with his theories of viral inactivation.

*Sanford S. Singer*

# Kenya's Mau Mau Uprising Creates Havoc

*After the Mau Mau began rebelling against British rule in Kenya, the British declared a state of emergency and gave their security forces special powers to hunt down the rebels.*

**What:** Civil strife
**When:** October 20, 1952
**Where:** Central Kenya
**Who:**
JOMO KENYATTA (1889-1978), a nationalist leader and head of the Kenya African Union
SIR EVELYN BARING (1903-1973), governor of Kenya from 1952 to 1959
FRED KUBAI (1915-    ) and
BILDAD MWANGANU KAGGIA (1922-    ), cofounders of Mau Mau
DEDAN KIMATHI (1920-1957), a Mau Mau field commander

## The Kenyans Want Land

Beginning in the early twentieth century, Europeans settled much of the best land in Kenya, a British colony. The European settlers were never more than 1 percent of Kenya's population, but many of their farms were very large, and they produced most of Kenya's farm exports. The settlers had major influence in Kenya's government as well.

The African residents of the "White Highlands" were mostly from the Kikuyu tribe. Most of them worked on the white farms, and some of them lived there as tenants or squatters. Others had to live in reserve areas.

As the Kikuyu increased in numbers, they wished more and more for their own land. Many of them moved to Nairobi, the capital city, where they crowded into "locations"—neighborhoods which were set aside for Africans. The British government had declared in 1923 that "Kenya is an African territory," yet Africans were treated as second-class citizens. They did not have the right to travel freely in their country or to live in certain areas. They were subject to their traditional chiefs and did not have a voice in the national government.

In 1944, the Kenya African Union (KAU) was formed, bringing together nationalists from various tribes. Jomo Kenyatta, a Kikuyu, became chair of the KAU in 1947; he worked hard to gain greater representation for Africans in government.

Meanwhile, Kenyan soldiers were returning from service under British commanders in World War II. Back in Kenya, they faced unemployment and discrimination, and they had no land to call their own. Many of them began calling for an overthrow of British rule.

In the White Highlands, the 200,000 or more Kikuyu squatters were becoming poorer. The white farmers were forcing many of them to move off the farms. Now they had no place to grow their own food, but their wages did not rise. The government offered them little education and few social services.

## Rise of the Mau Mau

In 1950, police claimed to have found a secret "Mau Mau Association" that aimed to drive the whites out of Kenya. (The exact meaning of the term "Mau Mau" is not known.) Some people were arrested for leading others in secret oaths of membership, and the organization was quickly banned.

This "association" was actually composed of radicals from the KAU and an older nationalist organization, the KCA; these radicals were led by Fred Kubai and Bildad Kaggia. Both organizations used oaths and ceremonies that contained vows of unity, loyalty, obedience, and secrecy. By the time the government banned Mau Mau, Kubai and Kaggia had spread the oaths throughout Nairobi and the Kikuyu area.

In November, 1951, a white settler was killed. Some Kikuyu who publicly supported the govern-

**1156**

ment were murdered; for example, Senior Chief Waruhiu Kungu, who had spoken out strongly against violence, was shot in his car by men disguised as police on October 9, 1952.

Frightened, the white settlers armed themselves and demanded protection from the government. The governor, Sir Evelyn Baring—who had been in Kenya only three weeks—declared a state of emergency on October 20, 1952. That night, many political and trade-union leaders—including Kenyatta, Kubai, and Kaggia—were arrested.

Forest zones where Mau Mau were active were called "prohibited areas," and the army was given permission to fire at will there. In other areas, the army and police had the right to search houses without a warrant and keep suspects locked up as long as they wished. Political meetings were banned. People who carried weapons illegally could be executed, along with those who led others in saying certain oaths.

Many Kikuyu were forced to move to special villages controlled by the government, and fences were put up around the African locations in Nairobi. There were actually very few attacks on Europeans, but the settlers were terrified. Some of them took matters into their own hands, killing anyone they suspected of being a Mau Mau follower.

At the settlers' insistence, the government forced Kikuyu squatters to leave the white farms; they were replaced with workers from other ethnic groups. About 100,000 Kikuyu squatters were sent to the reserves, which were already too crowded.

Kenyatta was not involved in the spread of "oathing" and violence. Yet the government considered him dangerous, and to the Mau Mau he was a hero. Along with Kubai, Kaggia, and three others, he was convicted in 1953 of being associated with the Mau Mau and was sentenced to seven years of hard labor.

In the next few years, tens of thousands of other Kenyans were forced to join their leaders

*Jomo Kenyatta in 1962.*

AP/Wide World Photos

in the detention camps. There they were punished with beatings or solitary confinement, and the government tried to reeducate them so that they would abandon the Mau Mau. The Hola camp was investigated by the government in 1959 after eleven men died there; they had been beaten severely when they refused to dig a ditch.

Meanwhile, the rebels were forcing many Kenyans to take oaths. Tom Mbotela, vice president of the Kau, opposed the Mau Mau, and he was assassinated in 1952. In the Lari massacre of March 26, 1953, rebels killed at least ninety-seven people, mostly women and children. Lari was a settlement set up by the government in 1940 for a group of people who had been moved from the Highlands to make way for white settlers. The Mau Mau considered the people of Lari to be traitors for having agreed to leave their original homes.

In 1954, one of the most important Mau Mau leaders, known as "General China," was captured. Soon afterward, the security forces detained about 16,500 people—mostly Kikuyu—in Nairobi. This was another serious blow to the Mau Mau. In the countryside, the search for rebels continued, and Kikuyu who were loyal to the government were recruited to police their own areas.

## Consequences

By 1956, the British had been able to flush many Mau Mau out of the forest. Dedan Kimathi, a Mau Mau field commander, was captured in October, 1956, and hanged in 1957. This event marked the real end of the uprising, although the state of emergency was not lifted until 1960.

According to official figures, 11,503 Mau Mau rebels had been killed in the fighting. More than one thousand people were executed for murder, administering oaths, or possessing firearms. Only 5,299 rebels surrendered or were captured.

The security forces lost 590, and the government said that the Mau Mau had killed 1,875 civilians—only 32 Europeans among them.

In 1961 Kenyatta was released from prison, and his people welcomed him as a hero. He led his country to independence on December 12, 1963. Though the Mau Mau uprising had been crushed, there is no doubt that it had prodded the British to leave Kenya more quickly than they would have otherwise.

*T. K. Welliver*

# Eisenhower Is Elected U.S. President

*With the election of Dwight D. Eisenhower, twenty years of Democratic domination of the American presidency came to an end.*

**What:** National politics
**When:** November 4, 1952
**Where:** United States
**Who:**

DWIGHT D. EISENHOWER (1890-1969), the Republican presidential candidate in 1952

RICHARD M. NIXON (1913-1994), Eisenhower's running mate

ADLAI E. STEVENSON (1900-1965), the Democratic presidential candidate in 1952

HARRY S. TRUMAN (1884-1972), president of the United States from 1945 to 1953

JOSEPH MCCARTHY (1908-1957), senator from Wisconsin from 1947 to 1957

ROBERT A. TAFT (1889-1953), a contender for the Republican nomination in 1952

## Democrats in Trouble

As the United States moved into a new decade in 1950, the Democrats had been in national office for eighteen years, Cold War tensions seemed to be melting into a "hot war" in Korea, and nervous Americans suspected that communists and corruption were lurking within the Truman administration. Truman was also accused of making serious mistakes in foreign policy. Some believed that China had fallen to the communists because the administration had failed to give enough help to Chiang Kai-shek's Nationalist forces. The situation in Korea was tense and uncertain; the peace talks had bogged down, and Truman had dismissed General Douglas MacArthur, a popular war hero.

In general, Americans were very frightened about the advance of communism. Senator Joseph McCarthy had been insisting that the American government, especially the State Department, was filled with communist spies and traitors. Because the Democrats had been in office for nearly twenty years, many Americans held them responsible for this communist subversion of the government.

Another problem for the Democrats was the high cost of living. The Korean War had made it necessary for the government to spend heavily, so that inflation was likely to grow. Many Americans, especially Republicans, believed that it was time to balance the budget and to reduce government spending.

The leading candidates for the Democratic nomination included Vice President Alben Barkley and Senator Estes Kefauver of Tennessee. Governor Adlai Ewing Stevenson of Illinois announced that he would not run but would allow the party to draft him at the convention. That is what the party decided to do. After a struggle on the convention floor, Stevenson was nominated on the first ballot. Senator John J. Sparkman of Alabama was given the vice presidential slot to satisfy Democrats from the South.

## Republicans Bring Hope

The Republicans were not without their problems. As the Republican National Convention drew near, it seemed quite possible that the party would split.

Senator Robert A. Taft of Ohio and his followers wanted to return the country to isolationism and to turn back the reforms of the New Deal. Taft's foreign policy appealed to many Republicans, especially in the Midwest, where isolationism had always been popular. Yet many powerful Republican leaders feared that Taft had made

**1159**

EISENHOWER ELECTED PRESIDENT

AP/Wide World Photos

*Dwight D. Eisenhower (right) takes the oath of office on January 20, 1953.*

too many enemies within the party, so that he would not be supported wholeheartedly. Also, Taft had attached his name to the Taft-Hartley labor law and in so doing had come to be bitterly resented by labor leaders and union members.

Many Republicans wanted to find a less controversial person to be the presidential candidate. Dwight D. Eisenhower, who had led the heroic Allied invasion of Normandy during World War II and now served as president of Columbia University, seemed to fit the need perfectly. After the general at last declared that he was a Republican, Senator Henry Cabot Lodge II of Massachusetts encouraged his nomination.

After primary victories in New Hampshire and Minnesota, Eisenhower defeated Taft on the first convention ballot, 595 to 500. To satisfy the Taft wing of the party and to bring in representation from the West, the convention chose Senator Richard M. Nixon of California as the vice presidential candidate.

The Democrats had tried to keep their Southern supporters by avoiding a strong civil rights

platform and by declaring that states should have the right to control oil in their tideland areas. Nevertheless, the 1952 election campaign made a crack in the traditionally Democratic South: Two well-known Democratic governors, James F. Byrnes of South Carolina and Allen Shivers of Texas, declared their support for Eisenhower.

During the campaign, Republicans made it clear that they would keep certain New Deal reforms such as the Tennessee Valley Authority and social security. Eisenhower announced that his party was not out to change the way the economy was handled. He did promise to balance the budget, reduce federal spending, lower taxes, protect free enterprise, and lessen government interference in business and industry.

Yet it was the Korean War and communists in the federal government that became the campaign's hottest issues. On the subject of the war, Eisenhower declared that he would "go to Korea." He did not explain exactly what he expected to accomplish there, but his promise was reassuring to many Americans.

On the Communist issue, Eisenhower did not adopt McCarthy's "smear" method of attacking opponents by accusing them of sympathizing with communism. At the same time, Eisenhower probably benefited from the crusade to ferret out communist spies. No one could accuse him of disloyalty to American ideals.

As a city intellectual who called Americans to lofty ideals, Stevenson could not compete with Eisenhower's down-home appeal. Though the Democrats suggested that a depression would follow a Republican victory, they could not turn the tide. On Election Day, Eisenhower won thirty-nine states—including four in the South—and 442 electoral votes, while Stevenson took only nine states and 89 electoral votes.

## Consequences

Though Eisenhower won the election easily, the Republican Party as a whole did not do so well. Republicans only barely gained control of Congress, and they lost it after Eisenhower had been in office two years.

In foreign policy, an important result of Eisenhower's election was that Taft and other isolationist Republicans were not allowed to pull the United States out of its involvements overseas. Having been involved with forces of the North Atlantic Treaty Organization (NATO) in Europe, Eisenhower believed that it was his duty to keep this commitment to the well-being of American allies.

*Fredrick J. Dobney*

# De Vaucouleurs Identifies Local Supercluster of Galaxies

*Gérard Henri de Vaucouleurs observed the local supercluster of galaxies, sparking the study of the large-scale structure of the universe.*

**What:** Astronomy
**When:** 1953
**Where:** Mount Stromlo, Australia
**Who:**

GÉRARD HENRI DE VAUCOULEURS (1918-1995), a French-born American astronomer

GEORGE OGDEN ABELL (1927-1983), an American astronomer

CHARLES DONALD SHANE (1895-1983), an American astronomer and director of the Lick Observatory

CARL ALVAN WIRTANEN (1910-      ), an American astronomer

JERZY NEYMAN (1894-1981), an American statistician and educator

ELIZABETH LEONARD SCOTT (1917-      ), an American astronomer and statistician

## Clusters of Galaxies

In the early 1920's, the debate over the nature of "spiral nebulas" was resolved. One school of thought was that spiral nebulas were pieces of the Milky Way—relatively nearby, small objects. Another theory held that they were distant, very large, independent star systems. In 1924, the American astronomer Edwin Powell Hubble was able to settle the debate by determining the distance to the Andromeda nebula. The distance was found to be a large one, which indicated that the Andromeda nebula is, in fact, a huge system, independent of the Milky Way. The American astronomer Harlow Shapley, in the 1920's, discovered the dimensions and rough structure of the Milky Way galaxy, and the work of Hubble and Shapley brought about the beginning of the present picture of the universe: The Milky Way is a spiral galaxy in a universe that contains other galaxies of various shapes and sizes.

Sky surveys, in which large portions of the sky are photographed and galaxies are counted and positioned, revealed interesting information about the way that galaxies appear to be distributed in space. Even before the nature of the spiral nebulas was known, astronomers had noted that spiral nebulas appear in clusters. In 1922, a band of nebulas stretching nearly 40 degrees across the northern sky was observed by the English astronomer J. H. Reynolds. In addition to identifying the "local group" (a group of nearby galaxies, of which the Milky Way is a part), astronomers identified other groupings of galaxies. The Coma cluster and the Virgo cluster of galaxies were defined and named for the constellations in which they appear. Hubble photographed faint galaxies, so faint that he thought he was seeing as far into the universe as he could and that he was witnessing a limit to the phenomenon of clustering. An earlier scheme had suggested that there was a hierarchy of structure to the universe, with clusters of galaxies making up larger clusters of clusters, which in turn made up still larger structures. Hubble's observations seemed to indicate that this hierarchy was not likely to exist.

By 1950, the largest cluster known was the Coma cluster, which contained more than one thousand individual galaxies. The galaxies in clusters were mostly elliptical galaxies (rounded or oval in shape, with no distinguishing structural features) and spirals without much spiral arm structure. Astronomers had also identified so-called field objects: isolated galaxies, mostly spirals, that did not appear to belong to any cluster. It had been suggested that perhaps the group

**1162**

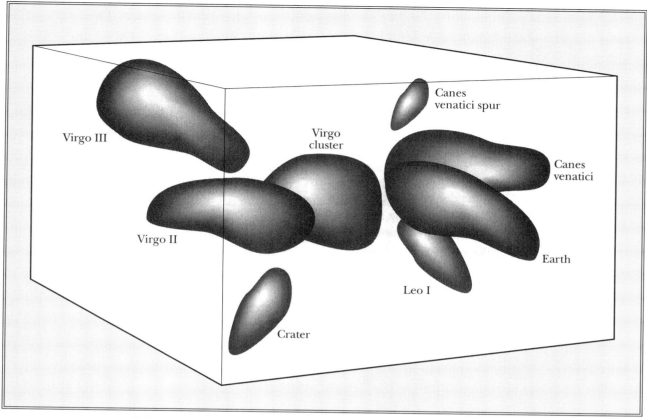

*Relative positions of the galaxy clusters that form the "local supercluster."*

of galaxies in the Virgo area of the sky might contain more than one cluster, but Hubble's work still seemed to rule out this possibility.

## Clusters of Clusters

It was on the basis of a sky survey, completed at Lick Observatory between 1947 and 1954 by Charles Donald Shane and Carl Alvan Wirtanen, that Elizabeth Leonard Scott and Jerzy Neyman applied the techniques of statistics to the question of the large-scale structure of the universe. Between 1952 and 1954, they published several papers regarding the laws that describe clustering, proposed that all galaxies belong to clusters, and mentioned the existence of "clouds" of galaxies. (That was their term for "superclusters"; a supercluster is simply a group of neighboring clusters of galaxies.) George Ogden Abell, at the University of California, Los Angeles, used plates taken at Mount Palomar Observatory to make a catalog of 2,712 clusters of galaxies, and his work indicated that many of the clusters seemed to be members of superclusters.

In the early 1950's, Gérard Henri de Vaucouleurs first defined and described what is called the "local supercluster." De Vaucouleurs had begun working at the Mount Stromlo Observatory in Australia to update the Shapley-Ames catalog of bright galaxies, a standard tool for astronomers. While doing this work, he observed that the local group was located at the edge of a much larger grouping of clusters of galaxies. He referred to this larger grouping as a "supergalaxy," and he further estimated that other supergalaxies might exist as well.

De Vaucouleurs estimated the local supercluster to be approximately fifty million light-years across and to be roughly disk shaped. The supercluster is centered on the Virgo cluster of galaxies, about fifty million light-years away. De Vaucouleurs also identified what appeared to be another supercluster, which he called the "southern supergalaxy"; he posited that the local supercluster is neither unique nor unusual. De Vaucouleurs would go on to conduct many studies of the superclustering phenomenon.

**1163**

## Consequences

Astronomers now estimate the local super-cluster to be about one hundred million light-years across and to have a total mass of about one thousand trillion times that of the sun. Astronomers also have discovered fine detail in the local supercluster and other superclusters. In addition to the local supercluster, others have been identified: the Hercules, Coma, and Perseus-Pisces superclusters. Common features of these superclusters have been identified. The hypothesis of de Vaucouleurs—that superclusters exist as organized structures—has been confirmed.

Fine structure has been discovered, such as filaments and streamers, or strings of galaxies that link the various parts of the superclusters. In addition, astronomers have discovered voids, or spaces, in which no bright galaxies appear.

The structure of superclusters may have much to teach astronomers about how the universe cooled and formed after the big bang theory. This field is at the forefront of cosmological research and may reveal even more startling information about the large-scale structure of the universe and how it came to be.

*Mary Hrovat*

# Du Vigneaud Synthesizes First Artificial Hormone

> *Vincent du Vigneaud's synthesis of oxytocin, a small polypeptide hormone from the pituitary gland, suggested that complex polypeptides and proteins could be synthesized and used in medicine.*

**What:** Medicine
**When:** 1953
**Where:** New York, New York
**Who:**

VINCENT DU VIGNEAUD (1901-1978), an American biochemist and winner of the 1955 Nobel Prize in Chemistry

OLIVER KAMM (1888-1965), an American biochemist

SIR EDWARD ALBERT SHARPEY-SCHAFER (1850-1935), an English physiologist

SIR HENRY HALLETT DALE (1875-1968), an English physiologist and winner of the 1936 Nobel Prize in Physiology or Medicine

JOHN JACOB ABEL (1857-1938), an American pharmacologist and biochemist

## Body-Function Special Effects

In England in 1895, physician George Oliver and physiologist Edward Albert Sharpey-Schafer reported that a hormonal extract from the pituitary gland of a cow produced a rise in blood pressure (a pressor effect) when it was injected into animals. In 1901, Rudolph Magnus and Sharpey-Schafer discovered that extracts from the pituitary also could restrict the flow of urine (an antidiuretic effect). This observation was related to the fact that when a certain section of the pituitary was removed surgically from an animal, the animal excreted an abnormally large amount of urine.

In addition to the pressor and antidiuretic activities in the pituitary, two additional effects were found in 1909. Sir Henry Hallett Dale, an English physiologist, was able to show that the extracts could cause the uterine muscle to contract (an oxytocic effect), and Isaac Ott and John C. Scott found that when lactating (milk-producing) animals were injected with the extracts, milk was released from the mammary gland.

Following the discovery of these various effects, attempts were made to concentrate and isolate the substance or substances that were responsible. John Jacob Abel was able to concentrate the pressor activity at The Johns Hopkins University using heavy metal salts and extraction with organic solvents. The results of the early work, however, were varied. Some investigators came to the conclusion that only one substance was responsible for all the activities, while others concluded that two or more substances were likely to be involved.

In 1928, Oliver Kamm and his coworkers at the drug firm of Parke, Davis and Company in Detroit reported a method for the separation of the four activities into two chemical fractions with high potency. One portion contained most of the pressor and antidiuretic activities, while the other contained the uterine-contracting and milk-releasing activities. Over the years, several names have been used for the two substances responsible for the effects. The generic name "vasopressin" generally has become the accepted term for the substance causing the pressor and antidiuretic effects, while the name "oxytocin" has been used for the other two effects. The two fractions that Kamm and his group had prepared were pure enough for the pharmaceutical firm to make them available for medical research related to obstetrics, surgical shock, and diabetes insipidus.

## A Complicated Synthesis

The problem of these hormones and their nature interested Vincent du Vigneaud at the

**1165**

George Washington University School of Medicine. Working with Kamm, he was able to show that the sulfur content of both the oxytocin and the vasopressin fractions was a result of the amino acid cystine. This helped to strengthen the concept that these hormones were polypeptide, or proteinlike, substances. Du Vigneaud and his coworkers next tried to find a way of purifying oxytocin and vasopressin. This required not only the separation of the hormones themselves but also the separation from other impurities present in the preparations.

During World War II (1939-1945) and shortly thereafter, other techniques were developed that would give du Vigneaud the tools he needed to complete the job of purifying and characterizing the two hormonal factors. One of the most important was the countercurrent distribution method of chemist Lyman C. Craig at the Rockefeller Institute. Craig had developed an apparatus that could do multiple extractions, making possible separations of substances with similar properties. Du Vigneaud had used this technique in purifying his synthetic penicillin, and when he returned to the study of oxytocin and vasopressin in 1946, he used it on his purest preparations. The procedure worked well, and milligram quantities of pure oxytocin were available in 1949 for chemical characterization.

Using the available techniques, Vigneaud and his coworkers were able to determine the structure of oxytocin. It was du Vigneaud's goal to make synthetic oxytocin by duplicating the structure his group had worked out. Eventually, du Vigneaud's synthetic oxytocin was obtained and the method published in the *Journal of the American Chemical Society* in 1953.

Du Vigneaud's oxytocin was next tested against naturally occurring oxytocin, and the two forms were found to act identically in every respect. In the final test, the synthetic form was found to induce labor when given intravenously to women about to give birth. Also, when microgram quantities of oxytocin were given intravenously to women who had recently given birth, milk was released from the mammary gland in less than a minute.

**Consequences**

The work of du Vigneaud and his associates demonstrated for the first time that it was possible to synthesize peptides that have properties identical to the natural ones and that these can be useful in certain medical conditions. Oxytocin has been used in the last stages of labor during childbirth. Vasopressin has been used in the treatment of diabetes insipidus, when an individual has an insufficiency in the natural hormone, much as insulin is used by persons having diabetes mellitus.

After receiving the Nobel Prize in Chemistry in 1955, du Vigneaud continued his work on synthesizing chemical variations of the two hormones. By making peptides that differed from the oxytocin and vasopressin by one or more amino acids, it was possible to study how the structure of the peptide was related to its physiological activity.

After the structure of insulin and some of the smaller proteins were determined, they, too, were synthesized, although with greater difficulty. Other methods of carrying out the synthesis of peptides and proteins have been developed and are used at the beginning of the twenty-first century. The production of biologically active proteins, such as insulin and growth hormone, has been made possible by efficient methods of biotechnology. The genes for these proteins can be put inside microorganisms, which then make them in addition to their own proteins. The microorganisms are then harvested and the useful protein hormones isolated and purified.

*Richard A. Hendry*

# Miller Synthesizes Amino Acids

*Stanley Lloyd Miller synthesized amino acids by combining a mixture of water, hydrogen, methane, and ammonia and exposing it to an electric spark.*

**What:** Biology
**When:** 1953
**Where:** Chicago, Illinois
**Who:**

STANLEY LLOYD MILLER (1930-      ), an
   American professor of chemistry
HAROLD CLAYTON UREY (1893-1981), an
   American chemist who won the 1934
   Nobel Prize in Chemistry
ALEKSANDR IVANOVICH OPARIN (1894-1980),
   a Russian biochemist
JOHN BURDON SANDERSON HALDANE (1892-
   1964), a British biochemist

## Prebiological Evolution

The origin of life on Earth has long been a tough problem for scientists to solve. While most scientists can envision the development of life through geologic time from simple single-cell bacteria to complex mammals by the processes of mutation and natural selection, they have found it difficult to develop a theory to define how organic materials were first formed and organized into life-forms. This stage in the development of life before biologic systems arose, which is called "chemical evolution," occurred between 4.5 and 3.5 billion years ago. Although great advances in genetics and biochemistry have shown the intricate workings of the cell, relatively little light has been shed on the origins of this intricate machinery of the cell. Some experiments, however, have provided important data from which to build a scientific theory of the origin of life. The first of these experiments was the classic work of Stanley Lloyd Miller.

Miller worked with Harold Clayton Urey, a Nobel laureate, on the environments of the early earth. John Burdon Sanderson Haldane, a British biochemist, had suggested in 1929 that the earth's early atmosphere was a reducing one—that it contained no free oxygen. In 1952, Urey published a seminal work in planetology, *The Planets,* in which he elaborated on Haldane's suggestion, and he postulated that the earth had formed from a cold stellar dust cloud. Urey thought that the earth's primordial atmosphere probably contained elements in the approximate relative abundances found in the solar system and the universe.

It had been discovered in 1929 that the sun is approximately 87 percent hydrogen, and by 1935 it was known that hydrogen encompassed the vast majority (92.8 percent) of atoms in the universe. Urey reasoned that the earth's early atmosphere contained mostly hydrogen, with the oxygen, nitrogen, and carbon atoms chemically bonded to hydrogen to form water, ammonia, and methane. Most important, free oxygen could not exist in the presence of such an abundance of hydrogen.

As early as the mid-1920's, Aleksandr Ivanovich Oparin, a Russian biochemist, had argued that the organic compounds necessary for life had been built up on the early earth by chemical combinations in a reducing atmosphere. The energy from the sun would have been sufficient to drive the reactions to produce life. Haldane later proposed that the organic compounds would accumulate in the oceans to produce a "dilute organic soup" and that life might have arisen by some unknown process from that mixture of organic compounds.

## Primordial Soup in a Bottle

Miller combined the ideas of Oparin and Urey and designed a simple, but elegant, experiment. He decided to mix the gases presumed to exist

**1167**

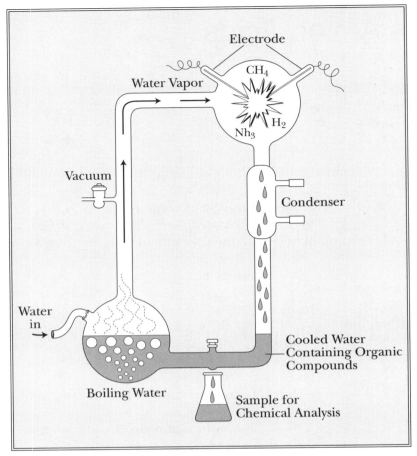

*The Miller-Urey experiment.*

in the early atmosphere (water vapor, hydrogen, ammonia, and methane) and expose them to an electrical spark to determine which, if any, organic compounds were formed. To do this, he constructed a relatively simple system, essentially consisting of two Pyrex flasks connected by tubing in a roughly circular pattern. The water and gases in the smaller flask were boiled and the resulting gas forced through the tubing into a larger flask that contained tungsten electrodes. As the gases passed the electrodes, an electrical spark was generated, and from this larger flask the gases and any other compounds were condensed. The gases were recycled through the system, whereas the organic compounds were trapped in the bottom of the system.

Miller was trying to simulate conditions that had prevailed on the early earth. During the one week of operation, Miller extracted and analyzed the residue of compounds at the bottom of the

system. The results were truly astounding. He found that numerous organic compounds had, indeed, been formed in only that one week. As much as 15 percent of the carbon (originally in the gas methane) had been combined into organic compounds, and at least 5 percent of the carbon was incorporated into biochemically important compounds. The most important compounds produced were some of the twenty amino acids essential to life on Earth.

The formation of amino acids is significant because they are the building blocks of proteins. Proteins consist of a specific sequence of amino acids assembled into a well-defined pattern. Proteins are necessary for life for two reasons. First, they are important structural materials used to build the cells of the body. Second, the enzymes that increase the rate of the multitude of biochemical reactions of life are also proteins.

Miller not only had produced proteins in the laboratory but also had shown clearly that the precursors of proteins—the amino acids—were easily formed in a reducing environment with the appropriate energy.

Perhaps the most important aspect of the experiment was the ease with which the amino acids were formed. Of all the thousands of organic compounds that are known to chemists, amino acids were among those that were formed by this simple experiment. This strongly implied that one of the first steps in chemical evolution was not only possible but also highly probable. All that was necessary for the synthesis of amino acids were the common gases of the solar system, a reducing environment, and an appropriate energy source, all of which were present on early Earth.

**Consequences**

Miller opened an entirely new field of research with his pioneering experiments. His results showed that much about chemical evolution could be learned by experimentation in the laboratory. As a result, Miller and many others tried variations on his original experiment by altering the combination of gases, using other gases, and trying other types of energy sources. Almost all the essential amino acids have been produced in these laboratory experiments.

Miller's work was based on the presumed composition of the primordial atmosphere of the earth. The composition of this atmosphere was calculated on the basis of the abundance of elements in the universe. If this reasoning is correct, then it is highly likely that there are many other bodies in the universe that have similar atmospheres and are near energy sources similar to the sun. Moreover, Miller's experiment strongly suggests that amino acids, and perhaps life as well, should have formed on other planets.

*Jay R. Yett*

# Ziegler Develops Low-Pressure Process for Making Polyethylene

*Karl Ziegler discovered that treating ethylene with certain catalysts caused the formation of high-density polyethylene (HDPE) at relatively low temperatures and pressures.*

**What:** Materials
**When:** 1953
**Where:** Mülheim, West Germany
**Who:**
KARL ZIEGLER (1898-1973), a German chemist
GIULIO NATTA (1903-1979), an Italian chemist
AUGUST WILHELM VON HOFMANN (1818-1892), a German chemist

## The Development of Synthetic Polymers

In 1841, August Wilhelm von Hofmann completed his Ph.D. with Justus von Liebig, a German chemist and founding father of organic chemistry. One of Hofmann's students, William Henry Perkin, discovered that coal tars could be used to produce brilliant dyes. The German chemical industry, under Hofmann's leadership, soon took the lead in this field, primarily because the discipline of organic chemistry was much more developed in Germany than elsewhere.

The realities of the early twentieth century found the chemical industry struggling to produce synthetic substitutes for natural materials that were in short supply, particularly rubber. Rubber is a natural polymer, a material composed of a long chain of small molecules that are linked chemically. An early synthetic rubber, neoprene, was one of many synthetic polymers (some others were Bakelite, polyvinyl chloride, and polystyrene) developed in the 1920's and 1930's. Another polymer, polyethylene, was developed in 1936 by Imperial Chemical Industries. Polyethylene was a tough, waxy material that was produced at high temperature and at pressures of about one thousand atmospheres. Its method of production made the material expensive, but it was useful as an insulating material.

World War II and the material shortages associated with it brought synthetic materials into the limelight. Many new uses for polymers were discovered, and after the war they were in demand for the production of a variety of consumer goods, although polyethylene was still too expensive to be used widely.

## Organometallics Provide the Key

Karl Ziegler, an organic chemist with an excellent international reputation, spent most of his career in Germany. With his international reputation and lack of political connections, he was a natural candidate to take charge of the Kaiser Wilhelm Institute for Coal Research (later renamed the Max Planck Institute) in 1943. Wise planners saw him as a director who would be favored by the conquering Allies. His appointment was a shrewd one, since he was allowed to retain his position after World War II ended. Ziegler thus played a key role in the resurgence of German chemical research after the war.

Before accepting the position at the Kaiser Wilhelm Institute, Ziegler made it clear that he would take the job only if he could pursue his own research interests in addition to conducting coal research. The location of the institute in the Ruhr Valley meant that abundant supplies of ethylene were available from the local coal industry, so it is not surprising that Ziegler began experimenting with that material.

Although Ziegler's placement as head of the institute was an important factor in his scientific breakthrough, his previous research was no less

**1170**

significant. Ziegler devoted much time to the field of organometallic compounds, which are compounds that contain a metal atom that is bonded to one or more carbon atoms. Ziegler was interested in organoaluminum compounds, which are compounds that contain aluminum-carbon bonds.

Ziegler was also interested in polymerization reactions, which involve the linking of thousands of smaller molecules into the single long chain of a polymer. Several synthetic polymers were known, but chemists could exert little control on the actual process. It was impossible to regulate the length of the polymer chain, and the extent of branching in the chain was unpredictable. It was as a result of studying the effect of organo-aluminum compounds on these chain formation reactions that the key discovery was made.

Ziegler and his coworkers already knew that ethylene would react with organoaluminum compounds to produce hydrocarbons, which are compounds that contain only carbon and hydrogen and that have varying chain lengths. Regulating the product chain length continued to be a problem.

At this point, fate intervened in the form of a trace of nickel left in a reactor from a previous experiment. The nickel caused the chain lengthening to stop after two ethylene molecules had been linked. Ziegler and his colleagues then tried to determine whether metals other than nickel caused a similar effect with a longer polymeric chain. Several metals were tested, and the most important finding was that a trace of titanium chloride in the reactor caused the deposition of large quantities of high-density polyethylene at low pressures.

Ziegler licensed the procedure, and within a year, Giulio Natta, an Italian chemist, had modified the catalysts to give high yields of polymers with highly ordered side chains branching from the main chain. This opened the door for the easy production of synthetic rubber. For their discovery of Ziegler-Natta catalysts, Ziegler and Natta shared the 1963 Nobel Prize in Chemistry.

## Consequences

Ziegler's process produced polyethylene that was much more rigid than the material produced at high pressure. His product also had a higher density and a higher softening temperature. Industrial exploitation of the process was unusually rapid, and within ten years more than twenty plants utilizing the process had been built throughout Europe, producing more than 120,000 metric tons of polyethylene. This rapid exploitation was one reason Ziegler and Natta

*A woman displays products wrapped in polyethylene, which was commonly used in packaging after the 1950's.*

**1171**

were awarded the Nobel Prize after such a relatively short time.

By the late 1980's, total production stood at roughly 18 billion pounds worldwide. Other polymeric materials, including polypropylene, can be produced by similar means. The ready availability and low cost of these versatile materials have radically transformed the packaging industry. Polyethylene bottles are far lighter than their glass counterparts; in addition, gases and liquids do not diffuse into polyethylene very easily, and it does not break easily. As a result, more and more products are bottled in containers made of polyethylene or other polymers. Other novel materials possessing properties unparalleled by any naturally occurring material (Kevlar, for example, which is used to make bulletproof vests) have also been an outgrowth of the availability of low-cost polymeric materials.

*Craig B. Lagrone*

# Cockerell Invents Hovercraft

Christopher Cockerell invented a vehicle that requires no surface contact for traction and moves freely over a variety of surfaces while supported on a self-generated cushion of air.

**What:** Transportation
**When:** 1953-1959
**Where:** Isle of Wight
**Who:**

CHRISTOPHER SYDNEY COCKERELL (1910-1999), a British engineer who built the first hovercraft

RONALD A. SHAW (1910-    ), an early pioneer in aerodynamics who experimented with hovercraft

SIR JOHN ISAAC THORNYCROFT (1843-1928), a Royal Navy architect who was the first to experiment with air-cushion theory

## Air-Cushion Travel

The air-cushion vehicle was first conceived by Sir John Isaac Thornycroft of Great Britain in the 1870's. He theorized that if a ship had a plenum chamber (a box open at the bottom) for a hull and it were pumped full of air, the ship would rise out of the water and move faster, because there would be less drag. The main problem was keeping the air from escaping from under the craft.

In the early 1950's, Christopher Sydney Cockerell was experimenting with ways to reduce both the wave-making and frictional resistance that craft had to water. In 1953, he constructed a punt with a fan that supplied air to the bottom of the craft, which could thus glide over the surface with very little friction. The air was contained under the craft by specially constructed side walls. In 1955, the first true "hovercraft," as Cockerell called it, was constructed of balsa wood. It weighed only 127 grams and traveled over water at a speed of 13 kilometers per hour.

On November 16, 1956, Cockerell successfully demonstrated his model hovercraft at the patent agent's office in London. It was immediately placed on the "secret" list, and Saunders-Roe Ltd. was given the first contract to build hovercraft in 1957. The first experimental piloted hovercraft, the SR.N1, which had a weight of 3,400 kilograms and could carry three people at the speed of 25 knots, was completed on May 28, 1959, and publicly demonstrated on June 11, 1959.

## Ground Effect Phenomenon

In a hovercraft, a jet airstream is directed downward through a hole in a metal disk, which forces the disk to rise. The jet of air has a reverse effect of its own that forces the disk away from the surface. Some of the air hitting the ground bounces back against the disk to add further lift. This is called the "ground effect." The ground effect is such that the greater the undersurface area of the hovercraft, the greater the reverse thrust of the air that bounces back. This makes the hovercraft a mechanically efficient machine because it provides three functions.

First, the ground effect reduces friction between the craft and the earth's surface. Second, it acts as a spring suspension to reduce some of the vertical acceleration effects that arise from travel over an uneven surface. Third, it provides a safe and comfortable ride at high speed, whatever the operating environment. The air cushion can distribute the weight of the hovercraft over almost its entire area so that the cushion pressure is low.

The basic elements of the air-cushion vehicle are a hull, a propulsion system, and a lift system. The hull, which accommodates the crew, passengers, and freight, contains both the propulsion and lift systems. The propulsion and lift systems can be driven by the same power plant or by separate power plants. Early designs used only one unit, but this proved to be a problem when adequate power was not achieved for movement and lift. Better results are achieved when two units

**1173**

are used, since far more power is used to lift the vehicle than to propel it.

For lift, high-speed centrifugal fans are used to drive the air through jets that are located under the craft. A redesigned aircraft propeller is used for propulsion. Rudderlike fins and an air fan that can be swiveled to provide direction are placed at the rear of the craft.

Several different air systems can be used, depending on whether a skirt system is used in the lift process. The plenum chamber system, the peripheral jet system, and several types of recirculating air systems have all been successfully tried without skirting. A variety of rigid and flexible skirts have also proved to be satisfactory, depending on the use of the vehicle.

Skirts are used to hold the air for lift. Skirts were once hung like curtains around hovercraft. Instead of simple curtains to contain the air, there are now complicated designs that contain the cushion, duct the air, and even provide a secondary suspension. The materials used in the skirting have also changed from a rubberized fabric to pure rubber and nylon and, finally, to neoprene, a lamination of nylon and plastic.

The three basic types of hovercraft are the amphibious, nonamphibious, and semiamphibious models. The amphibious type can travel over water and land, whereas the nonamphibious type is restricted to water travel. The semiamphibious model is also restricted to water travel but may terminate travel by nosing up on a prepared ramp or beach. All hovercraft contain built-in buoyancy tanks in the side skirting as a safety measure in the event that a hovercraft must settle on the water. Most hovercraft are equipped with gas turbines and use either propellers or water-jet propulsion.

**Consequences**

Hovercraft are used primarily for short passenger ferry services. Great Britain was the only nation to produce a large number of hovercraft. The British built larger and faster craft and pioneered their successful use as ferries across the English Channel, where they could reach speeds of 111 kilometers per hour (60 knots) and carry more than four hundred passengers and almost one hundred vehicles. France and the former Soviet Union have also effectively demonstrated hovercraft river travel, and the Soviets have experimented with military applications as well.

The military adaptations of hovercraft have been more diversified. Beach landings have been performed effectively, and the United States used hovercraft for river patrols during the Vietnam War.

Other uses also exist for hovercraft. They can be used as harbor pilot vessels and for patrolling shores in a variety of police- and customs-related duties. Hovercraft can also serve as flood-rescue craft and fire-fighting vehicles. Even a hoverfreighter is being considered.

The air-cushion theory in transport systems developed rapidly after the hovercraft began operation. The principle spread to trains and smaller people movers in many countries. Their smooth, rapid, clean, and efficient operation makes hovercraft continue to be attractive to transportation designers around the world.

*Larry N. Sypolt*

AP/Wide World Photos

*Christopher Sydney Cockerell.*

# Stalin's Death Ends Oppressive Soviet Regime

*With the death of Soviet dictator Joseph Stalin, the Soviet Union was saved from a new series of bloody purges and began a period of political reform.*

**What:** Political reform
**When:** March 5, 1953
**Where:** Moscow
**Who:**
JOSEPH STALIN (1879-1953), dictator of the Soviet Union from 1924 to 1953
NIKITA S. KHRUSHCHEV (1894-1971), first secretary of the Communist Party from 1953 to 1964

## Hope and Fear

On March 4, 1953, Radio Moscow announced that on March 1, Joseph Stalin had suffered a stroke, which had led to partial paralysis with heart and breathing difficulties. On the morning of March 6, it reported that Stalin had died the night before at 9:30 P.M.

The news brought great concern throughout the country that Stalin had ruled for more than a quarter of a century. In spite of his great cruelties, he had become a familiar, fatherly figure who had "made" the Soviet Union what it was in 1953.

With his two Five-Year Plans (1928-1939), Stalin had transformed the Soviet Union from a mostly agricultural country into one of the leading industrial nations of the world. He had led his people victoriously through a war that had brought death and destruction the Soviets had never experienced before, and he seemed to be the center around which all Soviet life revolved.

Though Stalin's policies had brought terror, bloodshed, and suffering to Soviet citizens—from the poorest collective farmers to the top leaders of the Communist Party—he at least represented the safety of the familiar. Many people had hated and feared him, yet leaders of the

Communist Party, especially, feared the consequences of his death.

## Stalin's Purge Plans

In the last year of his life, Stalin had been preparing to plunge his people once again into the suffering and killing that he had brought about in the 1930's, when a great "purge" had led to the executions of many Soviet leaders. Signs of a new purge that would begin in the Communist Party's highest ranks and extend to other areas of society had been evident at the Nineteenth Party Congress in October, 1952.

Suspicious of everyone and growing more distrustful as he aged, the old dictator had been planning to do away with many of his leading associates, including secret police chief Lavrenti Beria, Nikita Khrushchev (first secretary of the All-Union Central Committee), and Deputy Premier Georgi Malenkov. Furthermore, a major element in Stalin's new purge was to be anti-Semitism—a clear threat to the entire Jewish population of the Soviet Union.

Among Stalin's intended victims were many Jews in the party and the government bureaucracy. He had often accused the American Central Intelligence Agency of recruiting Soviet Jews to work for "American-Zionist imperialism" within the Soviet Union. Most frightening of all was the announcement on January 13, 1953, that a "Doctors' Plot" had been uncovered. Nine prominent doctors, six of whom were Jews, had supposedly been guilty of a plot to kill leaders of the Communist Party and the armed forces through improper medical treatment.

All over the Soviet Union, Stalin had been piecing together accusations of assassination plots, economic crimes, sabotage, and dealings with foreign enemies. The prominent men of the

**1175**

AP/Wide World Photos

*Joseph Stalin in his funeral bier in Moscow. Among the mourners are Georgi Malenkov (far right) and Nikita S. Khrushchev (second from left).*

Politburo, along with many others, were to be the accused. Because of Stalin's anti-Semitic language and the fact that many of the "criminals" were Jews, it is thought that Stalin planned to appeal to the anti-Semitic feelings that already existed among many Soviet people. In this way he hoped to get public support for his purge, at least in the beginning.

Clearly, only Stalin's death had saved the Soviet Union from great bloodshed.

## Consequences

As they announced Stalin's death and made speeches in the following weeks, Communist Party leaders appealed to the people in the name of "collective leadership" to stay calm. They tried to tighten control over the party and other political and economic organizations by reducing the membership of governing bodies and placing themselves in key positions. Khrushchev, for example, became first secretary of the Communist Party, while Georgi Malenkov became premier.

These leaders—many of whom had been on Stalin's "hit list"—began almost immediately to diminish the power of the secret police, which had brought such terror into Soviet life; the army, which had always been popular among the Soviet people, was given control over the secret police. Articles in the Party newspaper, *Pravda,* assured the people that their rights under the Soviet constitution would be respected and that more consumer goods would be made available.

The powerful Party leaders carefully avoided suggesting that any one individual would come

to rule the Party as Stalin had done. Though Malenkov attempted in March, 1953, to establish himself in Stalin's position, he was quickly defeated in the inner Party councils.

At the Twentieth Party Congress in 1956, Khrushchev made a secret four-hour speech before the Central Committee, painting a sickening picture of the cruelty and madness that had come along with Stalin's rule. Even that hard-headed group of Communists were dismayed to hear the details of tortures and humiliations that loyal Party leaders had suffered at Stalin's hands.

Soon Khrushchev began the process of "de-Stalinization." Stalin's name stopped appearing in newspapers and was removed from public landmarks (the city of "Stalingrad," for example, became "Volgograd"). Finally, his corpse was removed from its place next to that of Vladimir Ilich Lenin.

These changes were clearly meant to show that the Soviet Union could no longer progress under a system of terror. Yet for many Soviet people, the reforms did not go nearly far enough. The masses of people did not gain political freedom, and literature, art, philosophy, and the writing of history continued to be censored. Those who criticized the Soviet government too publicly were still likely to be arrested and shipped off to labor camps. Domination by a powerful individual had simply been replaced by the domination of a group—the Communist Party.

*George F. Putnam*

# Gibbon Develops Heart-Lung Machine

*John H. Gibbon, Jr., developed and tested the first artificial device to oxygenate and circulate blood during surgery, thus beginning the era of open-heart surgery.*

**What:** Medicine
**When:** May 6, 1953
**Where:** Boston, Massachusetts
**Who:**

JOHN H. GIBBON, JR. (1903-1974), a cardiovascular surgeon

MARY HOPKINSON GIBBON, a research technician

THOMAS J. WATSON (1874-1956), chairman of the board of IBM

T. L. STOKES and

J. B. FLICK, researchers in Gibbon's laboratory

BERNARD J. MILLER, a cardiovascular surgeon and research associate

CECELIA BAVOLEK, the first human to undergo open-heart surgery successfully using the heart-lung machine

## A Young Woman's Death

In the first half of the twentieth century, cardiovascular medicine had many triumphs. Effective anesthesia, antiseptic conditions, and antibiotics made surgery safer. Blood-typing, anticlotting agents, and blood preservatives made blood transfusion practical. Cardiac catheterization (feeding a tube into the heart), electrocardiography, and fluoroscopy (visualizing living tissues with an X-ray machine) made the nonsurgical diagnosis of cardiovascular problems possible.

As of 1950, however, there was no safe way to treat damage or defects within the heart. To make such a correction, this vital organ's function had to be interrupted. The problem was to keep the body's tissues alive while working on the heart. While some surgeons practiced so-called blind surgery, in which they inserted a finger into the heart through a small incision without ob-

serving what they were attempting to correct, others tried to reduce the body's need for circulation by slowly chilling the patient until the heart stopped. Still other surgeons used "cross-circulation," in which the patient's circulation was connected to a donor's circulation. All these approaches carried profound risks of hemorrhage, tissue damage, and death.

In February of 1931, Gibbon witnessed the death of a young woman whose lung circulation was blocked by a blood clot. Because her blood could not pass through her lungs, she slowly lost consciousness from lack of oxygen. As he monitored her pulse and breathing, Gibbon thought about ways to circumvent the obstructed lungs and straining heart and provide the oxygen required. Because surgery to remove such a blood clot was often fatal, the woman's surgeons operated only as a last resort. Though the surgery took only six and one-half minutes, she never regained consciousness. This experience prompted Gibbon to pursue what few people then considered a practical line of research: a way to circulate and oxygenate blood outside the body.

## A Woman's Life Restored

Gibbon began the project in earnest in 1934, when he returned to the laboratory of Edward D. Churchill at Massachusetts General Hospital for his second surgical research fellowship. He was assisted by Mary Hopkinson Gibbon. Together, they developed, using cats, a surgical technique for removing blood from a vein, supplying the blood with oxygen, and returning it to an artery using tubes inserted into the blood vessels. Their objective was to create a device that would keep the blood moving, spread it over a very thin layer to pick up oxygen efficiently and remove carbon dioxide, and avoid both clotting and damaging blood cells. In 1939, they reported that prolonged survival after heart-lung bypass was possible in experimental animals.

World War II (1939-1945) interrupted the progress of this work; it was resumed by Gibbon at Jefferson Medical College in 1944. Shortly thereafter, he attracted the interest of Thomas J. Watson, chairman of the board of the International Business Machines (IBM) Corporation, who provided the services of IBM's experimental physics laboratory and model machine shop as well as the assistance of staff engineers. IBM constructed and modified two experimental machines over the next seven years, and IBM engineers contributed significantly to the evolution of a machine that would be practical in humans.

Gibbon's first attempt to use the pump-oxygenator in a human being was in a fifteen-month-old baby. This attempt failed, not because of a malfunction or a surgical mistake but because of a misdiagnosis. The child died following surgery because the real problem had not been corrected by the surgery.

On May 6, 1953, the heart-lung machine was first used successfully on Cecelia Bavolek. In the six months before surgery, Bavolek had been hospitalized three times for symptoms of heart failure when she tried to engage in normal activity. While her circulation was connected to the heart-lung machine for forty-five minutes, the surgical team headed by Gibbon was able to close an opening between her atria and establish normal heart function. Two months later, an examination of the defect revealed that it was fully closed; Bavolek resumed a normal life. The age of open-heart surgery had begun.

**Consequences**

The heart-lung bypass technique alone could not make open-heart surgery truly practical. When it was possible to keep tissues alive by diverting blood around the heart and oxygenating it, other questions already under investigation became even more critical: how to prolong the survival of bloodless organs, how to measure oxygen and carbon dioxide levels in the blood, and how to prolong anesthesia during complicated surgery. Thus, following the first successful use of the heart-lung machine, surgeons continued to refine the methods of open-heart surgery.

The heart-lung apparatus set the stage for the advent of "replacement parts" for many types of cardiovascular problems. Cardiac valve replacement was first successfully accomplished in 1960 by placing an artificial ball valve between the left atrium and ventricle. In 1957, doctors performed the first coronary bypass surgery, grafting sections of a leg vein into the heart's circulation system to divert blood around clogged coronary arteries. Likewise, the first successful heart transplant (1967) and the controversial Jarvik-7 artificial heart implantation (1982) required the ability to stop the heart and keep the body's tissues alive during time-consuming and delicate surgical procedures. Gibbon's heart-lung machine paved the way for all these developments.

# Hillary and Tenzing Are First to Reach Top of Mount Everest

*After thirty years of unsuccessful attempts by many different expeditions, New Zealand beekeeper Edmund Hillary and Nepalese Sherpa Tenzing Norgay reached the summit of Mount Everest on May 29, 1953, nearly one hundred years after the height of the mountain was first measured.*

**What:** Sports; International relations
**When:** May 29, 1953
**Where:** Mount Everest, Nepal
**Who:**
SIR EDMUND HILLARY (1919-        ), New Zealand beekeeper, founder of the Himalayan Trust and former New Zealand high commissioner to India
TENZING NORGAY (1914-1986), Nepalese Sherpa from the Everest region of Nepal and first field director of the Himalayan Mountaineering Institute

## Why Everest?

On May 29, 1953, Edmund Hillary and Nepalese Sherpa Tenzing Norgay reached the summit of Mount Everest, the world's highest peak at 29,035 feet (8,850 meters). For thirty years, many people had been attempting to reach the top of Everest, but the desire to summit the mountain probably first arose shortly after the mountain's identification as the world's highest peak nearly one hundred years earlier.

Everest was first noticed in 1803 by British army officers stationed in Nepal, a small country located between India and Tibet. Mount Everest, known in Nepal as Sagarmatha and in Tibet as Chomolungma, is located in the Himalayan mountain range, which runs along the Nepalese border with Tibet. Throughout the 1800's, the British surveyed India in an effort to accurately map their colony. Because the independent Nepalese would not allow British surveyors into Nepal, the Himalayan peaks were first measured from the India/Nepal border, more than one hundred miles away. Measurements taken dur-

ing 1847 and 1849 were analyzed in 1854, and a final height of 29,002 feet (8,840 meters) was announced in 1856. The British then named the peak for George Everest, surveyor-general of India from 1830 to 1843. Although measured from a great distance under difficult conditions, the height announced in 1856 is remarkable for the fact that it was within 33 feet (10 meters) of the actual height of 29,035 feet established by global positioning satellites in 1999.

## Climbing Mount Everest

Efforts to climb Mount Everest, which had been identified as the world's highest peak, began with the first expedition through Tibet in 1921. Additional expeditions continued without success until the start of World War II. Because of postwar unrest in China and Tibet, diplomatic efforts were made to gain permission from the Nepalese government for access through Nepal. These efforts were successful in 1951, the same year that China's occupation of Tibet closed the traditional northern route to the mountain.

The first expedition through Nepal was made by the British in 1951 and included New Zealand beekeeper Hillary as a member. Although they did not reach the summit, they did find a route through the Khumbu Icefall, the tangled and dangerous base of a glacier. Above this area, the path to the summit crossed a snow-and-ice-filled valley, the Western Cwm, ascended the Lhotse face, and finally climbed up the southern ridge to the summit. The Swiss received the next permit for 1952. They were also unsuccessful, although it was on this expedition that Tenzing Norgay climbed to within 783 feet (225 meters) of the summit.

**1179**

The British were back in Nepal for another attempt in March, 1953. Hillary was again a member of the team, as was Tenzing Norgay as head Sherpa. Following a seventeen-day hike from the city of Kathmandu to Tengboche Monastery involving more than eight hundred porter loads of equipment, the expedition was ready to climb Mount Everest. In April, the group made its way to Everest base camp and began to find a route through the Khumbu Icefall. By May 6, 1953, the team had started the ascent of the Lhotse face. This took two weeks, with poor weather and an altitude of more than 24,000 feet (7,315 meters) creating a very difficult environment in which to climb.

After more than two months of climbing, Hillary and Tenzing were ready for the summit. With other climbers establishing a route and carrying equipment, the last camp was set at 27,900 feet (8,504 meters). At 6:30 A.M. on May 29, 1953, Hillary and Tenzing began their final ascent, using bottled oxygen to supplement their breathing. Over the next five hours, they climbed through difficult snow and ice conditions, scaling a very steep rock and ice cliff. At 11:30 A.M., they found there was nowhere else to climb—they had reached the summit. After spending fifteen minutes on the top of Mount Everest, taking pictures and leaving small mementos, they began their descent.

When they returned to Everest base camp on May 31, 1953, they were met with great excitement. The news of their feat was received in En-gland in time for Queen Elizabeth's coronation on June 2. In one of her first acts, the queen made Hillary a knight of the British Empire. Tenzing was awarded the George Medal, Britain's highest civilian award for bravery. After their climb, questions have arisen about who actually set foot on the summit first. Both men's integrity and character is shown by the fact that neither claimed outright credit for being the first. Instead, both have agreed that they reached the summit together.

## Consequences

Since the first ascent in 1953, many people have climbed Mount Everest, with more than 1,300 reaching the summit. However, many of the expeditions have experienced tragedy, as more than 160 people have died on Mount Everest. Although Hillary and Tenzing's success in reaching the summit of Mount Everest is important, the most enduring legacy of their achievement can be found in Nepal itself.

Hillary's continued presence in Nepal after his Everest success led to the decision to provide assistance to the Nepalese people, especially those in the Everest region. This decision and the projects undertaken between the 1950's and the 1980's resulted in the establishment of the Sir Edmund Hillary Himalayan Trust. The efforts of this and other similar organizations have provided schools and health care facilities to the people of the Everest region. In addition, conservation efforts have improved, with a focus on regional reforestation. Tourism is now the largest segment of the Nepalese economy, providing foreign currency to the nation and jobs to many people throughout the country. It is a remarkable change for a country that was largely unknown and forgotten until a New Zealand beekeeper and a Nepalese Sherpa climbed the highest mountain in the world.

*Peter D. Lindquist*

# Rosenbergs Are First Americans Executed for Espionage During Peace Time

*Julius and Ethel Rosenberg, a married couple convicted of providing classified military information to the Soviet Union, became the first U.S. civilians to be executed for espionage amid objections by various liberal and civil rights groups, several members of the U.S. Supreme Court, and others.*

**What:** Human rights; Crime
**When:** June 19, 1953
**Where:** Sing Sing Prison, New York
**Who:**
JULIUS ROSENBERG (1918-1953), former U.S. Army soldier convicted of espionage
ETHEL ROSENBERG (1915-1953), wife of Julius, also convicted of espionage
DAVID GREENGLASS (1922-    ), soldier
MORTON SOBELL, codefendant of the Rosenbergs

## The Rosenberg Trial

The espionage trial of Julius and Ethel Rosenberg was a watershed in the Cold War experience in the United States. It occurred at the height of the Red Scare, as the United States faced an emerging nuclear threat from the Soviet Union, was engaged in the Korean War, and was beset by an unknown number of Communist spies in government positions. Depending on one's perspective, the trial became a symbol of either the excesses of anticommunist hysteria or the extent to which the Soviet Union had been able to infiltrate the U.S. government and military.

The Rosenbergs' arrest resulted from a chain of events that began with the deciphering of intercepted Soviet cables. The deciphering effort, known as the Venona project, led American authorities to David Greenglass, a soldier at a nuclear weapons laboratory in Los Alamos who had sold atomic bomb information to a Soviet agent in 1944 and 1945. This was at a time when the Americans, Soviets, and others were racing to develop an atomic bomb that presumably would provide victory in World War II and later an edge in the postwar balance of power. Greenglass confessed and identified his wife and his brother-in-law, Julius Rosenberg, as also involved in the spy ring.

Rosenberg was arrested in June, 1950, by Federal Bureau of Investigation (FBI) agents. Agents later arrested his wife, Ethel, on the basis of evidence that she had helped to type notes related to the stolen secrets. A third defendant and friend of the Rosenbergs, Morton Sobell, was arrested for involvement in the spy ring.

The trial began in March, 1951. The government prosecutor portrayed the charges gravely, asserting that the defendants' actions threatened to weaken the United States' defenses and to tip the balance of power in the Soviet Union's favor. The defense proclaimed the innocence of the defendants, arguing that the trial was so tainted by anticommunist hysteria that it could not provide them a fair hearing.

The prosecution brought forth a number of witnesses, including Greenglass, who were involved with the Rosenbergs and Sobell. The only witnesses called by the defense were the Rosenbergs. After a month, the jury returned a guilty verdict for all three defendants. The judge, calling the Rosenbergs' actions "Russian terrorism" and "worse than murder," sentenced the Rosenbergs to death. Sobell was sentenced to thirty years in prison.

## The Execution

Rather than settle the controversy surrounding the case, the sentence intensified it. Liberal, religious, and civil rights groups strenuously opposed the carrying out of the death sentences, arguing that the trial had not been fair and that

**1181**

the sentence was excessive. Ethel Rosenberg's plight received particular attention, given that the case against her was considered the weaker of the two. Some argued that the only reason the government sought to prosecute her at all was as a device to win her husband's confession.

The issue of the Rosenbergs' sentence was taken up by the international press, with Soviet bloc countries unsurprisingly making especially loud protests. However, others, including the Pope, also called for clemency. The Rosenbergs' lawyer, Emanuel Bloch, worked hard to prevent their execution, filing court appeals and seeking a meeting with President Dwight D. Eisenhower. None of these appeals succeeded. The Rosenbergs were executed in the electric chair at New York's Sing Sing Prison on June 19, 1953, almost exactly three years after their arrest. They thus became the first American citizens to be executed for espionage during peace time.

## Consequences

For decades afterward, the Rosenbergs' trial and execution were something of a *cause célèbre* among liberal groups. Most of these groups believed that the Rosenbergs were unfairly convicted, swept up by the anticommunist fervor of the day. The prosecution witnesses were portrayed as liars, key documents were dismissed as falsified, and the jury was thought to have been browbeaten and cajoled into returning a guilty verdict. The Rosenbergs themselves were seen as martyrs for various liberal causes.

However, in 1995, the Central Intelligence Agency released translations of the intercepted Venona cables that revealed that Julius Rosenberg was indeed involved in espionage for the Soviet Union. Also, newly opened archives in the Soviet Union corroborated many of the charges against Rosenberg. Moreover, Rosenberg's alleged Soviet contact, Aleksander Feklisov, confirmed that he met with Rosenberg in the 1940's. After the emergence of these pieces of evidence, the Rosenberg case was largely drained of its power to symbolize governmental overreaching during the Red Scare of the 1950's. To be sure, much of the rhetoric of the trial, including the judge's sentencing statement, was tinged with anti-Sovietism and Cold War anxiety. However, the basic accuracy of the verdict, if not necessarily the appropriateness of the sentence, has been validated. Aside from the Rosenbergs' two sons, most observers seem to accept this.

Moreover, the demise of the Soviet Union, the end of the Cold War, and the general rejection of Communism throughout Eastern Europe and Latin America in the 1990's made questions about anticommunist sentiment rather moot. The Rosenberg case has become more of a historical event illuminating a past era rather than a cautionary tale about anticommunist hysteria or the threat posed by espionage.

*Steve D. Boilard*

AP/Wide World Photos

*Ethel and Julius Rosenberg in 1951, during their trial.*

# Battle of Dien Bien Phu Ends European Domination in Southeast Asia

> *Through determination and careful strategy, Communist-led forces managed to besiege French fortifications at Dien Bien Phu—and eventually to gain control of North Vietnam.*

**What:** Military conflict
**When:** November 20, 1953-May 7, 1954
**Where:** Dien Bien Phu, Vietnam
**Who:**
Ho Chi Minh (1890-1969), Communist leader of the Viet Minh, and president of the Democratic Republic of Vietnam from 1945 to 1969
Vo Nguyen Giap (1911-    ), commander of the Viet Minh forces
Henri Navarre (1898-1983), commander in chief of French Union forces

## The French Return

After World War II, France returned its army to Indochina to reestablish its colonial power there. With the defeat of the Japanese, a power vacuum had developed. Various nationalist groups tried to assume power, but one by one they were defeated or absorbed by the well-organized Communist parties of Laos, Cambodia, and Vietnam.

The return of French forces marked the beginning of a bloody contest for political and economic control of Indochina. Ho Chi Minh, the dedicated leader of the Viet Minh force, proclaimed himself president of the Democratic Republic of Vietnam in 1945. As the Viet Minh began to fight the French, the struggle was indecisive. The Viet Minh harassed the French with hit-and-run tactics, while the French tried to bring the Viet Minh into open battle.

By 1954, the French Expeditionary Force had been fighting in Indochina for seven years. French forces totaled 240,000 persons, including Foreign Legion units, African colonial troops, and Indochinese recruits. They were assisted by the Vietnamese National Army of 211,000, 21,000 Laotians, and a Cambodian army of 16,000. Their enemies, the Viet Minh, numbered 300,000—about half of them regular army units and the remainder militia and guerrilla bands.

The French forces had the advantage of modern American equipment, and in most battles their numbers were greater. The Communists, however, were well organized and had the support of many people throughout the countryside.

As the war dragged on, people in France became discontented, wanting to see some decisive action. What was needed was a place to confront the Viet Minh forces in open conflict. Surely the French, with their superior firepower, manpower, and modern equipment, could then beat Ho Chi Minh's forces once and for all.

French intelligence discovered that General Vo Nguyen Giap, Commander of the Viet Minh, had begun to shift his forces out of Vietnam and was moving into Laos. This seemed the perfect opportunity to capture the Communists and defeat them before they could conquer Laos. General Henri Navarre, commander in chief of the French forces, planned to set up a base at Dien Bien Phu. From there he could launch attacks against the Viet Minh, force them into open conflict, and destroy them.

On November 20, 1953, Navarre ordered six paratroop battalions to jump into the bowl-shaped depression of Dien Bien Phu, two hundred miles northwest of Hanoi, Ho Chi Minh's capital. The valley, a flat, heart-shaped basin about twelve miles long and eight miles wide, contained a large village surrounded by paddy fields, which in turn were ringed by wooded hills. Only ten miles from the Laotian border, it lay at the crossroads of the major routes to China and Laos, over which the Viet Minh army would have to move.

**1183**

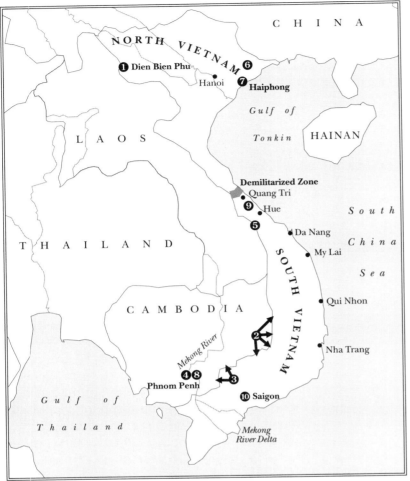

*(1) France falls, 1954. (2) Tet Offensive, January, 1968. (3) Cambodian invasion, April-May, 1970. (4) Sihanouk falls, April, 1970. (5) Laotian incursion, February, 1971. (6) Areas of U.S. bombing, 1972. (7) Mining of Haiphong Harbor, May, 1972. (8) Lon Nol falls, April, 1975. (9) North Vietnamese offensive, spring, 1975. (10) South Vietnam surrenders, April 20, 1975.*

Giap moved his Viet Minh army of fifty thousand into place, surrounding the garrison of fifteen thousand men.

By January, it was too late for the French to pull out of Dien Bien Phu. The best thing to do was to dig in and fight from a strongly fortified position.

Brigadier General Christian Marie-Ferdinand de la Croix de Castries and his deputy, Lieutenant Colonel Pierre-Charles Langlais, had an all but impossible task from the beginning. Most of the patrols they sent out were destroyed; they could not push far into enemy territory.

On January 31, 1954, the Communist artillery began to shell the airstrip and two of the French strong points in the valley. Unobserved, the Viet Minh had carried these heavy guns into place on their backs. They buried them in deep underground bunkers, hiding them so well that French counterfire hardly ever located them. Even aircraft attacks were unable to destroy them.

The French plan to interfere with the Communists' movement into Laos had failed. The Viet Minh pushed into Laos unopposed as they kept the French besieged at Dien Bien Phu. The French were hopelessly pinned down and unable to escape, except by air.

Meanwhile, the French had not received sufficient materials to fortify their strong points. Camouflage and concealment were impossible, so that it was easy for enemy gunners to find the most important French positions. Without enough aircraft, the French lacked firepower to protect their artillery and tanks.

Giap and his able chief of staff, General Hoang Van Thai, had dug a complicated network of siege trenches which completely surrounded the French strong points. They pulled the circle of trenches tighter and tighter, until the Viet

## The Dien Bien Phu Fight

Realizing that the French had not yet fortified their position at Dien Bien Phu, Giap sent some of his troops to tie down the French where they were; he planned to annihilate them by isolating them. Major General René Cogny, the fiery French commander in northern Vietnam, made up his mind to stay and fight.

Giap resisted the urge to strike swiftly, for he remembered the basic principle of revolutionary war: "Strike to win, strike only when success is certain; if it is not, then don't strike." He waited until the major French forces had reached their places in the valley before he sealed off the area. Between November 27 and December 7, 1953,

Minh forces were within grenade-throwing distance. Now they were ready to begin the attack.

On March 13, 1954, the Viet Minh overran the strong point Beatrice, northeast of the airfield. Gabrielle, the northernmost outpost, was taken the next night. The French defense caved in under heavy mortar, infantry, and artillery attacks.

By the middle of April, constant Viet Minh attacks forced the French completely off the airstrip and into an area little more than a mile across. Dropping supplies to the defenders became more and more difficult, so that eventually the Viet Minh were able to confiscate most of the dropped materials. After April 23, no more drops were possible.

During the last days of April, French resistance was slowly reduced to isolated pockets. With the coming of the rainy season, the French bunkers were flooded. The defenders were short of medicine, food, ammunition, and drinking water. General de Castries' headquarters fell on May 7, followed by the Isabelle garrison. The Viet Minh had succeeded in taking Dien Bien Phu.

**Consequences**

The Battle of Dien Bien Phu had a great influence on an international conference on Asian problems, which was already under way in Geneva. The Communist forces gained control of at least half of Vietnam, and the French people began demanding that their army move out of Indochina altogether.

Eventually the French parliament approved a settlement: Ho Chi Minh's Democratic Republic of Vietnam would govern the area north of the 17th parallel, while Bao Dai's Republic of Vietnam (which had been propped up by French and American support) would govern south of that line. Thus European domination in Asia began winding down, and the Communists were ready to take over. Soon Vietnam would be embroiled in yet another war—this one involving the United States.

# Ochoa Creates Synthetic RNA Molecules

> *Severo Ochoa discovered a method for synthesizing the biological molecule RNA, establishing that this process can occur outside the living cell.*

**What:** Biology; Genetics
**When:** Mid-1950's
**Where:** New York, New York
**Who:**
SEVERO OCHOA (1905-1993), a Spanish biochemist who shared the 1959 Nobel Prize in Physiology or Medicine
MARIANNE GRUNBERG-MANAGO, a French biochemist
MARSHALL W. NIRENBERG (1927-    ), an American biochemist who won the 1968 Nobel Prize in Physiology or Medicine
PETER LENGYEL (1929-    ), a Hungarian American biochemist

## RNA Outside the Cells

In the early decades of the twentieth century, genetics had not been experimentally united with biochemistry. This merging soon occurred, however, with work involving the mold *Neurospora crassa*. This Nobel award-winning work by biochemist Edward Lawrie Tatum and geneticist George Wells Beadle showed that genes control production of proteins, which are major functional molecules in cells. Yet no one knew the chemical composition of genes and chromosomes, or, rather, the molecules of heredity.

The American bacteriologist Oswald T. Avery and his colleagues at New York's Rockefeller Institute determined experimentally that the molecular basis of heredity was a large polymer known as deoxyribonucleic acid (DNA). Avery's discovery triggered a furious worldwide search for the particular structural characteristics of DNA, which allow for the known biological characteristics of genes.

One of the most famous studies in the history of science solved this problem in 1953. Scientists James D. Watson, Francis Crick, and Maurice H. F. Wilkins postulated that DNA exists as a double helix. That is, two long strands twist about each other in a predictable pattern, with each single strand held to the other by weak, reversible linkages known as "hydrogen bonds." About this time, researchers recognized also that a molecule closely related to DNA, ribonucleic acid (RNA), plays an important role in transcribing the genetic information as well as in other biological functions.

Severo Ochoa was born in Spain as the science of genetics was developing. In 1942, he moved to New York University, where he studied the bacterium *Azobacter vinelandii*. Specifically, Ochoa was focusing on the question of how cells process energy in the form of organic molecules such as the sugar glucose to provide usable biological energy in the form of adenosine triphosphate (ATP). With postdoctoral fellow Marianne Grunberg-Manago, he studied enzymatic reactions capable of incorporating inorganic phosphate (a compound consisting of one atom of phosphorus and four atoms of oxygen) into adenosine diphosphate (ADP) to form ATP.

One particularly interesting reaction was followed by monitoring the amount of radioactive phosphate reacting with ADP. Following separation of the reaction products, it was discovered that the main product was not ATP, but a much larger molecule. Chemical characterization demonstrated that this product was a polymer of adenosine monophosphate. When other nucleocide diphosphates, such as inosine diphosphate, were used in the reaction, the corresponding polymer of inosine monophosphate was formed. Thus, in each case, a polymer (a long string of building-block units) was

**1186**

formed. The polymers formed were synthetic RNAs, and the enzyme responsible for the conversion became known as "polynucleotide phosphorylase." This finding, once the early skepticism was resolved, was received by biochemists with great enthusiasm because no technique outside the cell had ever been discovered previously in which a nucleic acid similar to RNA could be synthesized.

## Learning the Language

Ochoa, Peter Lengyel, and Marshall W. Nirenberg at the National Institute of Health took advantage of this breakthrough to synthesize different RNAs useful in cracking the genetic code. Crick had postulated that the flow of information in biological systems is from DNA to RNA to protein. In other words, genetic information contained in the DNA structure is transcribed into complementary RNA structures, which, in turn, are translated into the protein. Protein synthesis, an extremely complex process, involves bringing a type of RNA, known as messenger RNA, together with amino acids and huge cellular organelles known as ribosomes.

Yet investigators did not know the nature of the nucleic acid alphabet—for example, how many single units of the RNA polymer code were needed for each amino acid, and the order that the units must be in to stand for a "word" in the nucleic acid language. In 1961, Nirenberg demonstrated that the polymer of synthetic RNA with multiple units of uracil (poly U) would "code" only for a protein containing the amino acid phenylalanine. Each three units (U's) gave one phenylalanine. Therefore, genetic words each contain three letters. UUU translates into phenylalanine. Poly A, the first polymer discovered with polynucleotide phosphorylase, was coded for a protein containing multiple lysines. That is, AAA translates into the amino acid lysine.

The words, containing combinations of letters, such as AUG, were not as easily studied, but Nirenberg, Ochoa, and Gobind Khorana of the

*Severo Ochoa.*

The Nobel Foundation

University of Wisconsin eventually uncovered the exact translation for each amino acid. In RNA, there are four possible letters (A, U, G, and C) and three letters in each word. Accordingly, there are sixty-four possible words. With only twenty amino acids, it became clear that more than one RNA word can translate into a given amino acid. Yet, no given word stands for any more than one amino acid. A few words do not translate into any amino acid; they are stop signals, telling the ribosome to cease translating RNA.

The question of which direction an RNA is translated is critical. For example, CAA codes for the amino acid glutamine, but the reverse, AAC, translates to the amino acid asparagine. Such a difference is critical because the exact sequence of a protein determines its activity—that is, what it will do in the body and therefore what genetic trait it will express.

**1187**

## Consequences

Synthetic RNAs provided the key to understanding the genetic code. The genetic code is universal; it operates in all organisms, simple or complex. It is used by viruses, which are near life but are not alive. Spelling out the genetic code was one of the top discoveries of the twentieth century. Nearly all work in molecular biology depends on this knowledge.

The availability of synthetic RNAs has provided hybridization tools for molecular geneticists. Hybridization is a technique in which an RNA is allowed to bind in a complementary fashion to DNA under investigation. The greater the similarity between RNA and DNA, the greater the amount of binding. The differential binding allows for seeking, finding, and ultimately isolating a target DNA from a large, diverse pool of DNA—in short, finding a needle in a haystack. Hybridization has become an indispensable aid in experimental molecular genetics as well as in applied sciences, such as forensics.

*Keith Krom Parker*

# Atomic-Powered Submarine *Nautilus* Is Launched

*The* Nautilus *was the first vessel powered by a nuclear reactor, allowing it to remain submerged for months, compared with the single day a diesel-powered submarine could operate under water.*

**What:** Engineering; Military; Technology
**When:** January 21, 1954
**Where:** Groton, Connecticut
**Who:**
HYMAN GEORGE RICKOVER (1900-1986), U.S. Navy captain
EUGENE PARKS WILKINSON (1918-     ), commander

## The First Nuclear-Powered Vessel

The USS *Nautilus* (SSN571), the first boat ever powered by a nuclear reactor, was launched at the Electric Boat Company shipyard in Groton, Connecticut, on January 21, 1954. Eight months later, on September 30, 1954, the *Nautilus* was commissioned into the U.S. Navy fleet. On January 17, 1955, under Commander Eugene Parks Wilkinson, the *Nautilus* sailed from its dock, becoming the first vessel powered by a nuclear reactor.

The *Nautilus* was 319 feet (97 meters) long, measured about 27 feet (8 meters) across, and could sail at a speed in excess of 20 knots when submerged. The *Nautilus* was armed with six 21-inch (53-centimeter) diameter torpedo tubes, all in the forward section of the boat. The *Nautilus* originally sailed with a crew of eleven officers and one hundred enlisted sailors.

On its shakedown cruise in May, 1955, the *Nautilus* sailed submerged from New London, Connecticut, to San Juan, Puerto Rico, traveling more than 1,300 miles (2,092 kilometers) in 84 hours. That was more than ten times farther than any submarine had ever traveled while submerged. Over the next several years, the *Nautilus* shattered all submerged speed and distance records for submarines. The *Nautilus* completed several missions under the polar ice cap and, on August 3, 1958, sailed beneath the Arctic icepack all the way to the North Pole.

The U.S. Navy's interest in nuclear power traces back to 1947, when U.S. Navy captain Hyman George Rickover requested that engineers at the nuclear research facility at Oak Ridge, Tennessee, determine if a high-pressure, water-cooled nuclear reactor could be used to power a submarine. In January, 1948, the Department of Defense requested the Atomic Energy Commission to design, develop, and build a nuclear reactor that could propel a submarine. The chief of naval operations, in August, 1949, established an operational requirement to develop a submarine nuclear propulsion plant with a ready-for-sea date of January, 1955.

## How the *Nautilus* Power System Works

Much of the success of the nuclear power system on the *Nautilus* is credited to the decision to develop a full land-based prototype, the Mark 1 Submarine Thermal Reactor, built at the National Reactor Testing Station in Idaho. Construction of the reactor began in April, 1950. It was built inside a submarine hull surrounded by a tank of water, simulating a seagoing power plant. The Mark 1 reactor was a full, operational power plant, with no engineering shortcuts or "breadboard systems," prototypes that are constructed to perform like the real system but that may be larger or use different materials from those used by commercial systems.

The Mark 1 reactor "went critical," developing power from nuclear reactions, on March 30, 1953. This was four years before the first commercial nuclear reactor, located in Shippingport, Pennsylvania, began producing commercial electric power. Therefore, the commissioning of the

*The atomic-powered USS* Nautilus *is lowered into the water near Groton, Connecticut, after being christened by First Lady Mamie Eisenhower.*

AP/Wide World Photos

Mark 1 reactor marked the first production of significant quantities of useful nuclear power in the world. About three months later, on June 25, 1953, Mark 1 commenced a 96-hour sustained full-power run, simulating a submerged crossing of the Atlantic and determining what modifications needed to be made to the systems on the *Nautilus.*

Before the *Nautilus,* submarines were basically surface ships that could submerge and operate under water for short periods of time. The diesel engines used to recharge the batteries that propel the submarine when submerged needed oxygen to operate, and so did the crew members. When the submarine submerged, the oxygen supply was cut off, and the boat depended for its power on electric batteries, while the crew depended on the air trapped within the hull or carried in compressed air bottles. The *Nautilus* was

different. Its nuclear reactor required no oxygen to operate. To accommodate the crew's need for oxygen, the *Nautilus* was equipped with a chemical system to purify the air, removing the carbon dioxide that humans exhale, and converting it to oxygen for breathing.

The heat generated by a nuclear reactor produces steam that drives the main propulsion turbine generators that produce electric power. Water circulates continuously through a primary system consisting of the reactor, loops of piping, coolant pumps, and steam generators. The heat produced in the reactor by nuclear fission, in which energy is released as a heavy atom breaks up into two lighter atoms, is transferred to the primary coolant water, which is pressurized to prevent boiling. This water is pumped through the steam generators, where it transfers the heat it is carrying to a secondary circulating system.

**1190**

The water in the primary system is pumped back into the reactor, where it is heated again. The secondary system, which is isolated from the primary system to minimize the possibility of radioactive contamination of the whole ship, carries steam to the engine room, where it runs turbine generators that produce electricity. The electricity produced by the generators powers the ship's electrical systems and the main propulsion system, which drives the propeller.

## Consequences

The historic voyage of the *Nautilus* to the North Pole demonstrated that submarines could hide for weeks or months under the polar icecap, making detection almost impossible. This capability proved extremely valuable when submarines carrying ballistic missiles, such as the Polaris and Poseidon, were deployed. These ballistic missile submarines can serve as remote launching platforms, whose locations are unknown to the enemy. They provide an almost indestructible second strike capability, deterring an enemy from attacking land-based missiles and aircraft because of the almost certain retaliation by the ballistic missile submarines.

The *Nautilus* was the world's first true submarine. All vessels previously called "submarines" were only submersible craft that had to surface and use diesel generators to recharge their batteries. The nuclear reactor enabled the *Nautilus* to remain submerged for months, freeing the boat from dependence on oxygen from the surface to run its engines. The *Nautilus* was refueled in April, 1957, after more than two years of operation. The *Nautilus* sailed 62,562 miles (100,662 kilometers) on its first reactor core. A diesel-powered submarine the size of the *Nautilus* would have required more than two million gallons of fuel oil to sail that distance. The *Nautilus* was retired to the Historic Ship Nautilus and Submarine Force Museum, in Groton, where it went on public display.

*George J. Flynn*

# Barghoorn and Tyler Discover Two-Billion-Year-Old Microfossils

Elso Barghoorn and Stanley Allen Tyler's fossil discoveries were the first in a series of discoveries crucial to understanding the origin and early development of life on Earth.

**What:** Earth science
**When:** April 30, 1954
**Where:** Madison, Wisconsin, and Cambridge, Massachusetts
**Who:**
ELSO BARGHOORN (1915-1984), an American paleontologist
STANLEY ALLEN TYLER (1906-1963), an American geologist
CHARLES DOOLITTLE WALCOTT (1850-1927), the head of the U.S. Geological Survey

## Preserved in Chert

On April 30, 1954, the American journal *Science* published a brief article by Elso Barghoorn and Stanley Allen Tyler describing five distinct types of fossil microorganisms. These included slender, unbranched filaments and spherical colonies made up of filaments, which were judged to resemble living blue-green algae, and branched filaments and spherical bodies, which were compared with living aquatic fungi.

These fossils had first come to the attention of Tyler, a geologist, while he was working with banded iron deposits of Precambrian age (more than 570 million years old) on the shores of Lake Superior. A puzzling circumstance was the occurrence of coal with the iron ore deposits. The iron ore was known to be approximately 2 billion years old—no life-forms that could have produced coal had been proved to have existed that long ago. Examination of the coal revealed what appeared to be microscopic plants. Tyler showed a specimen to William Schrock of the Massachusetts Institute of Technology, who thought it resembled living fungi. Schrock suggested that Tyler consult with Barghoorn in the Harvard University botany department.

Barghoorn agreed that the material appeared to be microbial. As a biologist, he recognized that convincing proof of life-forms so early in geologic time would be immensely significant. He suggested returning to the site and conducting a systematic search for life-forms in the coal-bearing rocks.

Following the coal seams into Canada, they collected samples of black gunflint chert, which were cut into thin slices with a diamond saw for microscopic observation. Chert is a sedimentary rock whose qualities make it an excellent preserver of organic remains. Organic cell walls remain intact, preserving cellular structures and traces of surface ornamentation.

When examined under the microscope, the thin sections revealed spheres and filaments. Both clearly were hollow and bounded by a sturdy wall of organic material; there was no doubt that these were microorganisms. In the 1954 publication, some specimens were identified as blue-green algae, with a bacterial level of cellular organization, and others as relatively more advanced aquatic fungi. (Subsequent detailed investigations have cast doubt on the identification of fungi in this or any other assemblage more than 1.5 billion years old.) In papers published in 1965 and 1971, Barghoorn attributed a bacterial or blue-green algal origin to all the gunflint organisms.

## The Bulk of Evolution

Tyler and Barghoorn were not the first to report microorganisms from Precambrian rocks of the Canadian shield. Between 1922 and 1925, John W. Gruner had described and illustrated filaments of algal specimens of about the same age.

Gruner's papers attracted less attention than Tyler and Barghoorn's for several reasons. First, the material was not as well preserved or illustrated and was not completely convincing. Second, prior to the routine use of radioactive decay as a means of dating rocks, the extreme antiquity of the specimens was not appreciated. Finally, although students of evolution and the geologic history of life are aware now that about five-sixths of biological evolution on Earth took place in the Precambrian era, few people in the 1920's were actively interested in the Precambrian.

By 1960, there was a small body of evidence for Precambrian life, but none of it was universally accepted or completely convincing. Radiometric dating had confirmed what had already been suspected from stratigraphic evidence: that the Precambrian encompassed a far longer time than the Phanerozoic (literally, the age of "evident life"—everything since the Precambrian). Estimates suggest that the earth was created 4.5 billion years ago.

**Consequences**

In the absence of a fossil record, scientists turned to the laboratory and to comparisons of living forms in an effort to formulate hypotheses about the origins and early evolution of life on Earth. American chemists Harold C. Urey and Stanley Miller postulated in the early 1950's that nonbiological processes early in the earth's history had created an "organic soup" in which electrical discharge produced complex molecules capable of replicating themselves and in which the precursors of the earliest cells developed.

It was recognized that there was a fundamental distinction among living organisms between those that lack nuclei (prokaryotes), including bacteria and blue-green algae, and those having nuclei (eukaryotes), including plants, animals, fungi, and most algae, and that the transition from prokaryote to eukaryote was a major evolutionary hurdle. Assuming that life as it is known in the twenty-first century evolved on Earth (a debatable assumption as long as there is no convincing Precambrian fossil record), then the most fundamental milestones in biological history, the evolution of cells and the evolution of eukaryotes, must have taken place in the enigmatic Precambrian.

It has been recognized for some time that living organisms have transformed the face of the earth in the Phanerozoic; the importance of their role in geological processes in the Precambrian was evident only when a usable fossil record became available. The fossil record provides information about the types and in some cases the abundance of microorganisms that were present at various stages in geologic time; the inorganic geologic record provides evidence for atmospheric and climatic changes, and comparison of the biochemistry of living forms provides clues as to probable conditions at the time various metabolic processes evolved.

*Martha Sherwood-Pike*

# Bell Labs Develops Photovoltaic Cell

> *Researchers at the Bell Telephone Labs constructed the first photovoltaic cell, providing a means of powering instruments on the early space satellites as well as hope for the future use of solar energy.*

**What:** Energy
**When:** May, 1954
**Where:** Murray Hill, New Jersey
**Who:**
Daryl M. Chapin (1906-1995), an American physicist
Calvin S. Fuller (1902-1994), an American chemist
Gerald L. Pearson (1905-1987), an American physicist

## Unlimited Energy Source

All the energy that the human race has at its disposal ultimately comes from the sun. Some of this solar energy was trapped millions of years ago in the form of vegetable and animal matter that became the coal, oil, and natural gas that the world relies upon for energy. Some of this fuel is used directly to heat homes and to power factories and gasoline vehicles. Much of this fossil fuel, however, is burned to produce the electricity on which modern society depends.

The amount of energy that is available from the sun is difficult to imagine, but some comparisons may be helpful. During each forty-hour period, the sun provides the earth with as much energy as the earth's total reserves of coal, oil, and natural gas. It has been estimated that the amount of energy provided by the sun's radiation matches the earth's reserves of nuclear fuel every forty days. The annual solar radiation that falls on about twelve hundred square miles of land in Arizona matches the world's estimated total annual energy requirement for 1960. Scientists have been searching for many decades for inexpensive, efficient means of converting this vast supply of solar radiation directly into electricity.

## The Bell Solar Cell

Throughout its history, Bell Systems has needed to be able to transmit, modulate, and amplify electrical signals. Until the 1930's, these tasks were accomplished by using insulators and metallic conductors. At that time, semiconductors, which have electrical properties that are between those of insulators and those of conductors, were developed. One of the most important semiconductor materials is silicon, which is one of the most common elements on the earth. Unfortunately, silicon is usually found in the form of compounds such as sand or quartz, and it must be refined and purified before it can be used in electrical circuits. This process required much initial research, and very pure silicon was not available until the early 1950's.

Electric conduction in silicon is the result of the movement of negative charges (electrons) or positive charges (holes). One way of accomplishing this is by deliberately adding to the silicon phosphorus or arsenic atoms, which have five outer electrons. This addition creates a type of semiconductor that has excess negative charges (an n-type semiconductor). Adding boron atoms, which have three outer electrons, creates a semiconductor that has excess positive charges (a p-type semiconductor). Calvin Fuller made an important study of the formation of p-n junctions, which are the points at which p-type and n-type semiconductors meet, by using the process of diffusing impurity atoms—that is, adding atoms of materials that would increase the level of positive or negative charges, as described above. Fuller's work stimulated interest in using the process of impurity diffusion to create cells that would turn solar energy into electricity. Fuller and Gerald Pearson made the first large-area p-n junction by using the diffusion process. Daryl Chapin, Fuller, and Pearson made a similar p-n junction very close to the surface of a silicon crystal, which was then exposed to sunlight.

The cell was constructed by first making an ingot of arsenic-doped silicon that was then cut into very thin slices. Then a very thin layer of p-type silicon was formed over the surface of the n-type wafer, providing a p-n junction close to the surface of the cell. Once the cell cooled, the p-type layer was removed from the back of the cell and lead wires were attached to the two surfaces. When light was absorbed at the p-n junction, electron-hole pairs were produced, and the electric field that was present at the junction forced the electrons to the n side and the holes to the p side. The recombination of the electrons and holes takes place after the electrons have traveled through the external wires, where they do useful work. Chapin, Fuller, and Pearson announced in 1954 that the resulting photovoltaic

cell was the most efficient (6 percent) means then available for converting sunlight into electricity.

The first experimental use of the silicon solar battery was in amplifiers for electrical telephone signals in rural areas. An array of 432 silicon cells, capable of supplying 9 watts of power in bright sunlight, was used to charge a nickel-cadmium storage battery. This, in turn, powered the amplifier for the telephone signal. The electrical energy derived from sunlight during the day was sufficient to keep the storage battery charged for continuous operation. The system was successfully tested for six months of continuous use in Americus, Georgia, in 1956. Although it was a technical success, the silicon solar cell was not ready to compete economically with conventional means of producing electrical power.

## Consequences

One of the immediate applications of the solar cell was to supply electrical energy for Telstar satellites. These cells are used extensively on all satellites to generate power. The success of the U.S. satellite program prompted serious suggestions in 1965 for the use of an orbiting power satellite. A large satellite could be placed into a synchronous orbit of the earth. It would collect sunlight, convert it to microwave radiation, and beam the energy to an Earth-based receiving station. Many technical problems must be solved, however, before this dream can become a reality.

Solar cells are used in small-scale applications such as power sources for calculators. Large-scale applications are still not economically competitive with more traditional means of generating electric power. The development of the world's poorer nations, however, may provide the incentive to search for less-expensive solar cells that can be used, for example, to provide energy in remote villages. As the standards of living in such areas improve, the need for electric power will grow. Solar cells may be able to provide the necessary energy while safeguarding the environment for future generations.

*Grace A. Banks*

PhotoDisc

*Parabolic mirrors at a modern solar power plant.*

**1195**

# U.S. Supreme Court Orders Desegregation of Public Schools

*With the* Brown v. Board of Education *case, the United States Supreme Court ruled that segregation in schools must be ended so that all American children could receive equal educational opportunity.*

**What:** Social reform; Education; Law
**When:** May 17, 1954
**Where:** Washington, D.C.
**Who:**
FREDERICK MOORE VINSON (1890-1953), chief justice of the United States from 1946 to 1953
EARL WARREN (1891-1974), chief justice of the United States from 1953 to 1969

## Growth of Segregation

In 1865, just after the Civil War ended, the United States adopted the Thirteenth Amendment to the Constitution, which outlawed slavery. In 1866, Congress passed the Civil Rights Act, which gave African Americans and other racial minorities equal treatment under the law. Two years later, the Fourteenth Amendment was ratified. The Fifteenth Amendment, ratified in 1879, guaranteed African American men the right to vote.

All these laws helped to secure civil rights for all Americans, especially those who had been treated unfairly in the past. But after these laws had been made and federal troops withdrew from the South, Northern concern for the well-being of African Americans diminished. The Southern states, run mainly by white men, began to pass laws that made black people sit in separate sections of trains and streetcars; they established poll taxes to prevent African Americans from voting.

In the *Plessy v. Ferguson* case (1896), the Supreme Court ruled that a Louisiana statute segre-gating transportation was constitutional. The majority of the Court declared that "the enforced segregation of the two races" did not stamp "the colored race with a badge of inferiority." As long as the separate facilities were "equal," they did not violate the Constitution. Justice John Marshall Harlan, however, dissented from this opinion, arguing that segregation would lead to "race hate."

With the support of the 1896 Supreme Court decision, segregation became common throughout the United States. In 1917, however, the Supreme court began to change its position on segregation: It declared unconstitutional a Louisville, Kentucky, law that enforced segregation in housing. In 1927, it held that laws keeping black Americans from voting in primaries violated the Fourteenth Amendment.

## The School Question

Beginning in 1938, the Supreme Court began to make rulings about segregation in public colleges and universities. By the fall of 1952, several cases challenging segregation in public schools, including one brought by a man named Oliver Brown, were before the Court. The school districts that were being challenged were in Kansas, Delaware, Virginia, South Carolina, and the District of Columbia. Because the Brown and Kansas names appeared first, the name of the case became famous as *Brown v. Board of Education of Topeka, Kansas.*

Some of the Supreme Court justices believed that Congress, rather than the Court, should take the responsibility for enforcing desegregation. The Court did not settle the segregation cases immediately; instead, the cases were reargued in December, 1953. By that time, Earl Warren had replaced Fred Vinson as chief justice of the United States.

Warren delivered the Court's momentous decision on May 17, 1954. "We conclude," he said, "that in the field of public education the doctrine of 'separate but equal' has no place. Separate educational facilities are inherently unequal."

Although the Fourteenth Amendment had said that states could not deprive American citizens of equal treatment under the law, Chief Justice Warren did not base the Court's decision solely on the history of the Fourteenth Amend-

ment. He considered that amendment "inconclusive" when it came to public education, because when the amendment was ratified, in the 1860's, public schools that were subsidized by government budgets had been just beginning.

"Today," said Warren, "education is perhaps the most important function of state and local governments. . . . In these days, it is doubtful that any child may reasonably be expected to succeed in life if he is denied the opportunity of an education."

The opportunity for an education, the Court said, must be open to all equally. The question, then, was whether segregation in schools de-

*Black and white fourth-graders head for the playground at St. Martin School in Washington, D.C., in September, 1954. The federally administered District of Columbia ended segregation in compliance with the Supreme Court's ruling.*

AP/Wide World Photos

**1197**

prived minority children—African Americans, Hispanic Americans, and others—of the same chance to learn what they needed to know to make a good life for themselves. The Supreme Court concluded that segregation blocked that chance, and that school segregation was therefore unconstitutional.

## Consequences

A year after the decision, on May 31, 1955, Warren revealed how important the issue of school desegregation was to the Supreme Court:

He announced that the process of desegregation of public schools must proceed "with all deliberate speed." This decision and the Court's commitment to it paved the way for the Civil Rights movement of the 1960's.

Many school districts, especially in the South, resisted the Court's decision. How desegregation should be carried out has continued to be a matter for debate—and for quite a few court cases. Even so, the principle of "separate but equal" was dealt a death blow in 1954.

*Michael R. Bradley*

# Southeast Asia Treaty Organization Is Formed

> *Bringing together eight nations to sign a treaty of mutual defense in Southeast Asia, American leaders hoped to contain Communist aggression in that part of the world.*

**What:** International relations; Military defense

**When:** September 8, 1954-February 19, 1955

**Where:** Manila, Philippine Islands; and Southeast Asia

**Who:**

DWIGHT D. EISENHOWER (1890-1969), president of the United States from 1953 to 1961

JOHN FOSTER DULLES (1888-1959), U.S. secretary of state from 1953 to 1959

## Regional Changes

Within Southeast Asia there are a variety of island, peninsular, and mainland countries whose peoples have different languages, religions, and cultures. Almost all these areas have tropical or subtropical climates; growing rice has been the traditional employment, but manufacturing (especially the clothing industry) has become more and more important. Except for Thailand (Siam), the entire area was under European or American rule in the late nineteenth and early twentieth centuries.

World War II brought enormous changes to all of Southeast Asia. Japanese conquests removed many colonial rulers. Nationalist movements grew; at first they were directed against the Japanese, but after the Japanese surrender, the nationalists were determined not to allow the old colonial governments to return either. Communists—some trained in Moscow, some inspired by China, some "home-grown"—played a part in the nationalist movements.

Having promised independence to the Philippines, the United States fulfilled that promise in 1946. Though the United States was rather unsympathetic to colonialism, alliances in World War II and the Cold War kept Americans bound to the colonial powers—Great Britain, France, and the Netherlands.

In parts of Southeast Asia, the end of colonial rule came peacefully. Yet other lands, such as Indonesia (the former Dutch East Indies) and French Indochina, had to fight for their independence. In Vietnam, a former French colony, the Viet Minh had their capital in the northern city of Hanoi. Led by Ho Chi Minh, a Communist nationalist, they fought with great determination and managed to defeat the French forces. At Geneva in July, 1954, the French and other world leaders agreed that France would withdraw from Vietnam and that the country would be temporarily divided between north and south.

In the estimations of American president Dwight D. Eisenhower, Secretary of State John Foster Dulles, and many other Americans, events in Vietnam showed that the Communists were ready to expand throughout all of Southeast Asia. To respond to this threat, the U.S. government proposed a regional defense treaty. This treaty would bring nations of Southeast Asia together in an alliance somewhat similar to the North Atlantic Treaty Organization (NATO), set up in 1949.

## The SEATO Treaty

The treaty for the collective defense of Southeast Asia was written at Manila and signed on September 8, 1954, by representatives of eight nations. In most cases, the delegates were the foreign ministers of their states. The U.S. delegation was made up of Secretary of State Dulles and two senators: H. Alexander Smith, a Republican, and Mike Mansfield, a Democrat.

**1199**

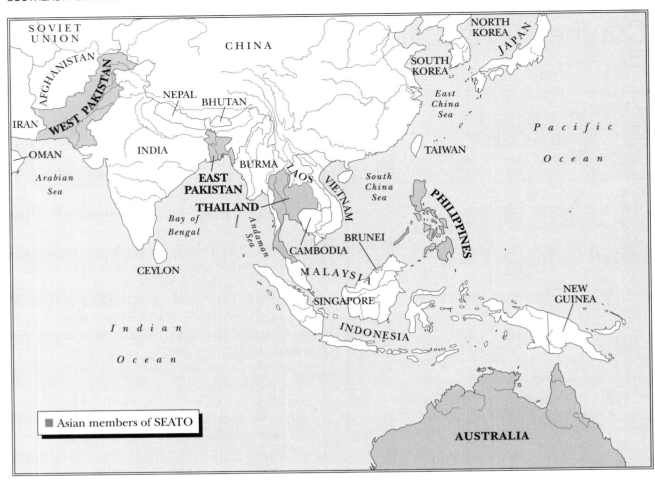

Asian members of SEATO

The Asian nations that joined the Southeast Asia Treaty Organization (SEATO) were Pakistan, the Philippine Republic, and Thailand. Australia, New Zealand, Great Britain, France, and the United States also signed.

Because of its policy of neutrality, and possibly also because Pakistan was participating, India did not become involved. Indonesia and Burma were neutral or antagonistic, and Singapore and Malaya did not join because they were still members of the British Commonwealth. The states of Indochina, according to the Geneva agreement, had to remain neutral.

According to the final version of the treaty, the participating nations agreed to abide by the Charter of the United Nations; they promised to settle disputes by peaceful means and to avoid the use of force. They also promised self-help and cooperation in resisting aggression as well as rebel movements that might be funded and directed by other governments. Each pledged to encourage free institutions and to cooperate in the development of business and trade.

Article 4, the core of the treaty, recognized that aggression against one member was an attack on all. Each member promised to resist aggression against the other members by taking steps that fit with its own constitution. If one member reported that there was a threat to its security (a threat that had not yet become actual aggression), all members would come together to decide what steps should be taken. The treaty also established a council and made provision for other states to join SEATO when they were ready.

The treaty defined its area as Southeast Asia, along with "the general area of the Southwest Pacific, not including the Pacific area north of 21 degrees 30 minutes north latitude." This definition excluded Taiwan, the Pescadores Islands, and Hong Kong. As soon as there were ratifications from a majority of the nations that had signed, the treaty would go into effect.

**1200**

The United States added a statement of "understanding" to the treaty, explaining that it was especially concerned to take a stand against Communist aggression. It is thought that the purpose of this statement was to allow the United States to stay out of any conflict between India and Pakistan. The signers of the treaty also added a "protocol" by which Laos, Cambodia, and "free Vietnam" (South Vietnam) came under the protection of the treaty, though they were not SEATO members.

At the Manila Conference in September, 1954, the participants also issued the Pacific Charter, promising to respect equal rights and self-determination and to work together toward higher living standards, economic progress, and the well-being of their people.

The required ratifications were deposited at Manila by February 19, 1955. Because of where it was signed, the treaty is sometimes called the Manila Pact; however, SEATO's headquarters were set up at Bangkok, Thailand. The organization was given a secretariat and permanent staff; the governing council was made up of the foreign ministers or their representatives. A group of military advisers was organized as well.

## Consequences

When conflicts developed in Laos and Vietnam, it was the United States alone, rather than SEATO, that took action. The treaty required its members to be in unanimity before any joint action could be taken, and this sort of agreement was hard to come by. For this reason, some U.S. citizens complained that the treaty and SEATO itself were not worth much, except in encouraging trade and cultural cooperation.

Those who supported SEATO, however, insisted that it was an important part of the American effort to contain Communist aggression in Southeast Asia. Though SEATO as a whole took no action in the Vietnam War, Australia, New Zealand, Thailand, and the Philippines all supplied troops to help the United States during that war.

In 1976, a year after the Vietnam War ended, SEATO troops engaged in their last exercise. A year later the organization disbanded.

*George J. Fleming*

# Ryle Constructs First Radio Interferometer

*Sir Martin Ryle developed the first radio interferometer, an astronomical instrument involving several radio telescopes controlled by computer.*

**What:** Astronomy
**When:** 1955
**Where:** Cambridge, England
**Who:**
SIR MARTIN RYLE (1918-1984), an English astronomer
KARL JANSKY (1905-1950), an American radio engineer
HENDRIK CHRISTOFFEL VAN DE HULST (1918-2000), a Dutch radio astronomer
HAROLD IRVING EWAN (1922-      ), an American astrophysicist
EDWARD MILLS PURCELL (1912-1997), an American physicist

## Seeing with Radio

Since the early 1600's, astronomers have relied on optical telescopes for viewing stellar objects. Optical telescopes detect the visible light from stars, galaxies, quasars, and other astronomical objects. Throughout the late twentieth century, astronomers developed more powerful optical telescopes for peering deeper into the cosmos and viewing objects located hundreds of millions of light-years away from the earth.

In 1933, Karl Jansky, an American radio engineer with Bell Telephone Laboratories, constructed a radio antenna receiver for locating sources of telephone interference. Jansky discovered a daily radio burst that he was able to trace to the center of the Milky Way galaxy.

In 1935, Grote Reber, another American radio engineer, followed up Jansky's work with the construction of the first dish-shaped "radio" telescope. Reber used his 9-meter-diameter radio telescope to repeat Jansky's experiments and to locate other radio sources in space. He was able to map precisely the locations of various radio sources in space, some of which later were identified as galaxies and quasars.

Following World War II (that is, after 1945), radio astronomy blossomed with the help of surplus radar equipment. Radio astronomy tries to locate objects in space by picking up the radio waves that they emit. In 1944, the Dutch astronomer Hendrik Christoffel van de Hulst had proposed that hydrogen atoms emit radio waves with a 21-centimeter wavelength. Because hydrogen is the most abundant element in the universe, van de Hulst's discovery had explained the nature of extraterrestrial radio waves. His theory later was confirmed by the American radio astronomers Harold Irving Ewen and Edward Mills Purcell of Harvard University.

By coupling the newly invented computer technology with radio telescopes, astronomers were able to generate a radio image of a star almost identical to the star's optical image. A major advantage of radio telescopes over optical telescopes is the ability of radio telescopes to detect extraterrestrial radio emissions day or night, as well as their ability to bypass the cosmic dust that dims or blocks visible light.

## More with Less

After 1945, major research groups were formed in England, Australia, and the Netherlands. Sir Martin Ryle was head of the Mullard Radio Astronomy Observatory of the Cavendish Laboratory, University of Cambridge. He had worked with radar for the Telecommunications Research Establishment during World War II.

The radio telescopes developed by Ryle and other astronomers operate on the same basic principle as satellite television receivers. A constant stream of radio waves strikes the parabolic-shaped reflector dish, which aims all the radio waves at a focusing point above the dish. The fo-

AP/Wide World Photos

*Sir Martin Ryle.*

cusing point directs the concentrated radio beam to the center of the dish, where it is sent to a radio receiver, then an amplifier, and finally to a chart recorder or computer.

With large-diameter radio telescopes, astronomers can locate stars and galaxies that cannot be seen with optical telescopes. This ability to detect more distant objects is called "resolution." Like optical telescopes, large-diameter radio telescopes have better resolution than smaller ones. Very large radio telescopes were constructed in the late 1950's and early 1960's (Jodrell Bank, England; Green Bank, West Virginia; Arecibo, Puerto Rico). Instead of just building larger radio telescopes to achieve greater resolution, however, Ryle developed a method called "interferometry." In Ryle's method, a computer is used to combine the incoming radio waves of two or more movable radio telescopes pointed at the same stellar object.

Suppose that one had a 30-meter-diameter radio telescope. Its radio wave-collecting area would be limited to its diameter. If a second identical 30-meter-diameter radio telescope was linked with the first, then one would have an interferometer. The two radio telescopes would point exactly at the same stellar object, and the radio emissions from this object captured by the two telescopes would be combined by computer to produce a higher-resolution image. If the two radio telescopes were located 1.6 kilometers apart, then their combined resolution would be equivalent to that of a single radio telescope dish 1.6 kilometers in diameter.

Ryle constructed the first true radio telescope interferometer at the Mullard Radio Astronomy Observatory in 1955. He used combinations of radio telescopes to produce interferometers containing about twelve radio receivers. Ryle's interferometer greatly improved radio telescope resolution for detecting stellar radio sources, mapping the locations of stars and galaxies, assisting in the discovery of "quasars" (quasi-stellar radio sources), measuring the earth's rotation around the sun, and measuring the motion of the solar system through space.

## Consequences

Following Ryle's discovery, interferometers were constructed at radio astronomy observatories throughout the world. The United States established the National Radio Astronomy Observatory (NRAO) in rural Green Bank, West Virginia. The NRAO is operated by nine eastern universities and is funded by the National Science Foundation. At Green Bank, a three-telescope interferometer was constructed, with each radio telescope having a 26-meter-diameter dish. During the late 1970's, the NRAO constructed the largest radio interferometer in the world, the Very Large Array (VLA). The VLA, located approximately 80 kilometers west of Socorro, New Mexico, consists of twenty-seven 25-meter-diameter radio telescopes linked by a supercomputer. The VLA has a resolution equivalent to that of a single radio telescope 32 kilometers in diameter.

Even larger radio telescope interferometers can be created with a technique known as "very long baseline interferometry" (VLBI). VLBI has been used to construct a radio telescope having an effective diameter of several thousand kilometers. Such an arrangement involves the precise synchronization of radio telescopes located in several different parts of the world. Supernova 1987A in the Large Magellanic Cloud was studied using a VLBI arrangement between observa-

tories located in Australia, South America, and South Africa.

Launching radio telescopes into orbit and linking them with ground-based radio telescopes could produce a radio telescope whose effective diameter would be larger than that of the earth. Such instruments will enable astronomers to map the distribution of galaxies, quasars, and other cosmic objects, to understand the origin and evolution of the universe, and possibly to detect meaningful radio signals from extraterrestrial civilizations.

*David Wason Hollar, Jr.*

# Backus Invents FORTRAN Computer Language

*John Backus led the effort to design and develop FORTRAN, the first major computer programming language.*

**What:** Computer science
**When:** 1955-1957
**Where:** New York, New York
**Who:**

JOHN BACKUS (1924-      ), an American software engineer and manager

JOHN W. MAUCHLY (1907-1980), an American physicist and engineer

HERMAN HEINE GOLDSTINE (1913-      ), a mathematician and computer scientist

JOHN VON NEUMANN (1903-1957), a Hungarian American mathematician and physicist

## Talking to Machines

Formula Translation, or FORTRAN—the first widely accepted high-level computer language—was completed by John Backus and his coworkers at the International Business Machines (IBM) Corporation in April, 1957. Designed to support programming in a mathematical language that was natural to scientists and engineers, FORTRAN achieved unsurpassed success in scientific computation.

Computer languages are means of specifying the instructions that a computer should execute and the order of those instructions. Computer languages can be divided into categories of progressively higher degrees of abstraction. At the lowest level is binary code, or machine code: Binary digits, or "bits," specify in complete detail every instruction that the machine will execute. This was the only language available in the early days of computers, when such machines as the ENIAC (Electronic Numerical Integrator and Calculator) required hand-operated switches and plugboard connections. All higher levels of language are implemented by having a program translate instructions written in the higher language into binary machine language (also called "object code"). High-level languages (also called "programming languages") are largely or entirely independent of the underlying machine structure. FORTRAN was the first language of this type to win widespread acceptance.

The emergence of machine-independent programming languages was a gradual process that spanned the first decade of electronic computation. One of the earliest developments was the invention of "flowcharts," or "flow diagrams," by Herman Heine Goldstine and John von Neumann in 1947. Flowcharting became the most influential software methodology during the first twenty years of computing.

Short Code was the first language to be implemented that contained some high-level features, such as the ability to use mathematical equations. The idea came from John W. Mauchly, and it was implemented on the BINAC (Binary Automatic Computer) in 1949 with an "interpreter"; later, it was carried over to the UNIVAC (Universal Automatic Computer) I. Interpreters are programs that do not translate commands into a series of object-code instructions; instead, they directly execute (interpret) those commands. Every time the interpreter encounters a command, that command must be interpreted again. "Compilers," however, convert the entire command into object code before it is executed.

Much early effort went into creating ways to handle commonly encountered problems—particularly scientific mathematical calculations. A number of interpretive languages arose to support these features. As long as such complex operations had to be performed by software (computer programs), however, scientific computation would be relatively slow. Therefore, Backus lobbied successfully for a direct hardware implementation of these operations on IBM's new scientific computer, the 704. Backus then started the Programming Research Group at IBM in order to develop a compiler that would allow programs to be written in a mathematically oriented language rather than a machine-oriented language. In November of 1954, the group defined an initial version of FORTRAN.

### A More Accessible Language

Before FORTRAN was developed, a computer had to perform a whole series of tasks to make certain types of mathematical calculations. FORTRAN made it possible for the same calculations to be performed much more easily. In general, FORTRAN supported constructs with which scientists were already acquainted, such as functions and multidimensional arrays. In defining a powerful notation that was accessible to scientists and engineers, FORTRAN opened up programming to a much wider community.

Backus's success in getting the IBM 704's hardware to support scientific computation directly, however, posed a major challenge: Because such computation would be much faster, the object code produced by FORTRAN would also have to be much faster. The lower-level compilers preceding FORTRAN produced programs that were usually five to ten times slower than their hand-coded counterparts; therefore, efficiency became the primary design objective for Backus. The highly publicized claims for FORTRAN met with widespread skepticism among programmers. Much of the team's efforts, therefore, went into discovering ways to produce the most efficient object code.

The efficiency of the compiler produced by Backus, combined with its clarity and ease of use, guaranteed the system's success. By 1959, many IBM 704 users programmed exclusively in FORTRAN. By 1963, virtually every computer

manufacturer either had delivered or had promised a version of FORTRAN.

Incompatibilities among manufacturers were minimized by the popularity of IBM's version of FORTRAN; every company wanted to be able to support IBM programs on its own equipment. Nevertheless, there was sufficient interest in obtaining a standard for FORTRAN that the American National Standards Institute adopted a formal standard for it in 1966. A revised standard was adopted in 1978, yielding FORTRAN 77.

### Consequences

In demonstrating the feasibility of efficient high-level languages, FORTRAN inaugurated a period of great proliferation of programming languages. Most of these languages attempted to provide similar or better high-level programming constructs oriented toward a different, nonscientific programming environment. COBOL, for example, stands for "Common Business Oriented Language."

FORTRAN, while remaining the dominant language for scientific programming, has not found general acceptance among nonscientists. An IBM project established in 1963 to extend FORTRAN found the task too unwieldy and instead ended up producing an entirely different language, PL/I, which was delivered in 1966. In the beginning, Backus and his coworkers believed that their revolutionary language would virtually eliminate the burdens of coding and debugging. Instead, FORTRAN launched software as a field of study and an industry in its own right.

In addition to stimulating the introduction of new languages, FORTRAN encouraged the development of operating systems. Programming languages had already grown into simple operating systems called "monitors." Operating systems since then have been greatly improved so that they support, for example, simultaneously active programs (multiprogramming) and the networking (combining) of multiple computers.

*Kevin B. Korb*

**1206**

# Congress Passes Formosa Resolution

> *With a resolution stating that the president of the United States had the power to send armed forces to defend Formosa (Taiwan) and the Pescadores Islands, Congress backed the policy of "containing" Communism.*

**What:** International relations
**When:** January 24-29, 1955
**Where:** Washington, D.C.; Taiwan; Quemo; and Matsu
**Who:**
DWIGHT D. EISENHOWER (1890-1969), president of the United States from 1953 to 1961
CHIANG KAI-SHEK (1887-1975), president of the Republic of China from 1950 to 1975
ZHOU ENLAI (CHOU EN-LAI; 1898-1976), premier of the People's Republic of China from 1949 to 1976

## Attacks and Debates

In 1949, as the Chinese civil war came to an end, the Communists took over mainland China. Chiang Kai-shek, president of the Republic of China, fled with part of his government and army to the island of Taiwan (which many Americans called Formosa) and the nearby Pescadores Islands. Claiming that his Nationalist government was still the real government of China, Chiang announced that he would return to the mainland to regain power as soon as possible.

Taiwan and the Pescadores had been held by the Japanese from 1895 until 1945, when these territories were returned to China at the end of World War II. Chiang's troops also held other islands off the China coast, including Quemoy (which lay close to the Chinese port of Amoy, or Xiamen; Hsia-men), Matsu (off the coast of Fuzhou; Foochow), and the Tachens, about two hundred miles north of Matsu.

Chiang and the Chinese Communists agreed that Taiwan was a province of China and Quemoy and Matsu were part of the mainland province of Fujian (Fukien). Though Quemoy and Matsu were small, both sides saw them as important stepping-stones. Chiang planned to use them to help him return to the mainland, while the Chinese wanted to use them for moving into Taiwan.

Having supported Chiang in the civil war, the United States still insisted that his regime was the legitimate government for all China. After the Chinese Communists became involved in the Korean War (1950-1953), Americans became even more opposed to the People's Republic of China. Military and economic aid went to Taiwan, and the U.S. Seventh Fleet patrolled the Formosa Strait to prevent an invasion of Taiwan.

Against the advice of U.S. military leaders, Chiang increased his forces on Quemoy and Matsu; meanwhile, the Communists put even larger forces on the shore facing the islands. In August and September, 1954, the Communists began to bombard the islands, and two U.S. military advisers were killed.

Throughout that autumn, the United States and its allies argued about how best to protect Taiwan. Some of the Joint Chiefs of Staff and some members of Congress wanted to help Chiang return to the mainland and to give U.S. support to his forces on Quemoy and Matsu. Others argued, however, that Chiang could not hope to win, and that greater American involvement would mean a major war in Asia, perhaps even World War III.

Secretary of State John Foster Dulles believed that helping the Chinese Nationalists to remain strong would keep the Chinese Communist government from feeling secure—and would also give some hope to those who wanted to see the Communists overthrown.

**1207**

On December 2, 1954, the United States and Nationalist China signed a treaty of mutual defense—but it made no specific mention of the offshore islands. A month later, the Chinese Communists began to bombard the islands again.

## The Resolution

As the attacks continued, United States president Dwight D. Eisenhower sent a special message to Congress on January 24, 1955, asking for authority to use American armed forces to protect Taiwan, the Pescadores, and certain "closely related localities." Though Eisenhower realized that he already had this authority as commander in chief of the Armed Forces, he wanted Congress to support him and to make a show of unity against the Communists. In Communist China, Premier Zhou Enlai called Eisenhower's message a "war message."

The new Eighty-fourth Congress had a Democratic majority in both houses. Democrats Walter George in the Senate and James P. Richards in the House introduced a joint resolution, which became known as the Formosa Resolution, stating that peace in the western Pacific was important to the United States. Communist attacks were a danger to peace; therefore, the resolution made it clear that the president of the United States was authorized to use American forces to protect Taiwan and the Pescadores against "armed attack." The president would also have the right to protect "related possessions and territories of that area now in friendly hands."

The House passed the resolution on January 25, 1955, by a vote of 410 to 3. In a Senate committee, several amendments were introduced to change the resolution. Some senators wanted to turn Taiwan and the Pescadores over to the United Nations and to give the president authority only until the United Nations took action. Other amendments would have specifically omitted Quemoy and Matsu from the president's

*Chinese family from the Tachen Islands being evacuated to Keelung, Formosa.*

AP/Wide World Photos

authority. But none of these amendments was passed, and on January 28, 1955, the Senate passed the Formosa Resolution 85 to 3. President Eisenhower signed it the next day.

## Consequences

The mutual security treaty with Nationalist China was ratified in February. In that same month, Chiang evacuated the Tachen Islands, though he refused to reduce his forces on Quemoy and Matsu. The Communist government quickly seized the Tachens.

In April, Zhou Enlai told the Afro-Asian Conference in Bandung, Indonesia, that his country did not want war with the United States. He said that he would be willing to negotiate on Far Eastern issues, including Taiwan. There was never any formal statement or agreement to a cease-fire, but the Communists did stop bombarding Quemoy and Matsu in May, 1955.

Eisenhower saw Taiwan and Korea as the United States' line of defense against Communism in Asia. Losing Taiwan would be a major break in that line. Chiang's forces needed to be kept strong and to be allowed to keep hoping for a return to the mainland. That was why the United States had to help the Nationalists keep control of the Formosa Straits. By voting strongly in favor of the Formosa Resolution, Congress agreed that Communism must be stopped, but

that the United States should avoid becoming involved in war.

At the time of the Vietnam War, the 1964 Tonkin Gulf Resolution gave the president authority similar to that given by the Formosa Resolution. Those who opposed the Vietnam War argued that both of these resolutions gave the president too much power. Others insisted, however, that the resolutions were important expressions of national unity against the Communist threat.

*George J. Fleming*

# East European Nations Sign Warsaw Pact

*The Warsaw Pact made the Soviet Union's control over most of Eastern Europe official.*

**What:** International relations
**When:** May 14, 1955
**Where:** Warsaw, Poland
**Who:**
NIKITA S. KHRUSHCHEV (1894-1971), first secretary of the Communist Party of the Soviet Union from 1953 to 1964, and premier of the Soviet Union from 1958 to 1964
NIKOLAI ALEKSANDROVICH BULGANIN (1895-1975), premier of the Soviet Union from 1955 to 1958
VYACHESLAV MIKHAILOVICH MOLOTOV (1890-1986), Soviet foreign minister from 1953 to 1956
IVAN STEPANOVICH KONEV (1897-1973), commander in chief of the Warsaw Pact armed forces

## Response to NATO

After World War II, because the Western Allies were not able to agree with the Soviet Union about the fate of Germany, the German nation eventually became two countries. The Soviet Union had wanted to punish Germany by keeping it weak, while the United States, Great Britain, and France wanted a strong Germany to stand as an obstacle to Soviet communist expansion. In the 1950's, Soviet leaders became more and more concerned about the fact that West Germany was rearming itself. The North Atlantic Treaty Organization (NATO), made up of North American and Western European nations, also admitted West Germany to membership.

Representatives from the Soviet Union and various Eastern European nations gathered in Moscow in November, 1954, and Soviet foreign minister Vyacheslav M. Molotov announced that West Germany's "remilitarization" would lead to "special vigilance"—and new policies for defending Eastern Europe. Molotov was quite concerned that the Eastern Bloc countries be well prepared for war with the West.

First Secretary Nikita Khrushchev and Premier Nikolai Bulganin, however, were not quite as hard-nosed toward the West as Molotov was. Khrushchev and Bulganin's ideas won at the Warsaw Conference in May, 1955, when Soviets and East Europeans gathered to sign the Treaty of Friendship, Cooperation, and Mutual Assistance, or the Warsaw Pact. Khrushchev saw the Warsaw Pact mostly as a political device to help keep the balance of power during the Cold War.

The Geneva Summit Conference was coming up in July, 1955, and Khrushchev wanted the Warsaw Pact mostly to strengthen the Soviet position at that international conference. He hoped that all European nations would agree to an overall treaty for collective security; then NATO and the Warsaw Pact could be disbanded. If that goal could not be achieved, then perhaps each side would at least sign a treaty of non-aggression.

Soviet leaders had another concern. Since the death of Soviet dictator Joseph Stalin in 1953, Khrushchev had been leading the Soviet Union in "de-Stalinization"—rejecting some of Stalin's oppressive methods and creating a more positive feeling in the country. In Eastern Europe, however, some communist nations were interpreting de-Stalinization as an opportunity to become more independent from the Soviet Union. This was not what Khrushchev wanted, and the Warsaw Pact provided an opportunity for the Soviets to regain control over their East European "satellites."

## The Pact

The nations that signed the Warsaw Pact on May 14, 1955, included Albania, Bulgaria, Hungary, East Germany, Poland, Romania, the Soviet Union, and Czechoslovakia. The treaty's eleven articles called Warsaw Pact members to consult with one another on all issues of common interest. Conflicts would be settled peacefully, and defense would be a joint concern. A "political consultative committee" was established so that policy decisions could be made in harmony.

The military agreement was the most important part of the treaty, for the Warsaw Pact nations were to contribute troops to a joint command for mutual defense. Soviet Marshal I. S. Konev was appointed commander in chief of the joint forces, and the ministers of defense of the other member states became his deputies. Each of these deputies was put in charge of the troops contributed by his own state. The Warsaw Pact's military headquarters was located in Moscow.

Khrushchev's hopes for the Geneva Summit Conference came to nothing, and the Warsaw Pact gave the Soviets a justification for quickly moving large numbers of troops into Eastern European countries. Military equipment in the Warsaw Pact nations was standardized, and officers and soldiers alike were indoctrinated with the communist message. The most important positions in all Warsaw Pact armies went to officers who had been trained in the Soviet Union.

At first, the German Democratic Republic was not made a part of the joint command. At a meeting of the political consultative committee in January, 1956, however, East Germany was given equal status. To balance West Germany, the East German National People's Army was created and made part of the joint command.

## Consequences

Though the Soviets had at first seen the Warsaw Pact simply as a political bargaining tool, it eventually came to be an important means of military control as well. Warsaw Pact nations came to have a single, Soviet-controlled policy toward noncommunist countries. Within the Eastern Bloc itself, the pact allowed Moscow to promote "fraternal bloc solidarity"—which meant that any member that moved toward independence was quickly squelched.

In the fall of 1956, riots and large public demonstrations flared up in Poland and Hungary. The

### WARSAW PACT NATIONS

**1211**

governments of these countries gave in to some of the protesters' demands, though Moscow had told them to take a hard line. In October, 1956, the Soviet Union decided to intervene. A group of Soviet leaders went to Warsaw; with Soviet troops stationed nearby, they were quickly able to bring the Polish government under control.

In Hungary, however, the demonstrations were not stopped so easily. After Premier Imre Nagy announced that his nation would withdraw from the Warsaw Pact, Soviet troops flooded Hungary, and the uprising was crushed. In 1968, a similar rebellion in Czechoslovakia was defeated when Soviet tanks rolled into Prague. By this time, the Warsaw Pact's most important purpose was simply keeping the Eastern European governments in line.

*Manfred Grote*

# Geneva Summit Conference Opens East-West Communications

> *Though Western leaders did not manage to form any real agreements with leaders of the Soviet Union, the Geneva Summit Conference brought about what some called a "thaw" in the Cold War—a new willingness to discuss East-West problems.*

**What:** International relations
**When:** July 18-23, 1955
**Where:** Geneva, Switzerland
**Who:**

DWIGHT D. EISENHOWER (1890-1969), president of the United States from 1953 to 1961

NIKOLAI ALEKSANDROVICH BULGANIN (1895-1975), premier of the Soviet Union from 1955 to 1958

NIKITA S. KHRUSHCHEV (1894-1971), first secretary of the Communist Party of the Soviet Union from 1953 to 1964, and premier of the Soviet Union from 1958 to 1964

ROBERT ANTHONY EDEN (1897-1977), prime minister of Great Britain from 1955 to 1957

EDGAR FAURE (1908-1988), president of France in 1952 and from 1955 to 1956

## Steps to Summit

The year 1953 was an important turning point in the ongoing Cold War. In January, Republican Dwight D. Eisenhower became president of the United States, ending twenty years of Democratic administration. In the Soviet Union, the death of Joseph Stalin in March was followed by a new era of "collective leadership." In July, the Korean War came to an end without a real victory for either side.

As a result of these events, some world leaders began to hope that the Cold War between the Soviet Union and the West, which had begun in 1946, might become less intense. Perhaps there could even be solutions to some of the issues that had divided East from West. Only one month after Stalin's death, Winston Churchill, who had once again become prime minister of Great Britain, suggested that the time for improving East-West relations had arrived. World leaders should plan a gathering, he said, in order to resolve some of the major issues dividing them.

The United States was not quick to agree to the idea. Secretary of State John Foster Dulles said that the Eisenhower administration would not consider direct talks with Soviet leaders until they had shown their "good faith." The Soviets, meanwhile, liked Churchill's idea but did not want to accept any preconditions for such a meeting.

In the fall of 1954, however, Eisenhower implied that a gathering of government leaders, or "summit meeting," might be possible once the Paris Accords had been ratified. The Paris agreements, signed in October, 1954, made West Germany a completely independent country, with membership in the North Atlantic Treaty Organization (NATO).

The Soviets had been opposed to West Germany's entrance into NATO, but once it was clear that the Paris Accords would be ratified, they began to soften their position. No final peace treaty for Austria had ever been signed after World War II, for the Soviets had insisted that East and West Germany would have to be reunified first. Yet now the Soviet government opened negotiations for the Austrian treaty.

On May 15, 1955, the foreign ministers of France, Great Britain, the United States, and the Soviet Union met in Vienna and signed a treaty with the Austrian government, putting an end to Allied occupation and guaranteeing that Austria would be neutral and completely independent.

**1213**

Austria's neutrality was especially important to the Soviets, who also wanted Germany to be neutral.

The Austrian negotiations were the act of good faith for which the American government had been waiting. On May 10, a joint note from U.S. president Eisenhower, British prime minister Anthony Eden, and French president Edgar Faure invited the Soviet Union to join "in an effort to remove the sources of conflict between us." On May 26, the Soviet government accepted the invitation.

### Talking in Geneva

The Geneva Summit Conference opened on July 18 in the Palais des Nations, an old League of Nations building. Nearly twelve hundred representatives of the press had come to cover the conference—the first summit meeting since the Potsdam Conference in the summer of 1945.

From the beginning, it was obvious that this would not be an informal, personal gathering of leaders, as Churchill had hoped. Forty men sat down around four tables arranged in a square, with rows of assistants and observers stretching out behind them.

As the only "chief of state" present, Eisenhower was the first chairman and the first to address the group. He stated that the conference should give most of its attention to reunifying Germany under a freely elected government, the question of disarmament with supervision and inspection, and improved communication between East and West.

Faure followed, emphasizing the need for a peace treaty with Germany and disarmament

*From left to right, Nikolai Aleksandrovich Bulganin, Dwight D. Eisenhower, Edgar Faure, and Robert Anthony Eden in Geneva.*

**1214**

through some kind of international organization.

Eden also stressed German reunification; he proposed a five-power security pact to include Germany and suggested that a demilitarized zone be created in Europe between East and West.

Last of the four leaders to speak was Soviet premier Nikolai Bulganin. Calling for an end to the arms race and a ban on nuclear weapons, Bulganin also suggested that all European states and the United States join in a system of "collective security." This would make NATO, the Paris Accords, and the Warsaw Pact no longer necessary. Germany should be reunified, Bulganin said, but not as a member of NATO. He asked that there be more cultural and scientific exchange between East and West.

On the basis of these statements, the four foreign ministers then made a list of topics for discussion: German unification, European security, disarmament, East-West contacts, neutrality, and (because the Soviets insisted) Far Eastern problems. The Soviets did not want the governments of Eastern Europe to be discussed.

The next day, Eisenhower agreed not to discuss Eastern European governments but added that the Communist-Nationalist conflict over Taiwan could not be discussed either. Discussions proceeded from this point, with each of the four leaders expressing views on the various topics that had been listed. Unfortunately, there was not a spirit of compromise; instead, each side seemed stubborn, praising its own ideas and criticizing the ideas of the other side.

On July 20, the Soviet delegation presented a specific proposal for a treaty of European collective security. Failing to get support for this proposal, they came back the next day to suggest a nonaggression pact between NATO and the Warsaw Pact until a collective security agreement could be made.

The Soviet leaders also suggested a disarmament agreement that would reduce the sizes of armed forces; those who signed would also agree not to use nuclear weapons and to work toward banning them completely. This proposal opened the door for Eisenhower to bring a proposal of his own.

Because effective inspection and supervision were absolutely necessary in disarmament, Eisenhower said, there should be "open skies" around the world. As a demonstration of the United States' goodwill, he proposed that the four powers exchange "a complete blueprint of our military establishments, from beginning to end, from one end of our countries to the other." Each nation would be allowed to make aerial photographs of the others' military facilities as well.

It was a startling proposal. At first Bulganin was positive, but during the first recess Khrushchev approached Eisenhower to declare that he disagreed with Bulganin. According to Khrushchev, "the idea was nothing more than a bald espionage plot against the U.S.S.R." During the last two days of the talks, there were no important proposals, only general discussions about establishing better communication, travel, and study.

## Consequences

The Geneva Summit Conference actually did not bring about any new decisions or agreements. Eisenhower had no private talks with Khrushchev, who had clearly become the most important leader of the Soviet Union. The "open skies" proposal had little chance of acceptance, and on August 4, Bulganin officially rejected it in a speech before the Supreme Soviet.

Bulganin said, however, that the Geneva Summit had been a success because it had reduced tensions between East and West. Eisenhower also expressed hope for future East-West relations because of the "Spirit of Geneva." Probably the most important result of the Geneva Summit was that both sides were talking to each other again, clearly stating their positions on important issues. Soon afterward, exchange visits between American and Soviet scientists, scholars, artists, and political leaders began to take place.

*Tyler Deierhoi*

# Sudanese Civil War Erupts

*Civil war began in the Sudan when the Southern Corps of the Sudanese army revolted against its northern officers.*

**What:** Civil war
**When:** August 18-30, 1955
**Where:** Torit, Equatoria Province, Sudan
**Who:**
ISMAIL AL-AZHARĪ (1902-1969), prime minister of the Sudan from 1954 to 1956
RENALDO LOLEYA (died 1956), a second lieutenant in the Southern Corps
IBRAHIM ABBOUD (1900-1983), commander in chief of the Sudanese armed forces and president of the Sudan from 1958 to 1964

## Inequality Breeds Mistrust

In the southern part of the Sudan—which includes the Bahr al-Ghazal, Upper Nile, and Equatoria provinces—thirty different languages are spoken, and there is a wide variety of religions, including Christianity. The people of the northern Sudan are mostly Arabic-speaking Muslims and tend to have lighter skin than the southerners do.

Great Britain and Egypt ruled the Sudan from 1898 to 1956. The British preferred to keep Islamic influence out of the south, so they closed the south to northerners, except for those on government business. Greek, Syrian, and Jewish traders were encouraged to set up business in the south, but northern Sudanese traders were expelled. Christian missionaries received government support, while Muslims were not allowed to proselytize in the south. Missionaries were given responsibility for educating the school-age children of the south.

Though the British had intended these rules to protect the cultures of the south, the result was inequality and mistrust between north and south. Education, economic development, and politics were not as developed in the south as in the north. The northerners believed that southerners were backward and uncivilized.

In 1946, the British changed their policy and tried to help the Sudan become more unified. By this time, nationalists in the north were demanding that the British leave their country. Negotiations for independence began, but few southerners were involved. Many of them feared that in an independent Sudan the south would be dominated by the north.

In 1954, the newly elected Sudanese government began taking control of the army, police, and administration. Most of the formerly British positions went to northerners—for example, of eight hundred administrative jobs, only six went to southerners. Even in the army's Southern Corps, which was made up of southern troops, the highest ranks (twenty-four positions) were filled by northerners; only nine less important posts were given to southerners.

In May, 1955, two southern members of the Sudan's cabinet were dismissed for disagreeing with the prime minister, Ismail al-Azharī. The southerners were trying to form a southern bloc in Parliament, calling for a federal constitution so that the south could have some control over its own affairs. In July, one southern member of Parliament was arrested, and five other southerners were arrested and given prison sentences. Rumors began flying that the government intended to persecute southerners.

## The Mutiny

On August 7, the government learned of a mutiny plot in the Southern Corps, and three days later, northern troops were brought into Juba, the capital of Equatoria Province. Civilians began to flee the city.

The government planned to move Number Two Company of the Southern Corps from Torit,

**1216**

*In August, 1955, crowds gather in front of the parliament building in Sudan to hear the prime minister, who proposed the evacuation of foreign troops from the country.*

Equatoria, to Khartoum, capital of the Sudan, on August 18. When that day came, however, the soldiers refused to obey their orders, for they had heard rumors that they were to be executed in the north. Soon they had taken over their base; they killed several northern officers and looted shops that were owned by northerners.

On August 19, Lieutenant Renaldo Loleya arrived from Juba, announcing that the northern troops there were shooting wildly at soldiers and civilians alike. He took command of the mutineers. Northerners who had taken shelter at the police station were rounded up by the mutineers on August 20, and several were killed. A total of seventy-eight northerners were killed at Torit.

Other garrisons of the Southern Corps in Equatoria soon joined in the mutiny, as did many police, civilians, and even some government officials. Northerners were attacked, and their property was looted. In Bahr al-Ghazal and Upper Nile provinces, however, the government authorities were able to prevent a similar mutiny.

Prime Minister al-Azharī asked the mutineers to surrender and promised them fair treatment, but he refused to pull the northern troops out of Juba or to invite the United Nations to come in and investigate the situation.

On August 27, the mutineers agreed to surrender, but when northern troops arrived in Torit on August 30, they found only Loleya and a few companions in the garrison. The others had fled, fearing that the northerners would punish them harshly. In all, only 461 southern soldiers surrendered, out of a total force of about 1,400.

On September 6, however, the government declared that the disturbance was over. The Southern Corps was disbanded and replaced with northern troops.

There were reports that northern troops used torture, mutilation, and execution without trial to punish southerners. Officially, 121 persons were convicted and executed for participating in the mutiny. One of them was Renaldo Loleya.

## Consequences

The Sudan proceeded to celebrate its independence on January 1, 1956. In the next few years, southern members of Parliament continued to push for a federation, but they had no success. On November 17, 1958, the Sudan's elected government was overthrown by Brigadier General Ibrahim Abboud. He was determined to force the Arabic language and the Islamic religion on the south. In 1960 he tried to arrest many southern political leaders, but they were warned and managed to leave the country. Many other educated southerners followed these leaders into exile.

Abboud also tried to get rid of the influence of Christian missionaries, who were blamed for the division between north and south. In 1960, the Sunday holiday was abolished, and a number of priests were arrested. In 1961, all religious gatherings outside churches were banned, and missionaries who left the country were not allowed to return. Finally, in 1964, all foreign missionaries were expelled from the south (although not from the north). As a result, nearly all the schools in the south were closed.

In 1962, southern politicians in exile formed an organization, later known as the Sudan African National Union (SANU), to push for complete independence for the southern Sudan. The next year, an organized guerrilla movement began to fight the government.

Abboud was overthrown in 1964, and the Sudan returned to civilian rule. Yet violence continued, and hundreds of thousands of southerners fled to neighboring countries or to the north. In 1972, President Jafir Nimeiri was able to negotiate a cease-fire with the guerrillas and the southern political leaders. The peace lasted for eleven years. Another mutiny broke out in 1983 in the Upper Nile Province, and the Sudan was once again plunged into a bitter civil war.

*T. K. Welliver*

# American Federation of Labor and Congress of Industrial Organizations Merge

> *The merger of the American Federation of Labor (AFL) and the Congress of Industrial Organizations (CIO) created the largest labor organization in American history.*

**What:** Labor; Economics
**When:** December 5, 1955
**Where:** New York City
**Who:**
GEORGE MEANY (1894-1980), president of the AFL
WALTER P. REUTHER (1907-1970), president of the CIO

## The Twenty-Year Split

From late in the nineteenth century until the 1930's, the American Federation of Labor (AFL) was the main body of organized American labor, although it was never the only labor group. Within this federation of national and international unions, the individual unions kept a considerable amount of independence. Most of the AFL unions were made up of skilled laborers who banded together according to their crafts or skills.

In the 1920's, and even more in the early Depression years (1929-1932), the AFL lost members and prestige. Yet with the inauguration of President Franklin D. Roosevelt in 1933, this downward trend stopped. New Deal laws—for example, section 7A of the National Industrial Recovery Act (1933) and the National Labor Relations Act, or Wagner Act (1934)—encouraged the formation of unions; free collective bargaining was guaranteed by law.

With this encouragement, there were new organizing drives for unions. Yet some AFL groups, especially the United Mine Workers, which was led by John L. Lewis, argued that the drives could not be successful unless they were based on "industrial unionism": organizing workers by the industry in which they worked instead of by craft (for example, there would be a union for auto workers rather than one for machinists). Many of the leaders of the older craft unions disliked this idea. Personal loyalties and antagonisms made the argument a bitter one.

In 1935, insisting that the AFL was not reaching the mass-production industries, Lewis joined with leaders of some other unions to form the Committee on Industrial Organization within the AFL. After angry debate, the AFL revoked the charters of the unions within the Committee. In 1938, these unions organized themselves as an independent federation, named the Congress of Industrial Unions (CIO).

Over the years that followed, the CIO created a number of industrial unions, especially in automobile manufacturing and steel. The CIO was more active in politics and social reform than the AFL had been. Some antagonism continued between the two groups, but sometimes they found themselves working toward the same goals, or even cooperating. During World War II, for example, both organizations were given opportunities to speak on behalf of working people. After the war, they were often on the same side in Washington, D.C., opposing laws that they considered harmful to labor.

Still, the division remained. Roosevelt had tried to bring unity, but to no avail.

## The Merger

By 1952, however, there were new faces in the leadership of both organizations. Both William Green, president of the AFL, and Philip Murray, president of the CIO, had died. Lewis, who had by now taken the United Mine Workers out of the CIO, was no longer involved in either of these organizations.

The new leaders, George Meany of the AFL and Walter Reuther of the CIO, began talks early

**1219**

in 1953. Later that year, committees from the two organizations met and appointed a subcommittee of three members from each federation to make a plan of unity. The subcommittees included Meany and Reuther, along with AFL secretary-treasurer William F. Schnitzler, CIO secretary-treasurer James B. Carey, Matthew Woll of the Photo-Engravers for the AFL, and David McDonald of the United Steelworkers for the CIO. (Harry Bates of the Bricklayers later replaced Woll.)

As a first step, the subcommittee worked out an agreement for "no raiding": AFL unions agreed not to recruit members from CIO unions, and vice versa. Though all union leaders were not pleased with this agreement, it was approved.

The next step, in February, 1955, was the writing of a merger agreement by the subcommittees, with the help of lawyers—Arthur Goldberg for the CIO and Joseph A. Woll (son of Matthew Woll) for the AFL. The draft soon went to the full committees, which approved it.

The proposed agreement went by what Meany called "the short route"—all affiliated unions of both federations would be automatically part of the merger. Yet each union's independence was to be respected. When there was a question of which union had jurisdiction in a particular workplace, it would be settled by negotiation. Craft and industrial unions would have equal status.

The agreement was approved by the governing boards, and the two lawyers were asked to write a constitution for the new federation. The constitution was revised and approved in May, 1955, by subcommittees, committees, and the two boards.

Separate conventions were scheduled in New York City on December 1, 1955, and the whole body gathered on December 5 under the name AFL-CIO. The constitution was approved by both groups. Upon Reuther's nomination, Meany was unanimously elected president of the AFL-CIO; Carey nominated Schnitzler to serve as secretary-treasurer, and he too was voted in. Vice-

*President George Meany (seated fifth from the right) and the other leaders of the newly merged AFL-CIO pose for a portrait.*

presidencies and places on the Executive Council and Executive Committee were assigned to various AFL and CIO leaders.

The AFL-CIO constitution made a strong statement against communism and other totalitarian systems; unions that were dominated by communists would be expelled from the federation. Also, the constitution stated that corruption and racketeering were problems in some unions and gave the council power to investigate these problems and to recommend solutions.

## Consequences

Twenty years of division had been brought to an end. Though some unions—most notably the United Mine Workers—remained outside the AFL-CIO, the new federation represented most of organized American labor. There were no exact figures for the membership of the AFL-CIO, but it was estimated at fifteen million, making it the largest single organization in the history of American labor.

*George J. Fleming*

# African Americans Boycott Montgomery, Alabama, Buses

*To protest segregation, African Americans organized a boycott of public buses in Montgomery, Alabama—and began the Civil Rights movement.*

**What:** Civil rights and liberties; Social reform; Economics

**When:** December 5, 1955-December 21, 1956

**Where:** Montgomery, Alabama

**Who:**

ROSA PARKS (1913-    ), a seamstress and former secretary of the local NAACP chapter

MARTIN LUTHER KING, JR. (1929-1968), pastor of Dexter Avenue Baptist Church

## The Boycott Begins

The *Brown v. Board of Education* case, in which the Supreme Court decided that racial segregation in public schools was unconstitutional, marked the beginning of a dramatic time of change for African Americans. There had been some important breakthroughs in the years before 1954—especially the desegregation of the armed forces. Yet up to this time African Americans were still a separate community, with fewer rights and opportunities and less legal protection than their white fellow citizens.

Especially in the "Deep South," the principle of "separate but equal" was used in everyday life. Hundreds of laws, many of them passed in the late nineteenth century, restricted Southern African Americans eating, traveling, studying, or worshiping together with whites.

The Supreme Court's ruling against school segregation did not bring any immediate changes to Montgomery, Alabama. This city of about 130,000 people—50,000 of them African Americans—had once been the capital of the Confed-

eracy, and its white leaders had no wish to leave the old ways.

Encouraged by the Court's decision, however, a group of African Americans in Montgomery began to organize themselves. Ministers, teachers, and local leaders of the National Association for the Advancement of Colored People (NAACP) banded together, but they did not find a way to make their voice heard until December 1, 1955.

On that day, Rosa Parks, a seamstress at a department store and former secretary of the local NAACP chapter, refused to give up her seat on the bus to a white passenger, as city ordinances required. She was arrested.

African American leaders in Montgomery began at once to organize a protest in Parks's behalf. E. D. Nixon, a porter on a Pullman train and long-time leader in the African American community, contacted two young Baptist ministers: Martin Luther King, Jr., and Ralph David Abernathy. The three men, together with leaders of the Women's Political Council (an African American group), made plans for all African Americans to boycott Montgomery buses beginning on Monday, December 5.

The first day of the boycott was almost a total success. At a meeting held that day, the Montgomery Improvement Association (MIA) was formed to coordinate the boycott; King was elected president.

## The Protest Succeeds

At first Montgomery's whites were simply amused or indifferent. Yet as the boycott continued into December and the bus company's income dropped more than 75 percent, they became concerned. There were a series of meetings between the city commissioners, representatives of the bus company, and MIA, but the whites were unwilling to satisfy the African Americans' demands: courteous treatment by bus drivers; first-come, first-served seating, with African Americans filling the back of the bus and whites the front; and employment of African American bus drivers for routes that served mostly African American passengers.

City police began to harass the car pools that the African Americans had set up, and some of their drivers were arrested. King himself was arrested for speeding, and on January 30, 1956, his house was blasted by dynamite. Within several weeks, the houses of two other boycott leaders had also been blasted.

These acts of violence and intimidation worked to bring Montgomery's African American community together, so that it was able to continue the boycott for more than a year. The violence also drew national attention to Montgomery, so that the boycott began to receive support from outside. King was able to bring his message of "nonviolent passive resistance" to a new and much larger audience. In one mass meeting after another, he urged people to ignore hostile attacks, to confront persecutors passively, and to refuse to fight back.

While the boycott continued, the legal issues were taken to the courts. On February 1, five Montgomery women filed suit in U.S. District Court to have Alabama's transportation laws declared unconstitutional. A hearing was set for May 1.

By that time, eighty-nine MIA members had been convicted for "conspiracy to interfere with normal business." Yet the court declared that Alabama's bus ordinance was, indeed, unconstitutional. The city appealed this ruling to

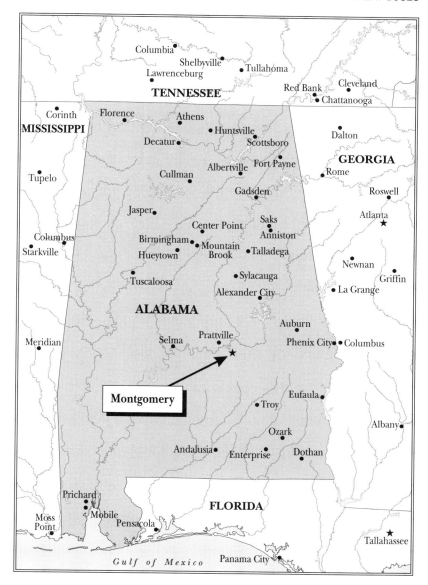

the Supreme Court, and new steps were taken against the boycott. In November, city officials were able to get an injunction against the car pool.

Nevertheless, King and other MIA leaders were able to keep the boycott going for another month, until the Supreme Court had upheld the lower court's ruling and forced the bus company to change its ways. On December 21, 1956, King, Abernathy, and Nixon marked the end of the boycott by boarding a bus in front of King's house. Two months of more bombings and arson followed before Montgomery's whites began to accept integration on city buses.

## Consequences

The boycott had succeeded in changing public transportation in Montgomery, but it had also achieved a greater success. As a result of the Montgomery struggle, Martin Luther King, Jr., became one of the most important spokespersons for African Americans. His methods of nonviolent passive resistance, which had been tested in Montgomery, were the major tool of the Civil Rights movement until the mid-1960's.

Soon after the boycott ended, King helped to found the Southern Christian Leadership Conference, which took the boycott method to other Southern states. Clearly, the Montgomery bus boycott was a turning point in the twentieth century, for it marked a new determination among African Americans to claim their rights.

*Courtney B. Ross*

# First Transatlantic Telephone Cable Starts Operation

*The first transatlantic telephone cable was designed to transmit calls between the continental United States and Europe.*

**What:** Communications
**When:** 1956
**Where:** Continental United States to Europe
**Who:**
SAMUEL F. B. MORSE (1791-1872), an American artist and inventor
CYRUS WEST FIELD (1819-1892), an American financier

## Repeated Frustration and Failure

The history of submarine cable design and development is largely a chronicle of repeated frustration and failure. From the time of the first telegraph communication, scientists had attempted to lay cable along the bottom of waterways in order to connect distant points in as direct a route as possible at minimal cost. Even before the first lines were constructed on land, telegraph inventor Samuel F. B. Morse had strung a telegraph cable across New York Harbor. In 1852, American financier Cyrus West Field organized a company to lay the first telegraph cable from the American continent to Europe. On August 5, 1858, the first cable was in place, and U.S. president James Buchanan communicated with England's Queen Victoria to inaugurate the first transoceanic electronic communication link. Unfortunately, the euphoria was short-lived. The cable failed a month later when its insulation was damaged by high-voltage transmission. In 1865, another company tried to lay a new cable but failed. Not until 1869 was the first truly successful effort completed and the first reliable transatlantic telegraph cable in place.

The time interval between the invention of the telegraph and the first reliable transoceanic telegraph cable transmissions was relatively short compared to the time it took to deploy a transoceanic telephone cable. The physical character of telephony requires much more of cable than does the relatively simple transmission mode used in telegraphy. As in land-line transmission, submarine cable must carry signals over long distances without significantly distorting them, a challenge whose solution proved elusive to researchers for almost three-quarters of a century.

In fairness, it must be pointed out that early interest in radiotelephony diverted the attention of researchers away from cable for much of the first half of the twentieth century. It was not until after World War II (1939-1945) that research on submarine cable design began in earnest. Radiotelephony proved somewhat unreliable for transoceanic communication, since shortwave broadcasts were highly susceptible to interference and noise caused by changes in weather patterns.

## The Plastic and Transistor Key

A key scientific discovery in the late 1940's, that of the transistor, made the problem of telephone circuit amplification appear solvable. Just as amplifiers, or repeaters, were needed in long-distance land-line transmission to maintain signal strength and quality, so were they necessary in submarine cable. In land lines, however, the maintenance of amplifiers was a much simpler task than it was in submarine cable. Early telephone repeaters were designed to incorporate vacuum tubes, which were devices used in early broadcasting to amplify and transmit radio waves. While research into their use in submarine cable actually occurred during the 1920's, the effort was regarded as futile even then.

Transistors held great promise in that they were compact, inexpensive, stable, long-lived,

**1225**

and energy-efficient. The first few years following their invention, however, proved frustrating to telephone engineers, who were unable to incorporate them successfully in facility designs. Eventually, refinements in the transistor concept led to the breakthrough that resulted in their ultimately successful application in submarine cable design.

Submarine cable had to be designed to withstand the unusual, often hostile environments found at the bottom of the ocean. Pressure, moisture, and cold were formidable opponents, and a variety of cable sheathing and insulating materials used in experiments that simulated such conditions had proved to be unreliable over long periods of time. Following World War II, polyethylene emerged as a primary material for encapsulating wire. After much experimentation, a formula of polyethylene and 5 percent butyl rubber was used in the manufacture of the SB submarine cable system, the design that would be used ultimately in the construction of the first transatlantic telephone cable.

The 1956 cable deployment and operation was a success. The cable sheathing and insulation proved to be reliable, and the quality of the circuits was acceptable. Soon, the new SD system was deployed. It carried 128 two-way circuits and utilized an improved polyethylene resin formula for improved quality and uniformity. Over the next decade, additional cables were laid, each an improvement over the last. By the early 1970's, yet another system, SG, had been put into service with forty-two hundred circuits.

## Consequences

The first transatlantic telephone cable was a significant achievement in telephony, but it also

*News reporters in New York talk to London using the new transatlantic telephone cable.*

AP/Wide World Photos

represented the first step in the establishment of a truly global, instantaneous electronic voice communication network. Its deployment meant that virtually every home with a telephone was connected to the transoceanic network. It was a significant achievement in the design of long-distance conductors, amplifiers, and submarine cable insulation and sheathing. For the first time, the reliability of submarine telephone circuitry had been demonstrated successfully, and efforts were under way to increase the number of circuits to accommodate growing demand for overseas voice communication.

Undersea telephone cable technology was challenged by another emergent technology soon after it was first deployed. By 1962, the American Telephone and Telegraph Company had begun to supplement underwater cable with the satellite transmission of telephone conversations. The company launched Telstar, a satellite that received signals from ground stations, amplified them, then relayed them back to down links around the globe. Telstar contained the equivalent of six hundred voice circuits. Over the years, many more satellites were launched, expanding the number of satellite voice circuits into the hundreds of thousands. As capacity increased, the cost of using satellites for international voice communication decreased substantially, and the transatlantic telephone cables, although they were still kept in service, became secondary transmission facilities.

As satellites were increasingly employed for long-distance telephone transmission, technology continued to improve. In the early 1990's, a new kind of submarine cable using fiber optics as the primary conductor was under development. Proponents believed that light-fiber submarine cable would be able to transmit a virtually unlimited number of voice, data, and video information exchanges at very low cost. By the 1990's, undersea cable using fiber optics in place of metallic transmission media was on the forefront of telephony, an emerging technology with great promise for the future.

*Richard L. Warms*

# Heezen and Ewing Discover Midoceanic Ridge

After leading an expedition to gather echograms of the ocean's floor, Maurice Ewing discovered the existence of the Midoceanic Ridge and postulated, with Bruce Charles Heezen, the existence of a midoceanic rift.

**What:** Earth science
**When:** 1956
**Where:** Atlantic Ocean
**Who:**
MAURICE EWING (1906-1974), an American geophysicist
BRUCE CHARLES HEEZEN (1924-1977), an American oceanographer and geologist
REGINALD AUBREY FESSENDEN (1866-1932), a Canadian physicist and engineer
HARRY HAMMOND HESS (1906-1969), an American geologist
ALFRED LOTHAR WEGENER (1880-1930), a German meteorologist and geophysicist

## Echolocation

Alfred Lothar Wegener was the multifaceted scientist responsible for proposing the theory of continental drift. He first began formulating his notion when he noticed that the coastlines of Africa and South America fit together as if they had once been part of one huge continent. Although Wegener was not the first to observe the jigsaw-like quality of continents, he was the first to try to figure out what all of this meant. His research spanned the sciences and culminated in the publication of *The Origin of Continents and Oceans*, in 1924. His hypothesis explained the mysteries of similar fossils found on opposite sides of the earth and the similarity of geographic detail in parts of the world. Unfortunately, his hypothesis neglected to explain how the continents could be in motion, constantly breaking and rejoining.

The explanation Wegener omitted was found several years later during World War II (1939-1945). In 1912, spurred by the *Titanic* disaster, Reginald Aubrey Fessenden completed a project aimed at creating an echo-detection device that would alert seagoing vessels to the presence of icebergs. About 1914, the former assistant to the famous inventor Thomas Alva Edison produced an invention that created soundwaves under water. By listening to the echoes picked up by an underwater microphone, Fessenden could detect objects in the same way that many ocean creatures, and even some land creatures, use echolocation in place of vision. Fessenden's early device became known later as "sonar" (an acronym for *sound navigation ranging*) when the French physicist Paul Langevin improved on his concept and introduced it to the seafaring public for general use.

During the war, the navy used a device called the "Fathometer," which was similar in working principle to sonar, to map the ocean floor around Iwo Jima. Aboard the USS *Cape Johnson* was Henry Hammond Hess, a geologist from Princeton University. He was the person who realized that with the technology of echo sounding, the seafloor could be mapped at last. While Hess's eventual contribution to the exploration of the seafloor was more along the lines of data interpretation than actual physical data gathering, the task he made possible was carried out by a colleague and friend from Columbia University, Maurice Ewing.

## The World Is a Baseball

Leading a group of scientists from the Lamont Geological Observatory in Palisades, New York, Ewing began his journey with the intent to map a string of undersea mountains called the Mid-atlantic Ridge. He found in the process, however, that merely mapping the terrain was not going to be enough. Sounding by echo with sonar and the Fathometer and utilizing data from oceanic earthquakes, the team members charted their progress, noting with interest that the underwater earthquakes reported by the seismologists Beno Gutenberg and Charles Richter occurred with increasing frequency along the ridge. Another exciting and inexplicable discovery was made during the process of dredging rock from the seafloor in the hope of finding ancient crustal rock from the formation of the earth. The oldest among the rocks found were only 150 million years old, and the others were even younger. Perhaps the greatest discovery of all—that of the Midoceanic Ridge—occurred after the actual data collection and during the interpretation of the echograms (records of echo soundings).

Columbia University graduate student Bruce Charles Heezen, in collaboration with Marie Tharp, also from Columbia University, began mapping the ocean floor based on data from the previous expeditions. Their examination of the echo-sounding passes revealed more about the topography. Much of the topography was very rugged terrain pierced by land jutting upward in the form of cliffs, crags, and, more important, mountains. The increasing detail of their mapping soon revealed that the ridge was not a regular mountain range or system.

The ridge was 1,829 to 3,048 meters high and 74,028 kilometers long, with a continuous, V-shaped, groovelike valley, later known as a "rift valley." Often compared to the laces of a baseball, the superlative mountain range crisscrossed the world. Heezen expanded on a suggestion from Tharp that the groove-shaped valley recorded by the echograms was a rift when it was compared to the median rift and the similar earthquake pattern found there. Although then-current echograms did not reveal the presence of a rift along the crest of the ridge, Heezen remained certain of this theory. He and Ewing developed a hypothesis about the existence of the rift and later made predictions about where the research team of the International Geophysical Year (1957-1958) might look. By 1958, the Heezen-Ewing theory that the ridge contained a rift was proved to be correct. The work of Ewing and Heezen sparked interest in further study of midoceanic ridges.

## Consequences

Harry Hammond Hess took the data used by Ewing and Heezen in the development of their theory and went one step further to propose the mechanism for continental drift. Hess proposed that hot rock rises from the earth's interior through the frequently recorded earthquakes surrounding the midoceanic ridges. The continuous upwelling of hot rock continually forces the rift to part and the seafloor to spread as the material from each eruption cools and is pushed away at the rapid pace of a few centimeters per year. When the spreading seafloor encounters an obstruction such as a continent base, it moves under the obstruction. Hess's theory, called "seafloor spreading," helped validate Wegener's idea of continents in drift. In addition to confirming a decades-old theory, the work of Heezen and Ewing opened a new venue of research to scientists as the world entered the International Geophysical Year.

*Earl G. Hoover*

# Pottery Suggests Early Contact Between Asia and South America

*Anthropologists asserted that design similarities between five-thousand-year-old pottery from Ecuador and pottery of the same age from Japan proved that contact had occurred between Asia and South America.*

**What:** Anthropology; Archaeology
**When:** 1956-1961
**Where:** Valdivia, Ecuador
**Who:**

BETTY J. MEGGERS (1921-    ), an American anthropologist, author, research associate, and expert in archaeology at the Smithsonian Institution

CLIFFORD EVANS (1920-1981), an anthropologist and curator of archaeology, later chairman of the anthropology division, at the Smithsonian Institution

EMILIO ESTRADA (1916-1961), a prominent Ecuadorian businessman, widely respected amateur archaeologist, and sportsman

## Diffusion Confusion

While there is disagreement over the earliest date at which humankind appeared in the Americas, it is generally agreed that both North and South America had been populated by about 10,000 B.C.E. These early inhabitants were of Asian origin and migrated to the New World across the Bering land bridge, a vast, subarctic plain connecting present-day Alaska and Siberia. The land bridge was exposed by a fall in ocean levels resulting from the Pleistocene ice age, which began about 2.5 million years ago and ended about 10,000 years ago. After the close of the last ice age, the sea level rose, submerging the land bridge and ending land contact between Asia and the Americas. Some anthropologists have argued, however, that occasional contacts between the Old and New Worlds continued and that these contacts had decisive impacts on the development of culture in America.

The possibility of transoceanic voyages predating those of Christopher Columbus, particularly voyages between Asia and South America, has been one of the most debated issues in anthropology. The issue is important because it plays a key role in the debate between anthropologists who argue for the theory of "diffusion" and those who favor the theory of "independent invention." Diffusionists believe that human beings are, by nature, very conservative and that new social practices and forms of society are rarely invented; such complex functions as agriculture, writing, and monumental architecture could not have been discovered more than once or twice in human history. Instead, the diffusionists theorize, such functions have spread around the world through contact between groups.

In the early twentieth century, most American anthropologists, following the work of Franz Boas, came to reject diffusionism in favor of independent invention. They were disturbed by scientific problems with diffusionism, as well as by its often racist and ethnocentric overtones. While they did not reject diffusionism altogether, they argued that humankind is equally inventive everywhere. They insisted that, unless contact between cultures can be scientifically proven, similarities between them must be assumed to be the result of convergent or parallel cultural development, or chance.

For both groups, the high civilizations of the Americas were important test cases. If all contact between Asia and the Americas ended with the disappearance of the Bering land bridge twelve thousand years ago, then complex civilizations

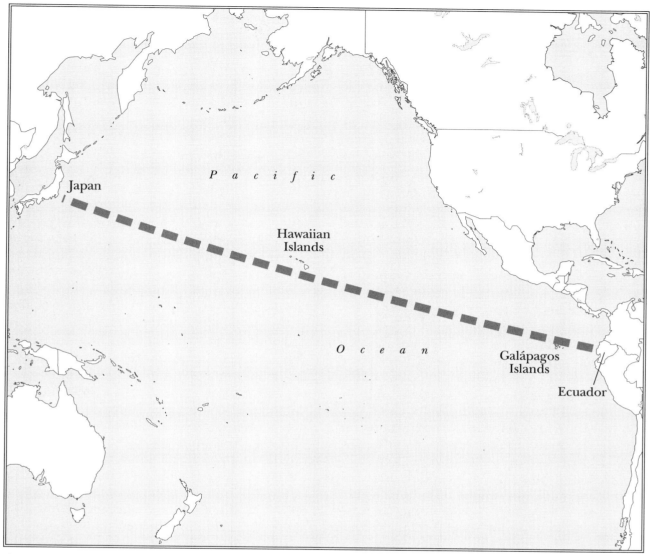

*Anthropologists who favor the theory of cultural diffusion believe that Japanese fishermen may have reached the area of modern Ecuador thousands of years ago.*

such as the Olmec, the Maya, and the Inca, which appeared only within the past few thousand years, must have been invented independently. If, however, transoceanic contact between Asia and the Americas continued, then the development of complex civilization in America may have been critically influenced by voyagers from Asia, thus supporting the diffusionist view.

### Seeming Proof

Despite the strength of academic opinion against diffusion, public interest in it remained strong. Numerous popular authors wrote about the similarities between American societies and Old World civilizations. Explorers such as Thor Heyerdahl captured the public imagination with dramatic demonstrations showing that transoceanic voyages were possible with simple technology. In 1956, Emilio Estrada, a well-respected Ecuadorian amateur archaeologist, discovered fragments of ancient pottery near Valdivia on the Ecuadorian coast.

Estrada was interested in the possibility of pre-Columbian transoceanic voyages and favored a diffusionist explanation of New World civilization. He enlisted the help of Smithsonian anthropologists Betty J. Meggers and Clifford Evans. Excavations at the Valdivia site continued

**1231**

from 1956 to 1961. Meggers and Evans published the results of their study in 1965 as the first volume of the prestigious *Smithsonian Contributions to Anthropology*. Meggers and Evans claimed that, at Valdivia, they had discovered the oldest pottery in the New World, dating to about 3000 B.C.E. The pottery was quite complex, and it bore an uncanny resemblance to pottery from Japan dating from about the same time period. Since no older or simpler forms of pottery had been found in the New World, it was assumed that the technology for making pottery must have been introduced from outside. Meggers and Evans hypothesized that a group of Japanese fishermen had been blown off course and eventually landed in Ecuador. The fishermen introduced pottery making at Valdivia, and from there it spread throughout the Americas.

James A. Ford, a prominent archaeologist and supporter of diffusionism, wrote that Meggers and Evans had clearly demonstrated that culture was diffused rather than reinvented and that "human culture is a single connected story." In 1966, Meggers and Evans received the Gold Medal of the International Congress of Americanists for outstanding Americanist studies and the Decoration of Merit from the government of Ecuador.

Other anthropologists were not as easily convinced and pointed to serious defects in Meggers and Evans's argument. David Collier of Chicago's Field Museum of Natural History noted that sailors adrift from Japan were very unlikely to reach Ecuador and would probably have ended their trip in Southern California or Mexico. Many argued that even if a boatload of fishermen had landed in Ecuador, it was unlikely that there would be specialists on board capable of

making pottery. The most telling blow against the diffusionists, however, was the discovery of older and simpler pottery in Colombia and Ecuador. This pottery bears no resemblance to Japanese work. Because of these flaws, the work did little to convince those who were not already supporters of diffusion.

## Consequences

Diffusionists have amassed a bewildering range of evidence showing the similarities between artifacts from New World cultures and those from China, Malaysia, the Mediterranean, Africa, India, and Southeast Asia. They argue that many of these similarities seem so unlikely that contact and diffusion are the best explanation of their presence. The sheer number of similarities, they say, makes any other interpretation unlikely.

Those who favor independent invention insist that even a very large number of design similarities between Old and New World artifacts would not prove contact. Startling similarities are frequently found in societies so greatly separated in time and space that no contact between them is possible. Author Nigel Davies, for example, points out that unusual bottles made in Africa in the 1930's are virtually indistinguishable from bottles made on the north coast of Peru in about 800 B.C.E. There is almost no chance of diffusion across more than twenty-five centuries.

Diffusionists have been unable to satisfy the demands of those who favor independent invention. Until they do, the idea that cultural advancements have been spread by diffusion will remain only a thought-provoking possibility.

# Soviet-Chinese Dispute Creates Communist Rift

---

*Denouncing Soviet leader Nikita Khrushchev for his program of "de-Stalinization," Chinese leader Mao Zedong helped create a major split in worldwide communism.*

---

**What:** International relations; Military capability
**When:** February, 1956-October, 1964
**Where:** Beijing and Moscow
**Who:**
NIKITA S. KHRUSHCHEV (1894-1971), first secretary of the Communist Party of the Soviet Union from 1953 to 1964, and premier of the Soviet Union from 1958 to 1964
MAO ZEDONG (1893-1976), chairman of the Chinese Communist Party from 1949 to 1976

## The Chinese Challenge

Though as leaders in the world communist movement the Soviet Union and the People's Republic of China were allies, they had differences—for example, a centuries-old conflict over Eastern Siberia. These differences did not erupt into open fights, however, until 1956.

In February, 1956, First Secretary Nikita Khrushchev made his famous de-Stalinization speech before the Soviet Twentieth Party Congress. It was a shocking condemnation of the policies of Soviet dictator Joseph Stalin, who had died in 1953, and the Chinese were dismayed. Why had they not been consulted ahead of time about such an important change?

Mao Zedong, chairman of the Chinese Communist Party, was pursuing policies similar to Stalin's, so an attack on Stalin was also a criticism of Maoism. Mao was especially unhappy with Khrushchev's call for "peaceful coexistence with America," for Mao believed that communists around the world should stand firm against the United States. Privately, Chinese leaders let the Soviets know that they were quite displeased.

Relations between Soviet and Chinese communists got much worse during 1957 and 1958. With the success of Sputnik, the world's first artificial satellite, Mao believed that the Soviets should become more aggressive and less compromising toward the West. Yet Khrushchev worked steadily toward a détente—relaxing of East-West tensions.

At the same time, the Soviets began giving more aid to "nonaligned" nations, such as India, while cutting back their economic aid to China. Since China could not count on large amounts of aid from the Soviet Union, Mao, in the fall of 1958, called his country into a "Great Leap Forward": Through will, hard work, and determination, the Chinese people would collectivize and industrialize their country on their own, making it as productive as the United States. This effort to set up giant agricultural collectives and to manufacture vast quantities of steel in "backyard furnaces" was publicized as "Chinese Communism"—a departure from the Soviet way.

Khrushchev was infuriated. Mao's move would surely encourage all the poorer communist and nonaligned nations to question the value of Soviet experience, advice, and leadership. If Chinese Communism were accepted, it would make China dominant in the world revolutionary movement that the Soviets had always claimed as their own.

## A Growing Split

There were some attempts at reconciliation, but between 1959 and 1964 an open division developed between the Soviets and the Chinese. In 1959 Khrushchev reneged on an earlier offer to help the Chinese develop atomic weapons. En-

**1233**

raging Mao even more, he supported members of the Chinese Communist Party who were pressing Mao to adopt more moderate policies.

In April, 1960, a Chinese article titled "Long Live Leninism" attacked Khrushchev's foreign policy. Within the international communist bloc, the Chinese began trying to form an anti-Soviet group. Khrushchev then withdrew all Soviet technical experts from China and stopped economic aid. Yet Mao became more determined than ever.

At international communist congresses—first at Bucharest in June, 1960, then at Moscow in November, 1960—the Soviets tried to get China condemned or expelled. These attempts failed, and Albania committed itself to the Chinese side.

When China came into conflict with India over Ladakh in October, 1962, the Soviets did not give China any help. Also, in the spring of 1962 the Soviets received hundreds of thousands of Kazakh and Uighur tribespeople who had first rebelled against Chinese rule in Xinjiang (Sinkiang) and then fled over the border to join their relatives in Soviet Central Asia. Meanwhile, the Chinese loudly criticized Khrushchev's decision to withdraw Soviet missiles from Cuba.

In late 1962 and in 1963, the Soviets organized a series of European Communist congresses at which the Chinese delegates and Chinese ideas were attacked with increasing fury. The Chinese continued about their business of trying to split the world's communist parties into a "truly revolutionary" group led by themselves and those that would follow the Soviet "revisionist" way.

Just before Soviet and Chinese leaders were to meet for talks in July, 1963, the Soviet Union signed a nuclear test-ban treaty with Great Britain and the United States. The Chinese responded fiercely, for the first time stating publicly their criticism of Khrushchev's de-Stalinization program, which they called treason to Marxism-Leninism. They also insisted that the test-ban treaty was treason against the communists of the world.

Early in 1964, the Chinese Communists attacked Khrushchev personally, ridiculing him with a variety of abusive names. They also publicly accused the Soviets of aggressive, hostile intentions for the use of atomic weapons and in Soviet-Chinese border disputes. Mao called Khrushchev a conspirator with "American imperialism" against Marxism-Leninism and against the People's Republic of China.

For his part, Khrushchev denounced Mao as a "Trotskyite" and a racist who had split the international communist movement. He tried to call a conference of communist leaders to expel China from the Party. In October, 1964, China successfully detonated an atomic bomb, proving that it was no longer helpless before the Soviet Union's nuclear might.

## Consequences

Naturally, leaders in the United States and other Western nations watched the Soviet-Chinese dispute with great interest. Communism, which had been feared as a monster that was trying to engulf the world, was now divided within itself, and many in the West breathed a sigh of relief.

The break between China and the Soviet Union continued for a number of years. In 1972 the United States began talks with Chinese leaders to begin opening up trade and diplomacy with that country. After Mao died in 1976, a process of economic reforms began in China, and the Soviets began moving to reopen relations with the Chinese.

# Soviet Premier Khrushchev Denounces Stalin

*In a speech before delegates to the Twentieth Congress of the Communist Party of the Soviet Union, Nikita Khrushchev harshly criticized the leadership of dictator Joseph Stalin, who had died in 1953.*

**What:** National politics
**When:** February 24-25, 1956
**Where:** Moscow
**Who:**

NIKITA S. KHRUSHCHEV (1894-1971), first secretary of the Communist Party of the Soviet Union from 1953 to 1964, and premier of the Soviet Union from 1958 to 1964

ANASTAS MIKOYAN (1905-1970), first deputy premier of the Presidium of the Soviet Communist Party from 1946 to 1964

JOSEPH STALIN (1879-1953), dictator of the Soviet Union from 1924 to 1953

LAVRENTI BERIA (1899-1953), head of the Soviet security forces, including the secret police, under Stalin

VLADIMIR ILICH LENIN (1870-1924), founder and head of the Soviet Union from 1918 to 1924

## The Party Convenes

The Twentieth Congress of the Communist Party of the Union of Soviet Socialist Republics convened in February, 1956, about six months before it had been originally scheduled. This was the first meeting of delegates from the Communist Party throughout the Soviet Union since the death of Joseph Stalin, the Soviet dictator, in March, 1953.

After his death, there had been a struggle for power among communist leaders. Nikita Khrushchev had become first secretary of the Party, and he and other leaders had announced that the Party would now benefit from "collective leadership."

In Eastern Europe, which was dominated by the Soviet Union, there had been a number of uprisings, protests, and strikes following Stalin's death, but by the end of 1953 they had been crushed. In the Soviet Union itself, the first reaction to Stalin's death had been sorrow and grief. As time passed, however, discontent and criticism of Stalinism had begun to appear. Within the Party leadership, First Deputy Premier Anastas Mikoyan had been asking for "de-Stalinization."

Beginning on February 14, the Party Congress began to go through the usual round of tedious business, economic reports, and speeches in praise of Marxism. On February 16, however, Mikoyan began stirring things up. Before the whole assembly, he made a speech criticizing Stalin's leadership, saying that it had been based on the "cult of personality"—in other words, Stalin had been seeking power for himself rather than the good of the Communist Party. Mikoyan also complained about the way history had been written during Stalin's regime: Stalin had been glorified as a great hero who did nothing wrong.

On February 17, Deputy Premier Georgi M. Malenkov made a speech that included criticisms of Stalin, and Premier Nikolai Bulganin followed suit on February 21. On February 24, First Secretary Khrushchev took the floor.

## Khrushchev's Speech

Clearly, Khrushchev had been preparing his speech for some time: It covered nearly one hundred typed pages. He delivered it in sections over a period of two days.

Khrushchev began by praising Vladimir Ilich Lenin, the Soviet Union's founder, who had helped build the Central Committee of the Communist Party as "a real expression of collective

**1235**

leadership." Lenin was an enlightened leader, Khrushchev said, and it was a shame that Stalin had not followed in his steps.

With long quotes from the "Last Testament" that Lenin had dictated shortly before his death in January, 1924, Khrushchev showed the assembled delegates that Lenin had not wanted Stalin to lead the nation. The Soviet founder had warned that Stalin was a proud man who intended to grab as much power as he could. To support his point, Khrushchev made sure that copies of the "Last Testament" were distributed to all the delegates at the congress.

Khrushchev then turned to the most devastating part of his speech—the story of the repression and executions that took place under Stalin's leadership in the 1930's. With these "purges," Stalin had claimed that he was cleansing the Party of those who were disloyal.

Leon Trotsky, a Bolshevik leader who opposed Party policy, had already been defeated by the time the purges began, Khrushchev said; so had a few others who wanted to depart from "pure" communism. Clearly, Stalin's purges were brutal and unnecessary; he had destroyed "many honest Communists" who had helped bring in the revolution.

In the next section of the speech, Khrushchev accused Stalin of having been a poor leader during World War II. Stalin's purges of Soviet military leaders were a "massacre of innocents," according to Khrushchev. Lavrenti Beria, who had been chief of Security Forces under Stalin, was also given much blame. (Beria had been arrested after Stalin's death and executed in December, 1953.)

The military purges, Khrushchev said, destroyed military commanders and political workers who could have helped prepare the Soviet Union for Germany's attack in 1941. It was clear that Stalin had misunderstood Nazi leader Adolf Hitler. Because of Stalin's stubbornness, the country had not been properly defended during the war.

Khrushchev then made a detailed criticism of the way Stalin had been glorified in Soviet history books and by the media in general. Finally, Khrushchev attacked Stalin's agricultural policies (Khrushchev claimed to be an expert in this field). To end his speech, he emphasized again that Stalin had destroyed "Leninist principles" by weakening the Party's leadership during the thirty years of his regime.

## Consequences

Though Khrushchev made his speech to a closed meeting, copies of it began appearing throughout the Soviet Union and abroad. Yet response from other countries was rather slow in coming. Within weeks, the communist parties of the Eastern European nations stated their agreement with Khrushchev's denunciation of Stalin.

During the next few months, however, Communist parties elsewhere criticized the speech. The leaders of the Italian Communist Party pointed out that several of those who had denounced Stalin shared responsibility for his policies, for they had been part of the Soviet Communist Party when he was in power. In China, the speech caused outright dismay among communist leaders.

What did the speech mean? Was the Soviet Union ready to leave behind its repressive policies and create a more open, democratic state? There were some changes in Soviet foreign policy; another speech at the Twentieth Party Congress called for "different paths to socialism" for different peoples and for "revolution without violence."

Yet as Khrushchev gained more power in the Soviet Union he continued to use some of Stalin's methods, including purges and labor camps. Perhaps the speech was mostly Khrushchev's way of affirming that he had risen to leadership and did not intend to live in Stalin's shadow.

*Deepa Mary Ollapally*

# First Birth Control Pills Are Tested

*Gregory Pincus directed the first large-scale test of the birth control pill in Puerto Rico, paving the way for worldwide use of a safe, effective contraceptive.*

**What:** Medicine
**When:** April-December, 1956
**Where:** Río Piedras, Puerto Rico
**Who:**
GREGORY PINCUS (1903-1967), an American biologist
MINJUE ZHANG (MIN-CHUEH CHANG; 1908-1991), a Chinese-born reproductive biologist
JOHN ROCK (1890-1984), an American gynecologist
CELSO-RAMON GARCIA (1921-    ), a physician
EDRIS RICE-WRAY (1904-    ), a physician
KATHERINE DEXTER MCCORMICK (1875-1967), an American millionaire
MARGARET SANGER (1879-1966), an American activist

## An Ardent Crusader

Margaret Sanger was an ardent crusader for birth control and family planning. Having decided that a foolproof contraceptive was necessary, Sanger met with her friend, the wealthy socialite Katherine Dexter McCormick. A 1904 graduate in biology from the Massachusetts Institute of Technology, McCormick had the knowledge and the vision to invest in biological research. Sanger arranged a meeting between McCormick and Gregory Pincus, head of the Worcester Institutes of Experimental Biology. After listening to Sanger's pleas for an effective contraceptive and McCormick's offer of financial backing, Pincus agreed to focus his energies on finding a pill that would prevent pregnancy.

Pincus organized a team to conduct research on both laboratory animals and humans. The laboratory studies were conducted under the direction of Minjue Zhang, a Chinese-born scientist who had been studying sperm biology, artificial insemination, and in vitro fertilization. The goal of his research was to see whether pregnancy might be prevented by manipulation of the hormones usually found in a woman.

It was already known that there was one time when a woman could not become pregnant—when she was already pregnant. In 1921, Ludwig Haberlandt, an Austrian physiologist, had transplanted the ovaries from a pregnant rabbit into a nonpregnant one. The latter failed to produce ripe eggs, showing that some substance from the ovaries of a pregnant female prevents ovulation. This substance was later identified as the hormone progesterone by George W. Corner, Jr., and Willard M. Allen in 1928.

If progesterone could inhibit ovulation during pregnancy, maybe progesterone treatment could prevent ovulation in nonpregnant females as well. In 1937, this was shown to be the case by scientists from the University of Pennsylvania, who prevented ovulation in rabbits with injections of progesterone. It was not until 1951, however, when Carl Djerassi and other chemists devised inexpensive ways of producing progesterone in the laboratory, that serious consideration was given to the medical use of progesterone. The synthetic version of progesterone was called "progestin."

## Testing the Pill

In the laboratory, Zhang tried more than two hundred different progesterone and progestin compounds, searching for one that would inhibit ovulation in rabbits and rats. Finally, two compounds were chosen: progestins derived from the root of a wild Mexican yam. Pincus arranged for clinical tests to be carried out by Celso-Ramon Garcia, a physician, and John Rock, a gynecologist.

Image Club Graphics

*Birth control pills became a widely used form of contraceptive quickly.*

Rock had already been conducting experiments with progesterone as a treatment for infertility. The treatment was effective in some women but required that large doses of expensive progesterone be injected daily. Rock was hopeful that the synthetic progestin that Zhang had found effective in animals would be helpful in infertile women as well. With Garcia and Pincus, Rock treated another group of fifty infertile women with the synthetic progestin. After treatment ended, seven of these previously infertile women became pregnant within half a year. Garcia, Pincus, and Rock also took several physiological measurements of the women while they were taking the progestin and were able to conclude that ovulation did not occur while the women were taking the progestin pill.

Having shown that the hormone could effectively prevent ovulation in both animals and humans, the investigators turned their attention back to birth control. They were faced with several problems: whether side effects might occur in women using progestins for a long time, and whether women would remember to take the pill day after day, for months or even years. To solve these problems, the birth control pill was tested on a large scale. Because of legal problems in the United States, Pincus decided to conduct the test in Puerto Rico.

The test started in April of 1956. Edris Rice-Wray, a physician, was responsible for the day-to-day management of the project. As director of the Puerto Rico Family Planning Association, she had seen firsthand the need for a cheap, reliable contraceptive. The women she recruited for the study were married women from a low-income population living in a housing development in Río Piedras, a suburb of San Juan. Word spread quickly, and soon women were volunteering to take the pill that would prevent pregnancy. In the first study, 221 women took a pill containing 10 milligrams of progestin and 0.15 milligram of estrogen. (The estrogen was added to help control breakthrough bleeding.)

Results of the test were reported in 1957. Overall, the pill proved highly effective in preventing conception. None of the women who took the pill according to directions became pregnant, and most women who wanted to get pregnant after stopping the pill had no difficulty. Nevertheless, 17 percent of the women had some unpleasant reactions, such as nausea or dizziness. The scientists believed that these mild side effects, as well as one death from congestive heart failure, were unrelated to the use of the pill.

Even before the final results were announced, additional field tests were begun. In 1960, the U.S. Food and Drug Administration (FDA) approved the use of the pill developed by Pincus and his collaborators as an oral contraceptive.

**1238**

## Consequences

Within two years of approval by the FDA, more than a million women in the United States were using the birth control pill. New contraceptives were developed in the 1960's and 1970's, but the birth control pill remains the most widely used method of preventing pregnancy. More than 60 million women use the pill worldwide.

The greatest impact of the pill has been in the social and political world. Before Sanger began the push for the pill, birth control was regarded often as socially immoral and often illegal as well. Women in those post-World War II years were ex-pected to have a lifelong career as a mother to their many children. With the advent of the pill, a radical change occurred in society's attitude toward women's work. Women had more freedom to work and enter careers previously closed to them because of fears that they might get pregnant. Women could control more precisely when they would get pregnant and how many children they would have. The women's movement of the 1960's—with its change to more liberal social and sexual values—gained much of its strength from the success of the birth control pill.

*Judith R. Gibber*

# Mao Delivers His "Speech of One Hundred Flowers"

> With the slogan "Let a hundred flowers bloom, let a hundred schools of thought contend," Mao Zedong invited Chinese intellectuals to criticize the Chinese Communist Party.

**What:** Civil rights and liberties
**When:** May, 1956
**Where:** Beijing and Shanghai, People's Republic of China
**Who:**
MAO ZEDONG (1893-1976), leader of the People's Republic of China from 1949 to 1976
HU FENG (1902-1985), a writer and member of the Chinese Communist Party (CCP)
FEI XIAOTONG (1910-     ), a sociologist
DING LING (1904-1986), a CCP member and feminist writer

## Bureaucracy and Control

China's First Five-Year Plan was scheduled to run from the beginning of 1953 to the end of 1957, and by 1956 many changes had been brought to Chinese society. Mao Zedong, China's leader, was eager to end private ownership of industry and farmlands, and by the end of 1956 about 88 percent of peasant families had become members of "advanced cooperatives," or collective farms.

The Chinese Communist Party (CCP), with its huge bureaucracy, rationed food, clothing, and other items for the people of China. Jobs were assigned by the state, and people did not always have free choice about where to live or travel. The Party also tried to control art and literature.

Hu Feng, an author and CCP member, was a leader in the Writers' Union and the National People's Congress. He disagreed with Mao's approach, however, saying that the Party should not use Marxist ideas to judge works of art. Mao had

him arrested and charged as an imperialist and a counterrevolutionary.

After Hu and other intellectuals were harassed and arrested, many Chinese intellectuals became discouraged. They had hoped to work with the CCP to build a better China, but it seemed that Mao was determined to keep them quiet.

Meanwhile, however, Mao was rethinking his strategy. In February, 1956, Soviet leader Nikita Khrushchev had publicly criticized Joseph Stalin for having oppressed the Soviet people and having dominated the Soviet Union through a "cult of personality." Khrushchev also called for "peaceful coexistence" with the West instead of constant conflict.

Mao estimated that, of China's five million intellectuals (high school graduates), only about 3 percent were actually hostile to the CCP. He decided that the other 97 percent could be used to help build socialism in China by bringing new, creative ideas into the CCP.

## A New Policy

In a speech before eighteen hundred communist and noncommunist delegates to China's Supreme State Conference on May 2, 1956, Mao presented an invitation: "Let a hundred flowers bloom, let a hundred schools of thought contend." Chinese intellectuals were encouraged to express their complaints against the CCP openly.

In closed meetings attended by CCP delegates, some pointed to corruption among Party leaders and abuse of workers in factories. Soon these criticisms became public, aired through the state-controlled press, wall posters, and street demonstrations. Critics claimed that the Chinese

**1240**

communist government had not respected human rights.

At Beijing University, students set up a "Democratic Wall," which was soon plastered with criticisms of the CCP. By the beginning of June, students had become so angry at the government's abuses that they began a protest movement in the large cities of China. They rioted, went on strike, raided files, demanded freedom of speech, and sometimes beat CCP leaders.

Meanwhile, workers began demanding better working conditions and higher wages, along with the right to form unions. They went on strike or slowed their rate of work in protest. Large numbers of peasants decided to leave the cooperatives, and many of them refused to pay taxes.

In June, 1957, sociologist Fei Xiaotong published his studies on a village in Jiangsu Province. He found that the villagers were less well-off in the 1950's than they had been in the 1930's, before the communists came to power.

## Consequences

Mao was overwhelmed by the flood of criticism by the very people he had been trying to please. To stop the rebellion without making himself look like a hypocrite, he revised his policy in a pamphlet published on June 18, 1957: "On the Correct Handling of Contradictions Among the People." According to Mao, all future criticisms must meet several requirements. The complaints must harmonize with the greater goal of socializing China and must strengthen the CCP.

With this move, Mao gained the approval of Party conservatives such as Liu Shaoqi, Zhu De, and Peng Zhen. They had not liked the Hundred Flowers Campaign because it had allowed outsiders to criticize the CCP.

Since the criticism could not be stopped with a few new rules, however, Mao and the hard-liners decided to begin the Antirightist Campaign in late June. During this campaign, which lasted until the end of 1958, more than 300,000 intellectuals were branded as "rightists." This label ruined their careers.

Fei Xiaotong was forced to deny the findings of his Jiangsu report and was forbidden to teach, do research, or publish any more articles or books about Chinese society. A famous feminist author, Ding Ling, was expelled from the CCP and sent to work on a farm in Manchuria.

敬祝毛主席万寿无疆

*Propaganda poster paying tribute to Mao Zedong.*

**1241**

Then there was the case of the Liang family. Mrs. Liang (whose maiden name was Yan Zhi-de) was a committed CCP member and was respected for her work as a public security officer. During the Hundred Flowers Campaign, the Changsha Public Security Bureau held meetings every night; everyone was encouraged to express criticisms. Mrs. Liang had no reason to criticize the CCP, but one of her supervisors insisted that she say something. So she said that her boss sometimes used bad language and was not always fair in assigning pay raises.

When the Antirightist Campaign started, Mrs. Liang's boss took revenge. He reduced her rank, cut her salary by nearly 75 percent, and sent her to work on a distant farm. Although her husband, Liang Yingqiu, was an important journalist and a trusted Party member, he had to divorce her for the sake of the children. "Rightist" children could not attend good schools and had no hope of career success.

During the Antirightist Campaign, about 1.7 million people were investigated for anti-CCP activities. More than half of them were rebuked, put on probation, or expelled from the Party. Those who survived the campaign learned to distrust the government, their fellow workers, their friends, and even their relatives.

*Peng-Khuan Chong*

# Egypt Seizes Suez Canal

*After Egyptian president Gamal Abdel Nasser declared that his country would nationalize the Suez Canal, the United Nations blocked Great Britain, France, and Israel from beginning an all-out war.*

**What:** International relations
**When:** July 26, 1956
**Where:** Egypt
**Who:**
GAMAL ABDEL NASSER (1918-1970),
    president of Egypt from 1956 to 1970
JOHN FOSTER DULLES (1888-1959), U.S.
    secretary of state from 1953 to 1959
ROBERT ANTHONY EDEN (1897-1977),
    prime minister of Great Britain from
    1955 to 1957
GUY MOLLET (1905-1975), premier of
    France from 1956 to 1957

## Taking the Canal

Great Britain, which had long been a colonial power in Egypt, had joined the United States in promising funds to finance Egyptian president Gamal Abdel Nasser's ambitious project: the High Aswan Dam. As Nasser began to speak out more and more against the policies of the Western Powers, however, the British and Americans cut off the promised funds. This was a serious problem for Nasser: Egypt desperately needed the dam to help develop its economy.

On July 24, 1956, Nasser bitterly attacked the United States, the U.S. State Department, and Secretary of State John Foster Dulles. The Americans, Nasser said, were either misinformed or lying. Nasser warned, "You may choke with rage but you will never succeed in ordering us about."

Two days later, Nasser made a major speech in Alexandria, Egypt, and announced that he was nationalizing the Suez Canal Company: "This canal is ours." Egypt would use the money collected from canal tolls, he said, for building the High Dam. "We shall rely on our strength, our own muscle, our own funds."

Reactions came immediately from the United States, Great Britain, and France. The United States protested Nasser's anti-American statements. Great Britain and France made an official protest; seizing the canal was an "arbitrary action," they said. Egyptian funds in British and French banks were frozen, and leaders of Great Britain and France began discussing military action to take the canal back.

Dulles advised against military moves and insisted that a conference be held in London to try to solve the problem. The conference's solution, an international control board to oversee the canal, was rejected by Nasser. A similar proposal was made in September, but Nasser again refused to accept "outside control." He did say, however, that Egypt would honor the 1888 agreement that allowed free passage through the canal.

On October 5, the United Nations Security Council met to debate the issue and hear British and French complaints. On October 10, the foreign ministers of several world powers met in New York, led by U.N. secretary general Dag Hammarskjöld. They came up with a set of six general principles for dealing with the problem, but this plan was defeated by the Soviet Union on October 13.

By now British and French leaders were becoming angry because the negotiations did not seem to be working. British prime minister Sir Anthony Eden, with the support of French premier Guy Mollet, stated openly that he was determined to "free" the canal, using force if necessary. Eden was disturbed by what he considered to be Nasser's "dictatorial" methods. Remembering how an earlier British prime minister, Neville Chamberlain, had failed to stop Adolf Hitler, Eden was determined to stop Nasser before things got out of control.

**1243**

## Israel Enters the Picture

Fearing Arab attacks, and with secret encouragement from Great Britain and France, an Israeli army led by Moshe Dayan invaded the Sinai Peninsula on October 29. Although Israel was not near the Suez Canal the next day, the British and French gave both Egypt and Israel an ultimatum: They were to withdraw ten miles away from the canal.

When the Egyptians rejected this ultimatum, the British and French decided that they had reason to take military action. On October 31, Egyptian airfields were attacked by French and British planes, and on November 5 and 6, British and French troops successfully landed at Port Said, on the northern end of the Suez Canal.

In the United Nations, France and Great Britain vetoed the possibility of a cease-fire. The U.N. Security Council met in emergency session and passed a second call for a cease-fire, with both United States and Soviet support.

On November 7, the hostilities ended, and it was agreed that a U.N. emergency force should enter the area. By this time, the British and French held one-fourth of the canal, and the main channel was blocked by wrecked ships.

Egypt demanded that all foreign troops be removed, but the British and French refused to move until the canal was cleared and a settlement was reached. Soviet premier Nikolai Bulganin, who had threatened earlier to send "volunteers" to help Egypt, called for the removal of the troops and payment of compensation to Egypt. Hoping that more negotiations could take place, American leaders asked that British, French, and Israeli troops withdraw unconditionally.

By November 15, the U.N. Emergency Force, led by Canadian major general E. L. M. Burns, entered northern Egypt, and by mid-December Burns had a force of four thousand men. On December 22, the British and French evacuated their forces from Egypt.

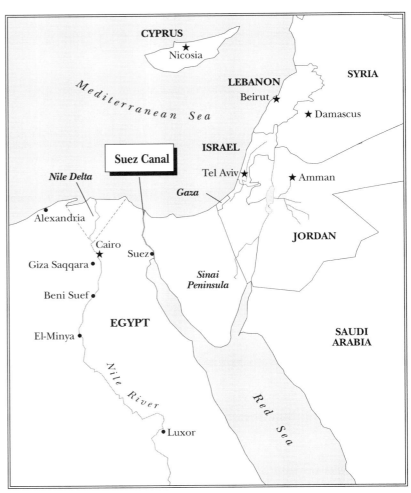

## Consequences

Egypt had faced down the European powers and won. People in other nations in Africa, Asia, and Latin America, tired of being dominated by the wealthy powers of the West, were inspired by Nasser's action. Because of Nasser's boldness, he also gained prestige throughout the Arab world, while Arab leaders became even more antagonistic to Israel than they had been before.

The Soviet Union began to give the Egyptians money to build the Aswan Dam, as well as other assistance and weapons; in this way, the Soviets became a new power in the Middle East.

Having lacked support from the United States during the crisis, French leaders decided that France would begin to act independently in foreign affairs. The British, on the other hand, had tried to act independently and failed, so they concluded that they must depend more on the

United States if they were to be effective. The Suez crisis was a defeat for British prime minister Eden, who resigned on January 9, 1957.

During the crisis and invasion, petroleum did not move through the canal or through Syrian pipelines. Although the United States provided temporary help, there were economic difficulties in Europe. The world had become deeply dependent on Middle Eastern oil to meet energy needs.

Through U.N. negotiations, a cease-fire was agreed upon, but nationalization of the canal was not completed until July 13, 1958. On that day, under the supervision of the World Bank, the Suez Canal Company signed an agreement with Egypt: The canal was left in Egyptian hands; the company kept its assets outside Egypt and paid Egypt for them.

*Thomas I. Dickson*

# Soviets Crush Hungarian Uprising

*A movement to create a new, independent kind of socialism in Hungary was stopped when Soviet leaders sent in troops.*

**What:** Civil strife; Political reform
**When:** October-November, 1956
**Where:** Hungary
**Who:**
MÁTYÁS RÁKOSI (1892-1971), first secretary of the Hungarian Communist Party from 1944 to 1956
IMRE NAGY (1896-1958), premier of Hungary from 1953 to 1955 and in 1956
ERNO GERÖ (1898-1980), first secretary of the Hungarian Communist Party in 1956
PAL MALÉTER, a Hungarian army officer and leader in the uprising
JÁNOS KÁDÁR (1912-1989), first secretary of the Hungarian Communist Party from 1956 to 1988
NIKITA S. KHRUSHCHEV (1894-1971), first secretary of the Soviet Communist Party from 1953 to 1964

## The New Course

After Joseph Stalin died in 1953, the new Soviet premier, Georgi Malenkov, began to force changes in Hungary, one of the Eastern European states that were kept under strict Soviet control. Mátyás Rákosi, first secretary of the Hungarian Communist Party, had been a Stalinist, but he was required to make a coalition with a group of reform-minded communists, led by Imre Nagy.

After Nagy became premier of Hungary on July 4, 1953, the Hungarian people began to reap the benefits of what he called his "new course." The power of the dreaded secret police was lessened, and political prisoners were released. Peasants were allowed to do away with their collective farms.

With his new policies, Nagy became one of the communist world's most popular leaders. Yet he had to struggle hard with Rákosi, for Stalinists still kept control of the Communist Party and the government bureaucracy. In April, 1955, party leaders expelled Nagy from the Communist Party in Hungary.

Rákosi was back in control, but he found himself in an awkward position. The Soviet Union supported Rákosi mainly because it needed strong, conservative leaders in Eastern Europe while the struggle for power continued in Moscow. Nikita Khrushchev, who had replaced Malenkov as the Soviet Union's most important leader, was starting his own "de-Stalinization" program, and Rákosi was again pressured to turn against his Stalinist beliefs.

After Khrushchev denounced Stalin's purges at the Twentieth Party Congress in February, 1956, a number of Hungarians made the point that Rákosi had also carried out political murders in the Stalinist style. During the spring and summer of 1956, Hungarian intellectuals—many of whom were communists—criticized Rákosi in magazine articles and in public discussions. Reform-minded communists called for a return to Nagy's "new course."

Between June 30 and July 21, Rákosi challenged Nagy's supporters and lost. He was quickly removed by Moscow. Unfortunately, he was replaced not by Nagy but by Erno Gerö, whom Hungarians despised as another Stalinist.

Gerö found that he could not stifle the loud calls for liberalization in Hungary. At a ceremony reburying a communist leader named Lazlo Rajk, whom Rákosi had executed, 300,000 people heard party leaders apologize for past wrongdoing. On October 14, the Party made Nagy a member once again. Gerö left for Yugoslavia, where he hoped to gain the support of Marshal Josip Broz Tito, a respected communist leader who had turned against Stalin in the 1940's.

## The Uprising Explodes

There were political meetings during Gerö's absence, and when he returned to Hungary late on October 23, enormous crowds filled the streets. By midnight, the crowds were fighting police and some Soviet troops.

Nagy was appointed to head a new government cabinet, but Gerö's followers released a false report that Nagy had requested help from the Soviet Union. Disappointed by Nagy's seeming treachery, the rebels continued the struggle with the help of some Hungarian soldiers.

Nagy appealed for order. He was in a tricky situation: The rebels no longer fully trusted him, yet he had to persuade them to stop the uprising. At the same time, he needed to act as a loyal communist; if he did not, the Soviets would remove him again and return Gerö or Rákosi to power.

By October 28, Nagy had succeeded in working out a cease-fire with the rebel leader, Colonel Pal Maléter. Nagy promised that the rebels would not be punished, that the secret police would be abolished, and that he would convince the Soviets to leave Hungary.

Soviet leaders Mikhail Suslov and Anastas Mikoyan had already arrived from Moscow to start negotiations. The Hungarian people once again gave their support to Nagy. Though the Soviets later claimed that Nagy intended to bring capitalism to Hungary, the Hungarian uprising was simply a movement of nationalist Hungarian communists against Soviet and Hungarian Stalinists. Maléter, who became minister of defense on October 31, spoke of establishing a "free, independent, and socialist" Hungary and said he would use force against anyone who tried to bring back capitalism.

At some point, probably about November 1, the Soviets decided that the Hungarian experiment could not continue. That day, Soviet troops reentered Hungary, and János Kádár formed a puppet government—a regime that was subser-

*In the main square in Budapest in November, 1956, Hungarians wave their flag from a Soviet tank they have captured.*

AP/Wide World Photos

**1247**

vient to Moscow. The Hungarian people rose up in arms and resisted ferociously, but they could not win. They appealed to the outside world to protect Hungarian neutrality, but there was little response.

On November 3, Maléter and his staff were arrested while they were negotiating at Soviet headquarters. Nagy fled to the Yugoslavian embassy in Budapest. By November 14, Kádár had settled in as Soviet puppet ruler with the help of Soviet troops and secret police. The Hungarian uprising had come to an end.

## Consequences

Having gotten a guarantee of protection from the Hungarian government, Nagy eventually came out of the Yugoslavian embassy—but he was quickly kidnapped. He and Maléter were shot in 1958. Rákosi never returned from Moscow; he died there in 1963. About 200,000 Hungarians fled to the West.

After he began to feel secure in his position, Kádár began to make reforms in Hungary. The Soviet leaders encouraged these reforms, though they made sure that liberalization did not happen too quickly. The Soviets had learned a lesson: If they were going to keep control of their "satellite" states in Eastern Europe, strong feelings of national pride would have to be crushed. Reforms could continue only as long as the Eastern European leaders stayed loyal to Moscow.

*Scott McElwain*

# Castro Takes Power in Cuba

*Driving out dictator Fulgencio Batista through a political and guerrilla campaign, Fidel Castro brought a Marxist regime to Cuba.*

**What:** Military conflict; Coups; Political reform

**When:** December 2, 1956-January 1, 1959

**Where:** Cuba

**Who:**

GERARDO MACHADO Y MORALES (1871-1939), an American-backed dictator who was overthrown by the revolution of 1933

RAMÓN GRAU SAN MARTÍN (1887-1969), president of Cuba from 1933 to 1934 and from 1944 to 1948

FULGENCIO BATISTA (1901-1973), dictator of Cuba from 1933 to 1944 and from 1952 to 1959

FIDEL CASTRO (1926 or 1927-        ), a revolutionary leader

## History of Revolutions

Cuba was one of the first Latin American nations to be colonized by the Spanish and it was the last to become politically independent. Although most of the other Latin American nations gained their independence in the 1820's, Cuba did not get rid of Spanish rule until 1898.

There were several reasons for this delay; one of them was that though Mexican and Colombian leaders tried to gain support for Cuban independence, the United States discouraged them. American leaders thought that an independent Cuba might too easily come under the control of a European nation far more powerful than Spain.

It was ironic, then, that Cuba finally became independent from Spain only after the United States intervened in 1898. The Cubans had fought bloody battles for three years and were close to winning the struggle when the United States won the Spanish-American War and suddenly gained control of Puerto Rico, the Philippines, and Cuba.

The Cubans felt robbed of their victory—and they discovered that they would have to pay a high price for the withdrawal of American troops from Cuba. In their new constitution of 1901, they were forced to include the Platt Amendment, which gave the United States the right to oversee Cuba's political affairs.

During the next third of a century, the American ambassador—and at times U.S. troops—were in control of Cuban politics. At the same time, American banks dominated the economy. Sugar cane had become Cuba's most valuable product, and American investors bought up the largest sugar plantations and mills. The United States was the most important customer for Cuban exports, and most products imported into Cuba came from the United States. It is not surprising, then, that when the Cuban people became unhappy with their political leaders, they blamed the United States.

In 1933, a revolution removed American-backed dictator Gerardo Machado from power. By this time, all reformist groups in Cuba were attacking what they called *plattistas* (people who supported the Platt Amendment) and *entreguistas* (people who were letting foreigners control Cuba's economy).

Yet once again, American intervention robbed the Cubans of a full victory. Before the revolution could get completely under way, the American ambassador persuaded Machado to resign. With the help of an army revolt led by Fulgencio Batista, the Cuban reformists were able to place Ramón Grau San Martín in the presidency. Yet the United States refused to recognize the Grau government, and U.S. ambassador Sumner Welles soon persuaded Batista to overthrow Grau.

**1249**

In 1934, the United States finally abolished the Platt Amendment. Even this action, however, was aimed mostly at preserving American influence in Cuba.

In the next twenty years, sugar production became less profitable. Americans invested more and more in other parts of Cuba's economy and brought tourism, gambling, prostitution, and organized crime. The governments dominated by Batista were eventually replaced by reformist governments, but there was much corruption and inefficiency. Another military coup returned Batista to power in 1952, and this time he became a very repressive dictator.

*Fidel Castro (center front) in the mountains of eastern Cuba in March, 1957.*

AP/Wide World Photos

## Castro's Movement

On July 26, 1953, Fidel Castro led an attack on the Moncada Army barracks in Santiago de Cuba—the second largest army barracks in the country. Although the attack was a miserable failure and landed Castro in prison, it was the first armed uprising against the dictatorship. Because of it, Castro became the leader of a general movement against Batista, and his group was named for the date of the uprising—the July 26 Movement.

An amnesty for political prisoners got Castro out of jail, and he spent some time doing political organizing. Then he left for Mexico to make preparations for another armed attack on the Batista government.

When Castro returned to Cuba on December 2, 1956, government troops were waiting for him. Only twenty or thirty of his small band of followers survived; together they made their way to the Sierra Maestra mountains and began a guerrilla struggle.

This war had no major battles and no impressive victories. Even at the end, Batista's army was much larger and better equipped than Castro's forces. But the government troops had lost the will to fight. All across Cuba, people despised Batista for his savage treatment of those who tried to oppose him.

On December 31, 1958, Batista fled the island, and in January, 1959, Castro entered Havana as a hero and the head of Cuba's new government.

## Consequences

Castro and his young followers wanted nothing to do with those who had served under Batista, and the reformists who had been involved in the revolution of 1933 had shown themselves to be corrupt. The "generation of 1953" was ready for radical changes in Cuban politics and society.

## CASTRO'S CUBAN REVOLUTION

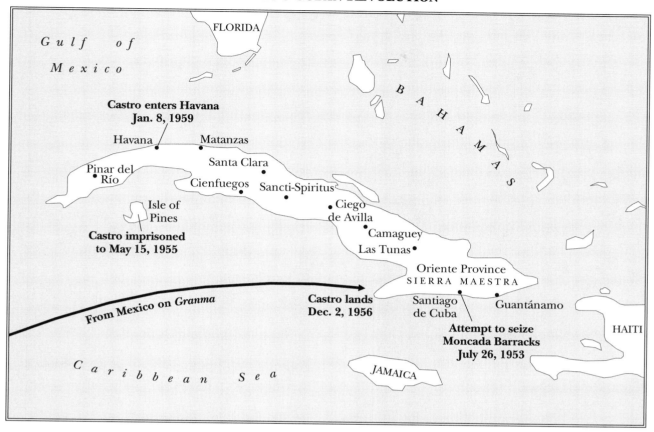

Castro was not interested in trying to make his country look like its giant neighbor to the north. Cuban nationalism meant defying the United States, and becoming economically independent meant turning away from American capitalism. So it was that Castro turned toward communism and friendship with the Soviet Union.

*John A. Britton*

# Calvin Explains Photosynthesis

*The 1961 Nobel Prize in Chemistry was awarded to Melvin Calvin for his studies of photosynthesis, which forms the basis of nutrition for all living things.*

**What:** Biology
**When:** 1957
**Where:** Berkeley, California
**Who:**
MELVIN CALVIN (1911-1997), an American chemist and biochemist who was the winner of the 1961 Nobel Prize in Chemistry for his work on photosynthesis
MICHAEL POLANYI (1891-1976), an English professor of chemistry

## Separating Light from Dark

Photosynthesis is the process in which plant cells use light energy, water, and carbon dioxide to produce oxygen and carbohydrates (sugars). In fact, photosynthesis is the source of most of the oxygen in the air. It is thus a means of capturing the sun's energy and converting it into a form useful to living things. For that reason, this process occupies a central place in the economy of life.

Yet it was not until Melvin Calvin published *The Path of Carbon in Photosynthesis* (1957) that photosynthesis was well understood. Calvin's interest in photosynthesis began while he was studying in England as a postdoctoral fellow with Michael Polanyi at the University of Manchester from 1935 to 1937. In 1937, Calvin left England and returned to the United States to teach at the University of California at Berkeley. From 1941 to 1945, during World War II, Calvin worked for the U.S. government. This included two years on the Manhattan Project (the U.S. effort to develop an atomic bomb).

In 1946, Calvin was appointed director of the bioorganic division of the Lawrence Radiation Laboratory at Berkeley. His work centered on photosynthesis.

Calvin had two goals in mind: determining how light energy is utilized, and learning how complex sugars and other molecules are formed. There had been speculation, even before 1940, that the conversion of carbon dioxide to carbohydrate might not require light, unlike the conversion of water to oxygen, which requires light. This had been proved in 1937, and Calvin confirmed it. When deprived of carbon dioxide while being illuminated, the plants were able to produce and store a high-energy reducing agent. When placed in a darkened environment, the plants were able to use this reducing agent to take up large amounts of carbon dioxide and convert them to sugar despite the darkness.

*Melvin Calvin.*

AP/Wide World Photos

**1252**

Thus, plant use of carbon dioxide was activated by the reducing agents produced by light energy, rather than by the light itself.

Calvin reasoned that if all chemicals involved in photosynthesis were identified, the sequence of reactions could be understood—but it would not be easy. Since all plant life is based on carbon, he needed to determine how carbon derived from carbon dioxide could be distinguished from carbon from other sources. This was made possible by the discovery of the long-lived radioactive isotope of carbon, carbon 14, by Samuel Ruben and Martin Kamen. Carbon dioxide could be produced with carbon 14, and the radioactive element would thus be identifiable at every stage of photosynthesis. Naturally occurring carbon would not be detected in this way, so it could be ignored.

### Calvin's Theory

In his Nobel address, Calvin described his strategy of investigating the chemicals into which the carbon is transformed during photosynthesis. If the reactions that transform the carbon into various chemicals could be stopped at some point, the identity of the molecules between carbon dioxide and sugars could be determined. The apparatus used for Calvin's studies was nicknamed the "lollipop" because of its shape. The alga *chlorella* was placed in the thin disklike apparatus and exposed to light and radioactive carbon dioxide. All chemical reactions were stopped by dropping the algae suspension into alcohol. This step also dissolved the organic molecules, separating them from the solid plant material, which could be analyzed for radioactivity.

It became obvious to Calvin that after only a few seconds the radioactive carbon had passed into a range of different chemical compounds. Using a special technique to separate the compounds and aid in analyzing them, Calvin came up with a complex sequence of reactions. Calvin was able to place each substance in the sequence in part by charting the amount of each substance present. A relatively small amount of carbon would have passed through to the last phase of photosynthesis, while progressively larger amounts of carbon compounds would distinguish earlier and earlier phases.

The photosynthesis reactions form a cycle, as Calvin learned. At the time of the Nobel award, Calvin attributed the idea for this convoluted pathway to a trip he took with his wife. Inspiration came to him as he sat waiting in his car.

### Consequences

Photosynthesis, as described by Calvin and his coworkers at Berkeley, is a complex series of chemical reactions. This series is referred to as the "Calvin cycle." He was able to show how this cycle—which occurs in a range of organisms from bacteria to higher plants—operates. The conversion of light energy to chemical energy serves to transform carbon dioxide into more complex organic molecules. The molecules that result from photosynthesis form the basis of nutrition for all living things. Moreover, these molecules are necessary to the formation of oil, fuel, petrochemicals, and all other materials derived from them, such as pharmaceuticals, plastics, feedstocks, and dyes. It is hard to imagine a world without the products of photosynthesis, since they touch almost every aspect of human life.

The work honored by the Nobel committee was characterized by its ambitious nature, coupled with enormous attention to detail. Efraim Racker of the Public Health Research Institute of New York, an expert in photosynthesis, said that "there was a lot of confusion in the field before Calvin. But he came up with the concept of the major cycle whereby carbon becomes sugar." In addition to the enormous contributions to the biochemical field, Calvin creatively applied a wide range of analytical techniques to the project.

Calvin proved to be a popular Nobel laureate. *Time* magazine referred to him as the "jolly biochemist" who had long been known as "Mr. Photosynthesis." Calvin's communication skills allowed his work to be widely understood and accepted.

# Isaacs and Lindenmann Discover Interferons

> *Alick Isaacs and Jean Lindenmann discovered interferons, proteins produced by body cells to fight viral diseases and possibly cancer.*

**What:** Medicine
**When:** 1957
**Where:** Mill Hill, England
**Who:**
ALICK ISAACS (1921-1967), a Scottish physician and microbiologist
JEAN LINDENMANN (1924-    ), a Swiss researcher
SIR MACFARLANE BURNET (1899-1985), an Australian microbiologist

## The Search for Virus Protection

It has been known for many years that people typically do not catch more than one virus at a time. This is thought to be caused by the biological action of proteins, called "interferons," secreted by virus-infected cells. There are three kinds of interferons: leukocytic, fibroepithelial, and immune types. The fibroepithelial and leukocytic interferons are believed to be most important in protection against viral infection. The interferon type produced in response to a viral invasion, however, depends on the kind of body cell that is infected and on the infecting virus involved.

Interferons are present in both plants and animals. In all the organisms that contain interferons, they are now believed to act as the primary cellular defense against viruses. Interferons not only destroy the original infecting viruses but also protect organisms from many other viruses after a specific virus attack. The discovery of interferons is credited to Alick Isaacs and Jean Lindenmann.

Isaacs, a physician and microbiologist, received his medical training at Glasgow University; however, he preferred basic research to clinical medicine. Therefore, he entered into graduate study of bacteriology. Isaacs did so well that he was granted a Medical Research Council Studentship (1947) and a Rockefeller Traveling Fellowship (1948); these awards allowed Isaacs to work with eminent microbiologists in England and Australia, such as Sir Macfarlane Burnet. He began to study viral influenzas (now known to be caused by a ribonucleic acid, or RNA, virus). Understanding influenza remained Isaacs's primary research goal throughout his career.

Isaacs's discovery of interferon arose from his examination of the so-called viral interference phenomenon—that is, the observation that any RNA virus usually can prevent the growth, in a cell, of other RNA viruses added to the cell. Isaacs's first efforts, carried out in Australia, showed that the interference caused by influenza viruses had nothing to do with the penetration of the cells by the viruses. Rather, it appeared to be caused by an event that occurred in the interior of the infected cell. In 1951, Isaacs returned to England to work in the National Institute for Medical Research at Mill Hill. His collaboration with Lindenmann—a visiting Swiss scientist—produced the crucial experiments in 1957 that led to their being credited with the discovery of interferons.

## Interfering with Viruses

Two very important initial discoveries that Isaacs and Lindenmann made were in "in vitro" experiments. The interfering agent was present in the culture fluid around the tissue samples being studied, and the infected tissues were stimulated to produce the interfering agent. They named the uncharacterized interfering agent "interferon." Subsequent study demonstrated that this interferon was a protein or a substance that required a protein for its action.

The discovery of interferon identified what appeared to be a new, important mechanism for defense against viral infection. It seemed likely

**1254**

that interferon would have exceptional therapeutic value because it worked against many different viruses, not only against the one that caused its production. In 1961, Isaacs was appointed to head the virus research division at Mill Hill. In ensuing years, he was awarded several honorary medical degrees in response to his work and was elected to the Royal Society of London. Regrettably, Isaacs did not live to see interferon research come to fuller fruition; he died of an intercranial hemorrhage on January 26, 1967.

Soon, however, the molecular basis for interferon action was identified. It was found that the entry of an RNA virus (actually its RNA core) into a cell causes the cell to produce interferon, which is then secreted from the cell or else is released from the cell if the RNA virus infection kills the cell. The interferon molecules secreted or released in these ways come into contact next with uninfected cells. They interact with these cells and confer upon them a great resistance to the original, infecting RNA virus and to a large number of other RNA viruses that would otherwise be infectious.

Interferons prevent viruses from killing protected cells by stopping the usual sequence of events that occurs when a virus infects a cell. When a virus interacts with a cell, it penetrates the cell membrane and releases its RNA core into the cell. The RNA core converts the cell into a virus factory that makes the viral RNA and the viral proteins that will become new viruses (viral progeny). These viral components are assembled and become hundreds of viral progeny. The viral progeny destroy the infected cell, enter other nearby cells, and start the process over again.

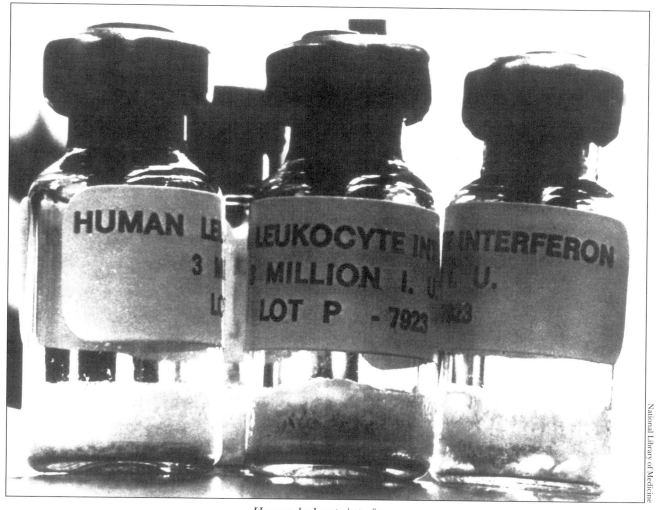

*Human leukocyte interferon.*

National Library of Medicine

The steps are repeated over and over. Interferons therefore act by stopping the synthesis of both viral proteins and viral RNA. Soon after a viral infection is under control, interferon production stops and, consequently, the antiviral proteins are no longer produced.

## Consequences

It is widely believed now that interferons are the cell's first line of defense against viral infection. Interferons are reported to be useful in the treatment of hepatitis, rabies, viral encephalitis, and eye infections. It has also been observed that many types of cancer cell tissue cultures are killed by the addition of interferons. These cancers are thought to be caused by RNA viruses. Osteogenic sarcoma, a bone cancer that attacks teenagers and usually kills 85 percent of those afflicted, is now treated successfully via interferons. Reports of other valuable interferon-based cancer therapy fill the literature of cancer research.

Until the mid-1980's development of genetically engineered interferons, interferons were very rare (for example, 25 gallons of blood yielded 1 milligram of pure interferon). Many early studies of interferon action were less than ideal. Consequently, there was considerable controversy about the overall efficacy of the interferons in cancer therapy. The results with genetically engineered interferons, however, were encouraging; these interferons promise to become part of the modern arsenal of cancer therapy and the treatment of serious viral diseases.

*Sanford S. Singer*

# Sabin Develops Oral Polio Vaccine

*Albert Bruce Sabin developed a polio vaccine that stimulated long-lasting immunity but did not cause paralytic disease.*

**What:** Medicine
**When:** 1957
**Where:** Cincinnati, Ohio
**Who:**
ALBERT BRUCE SABIN (1906-1993), a
    Russian-born American virologist
JONAS EDWARD SALK (1914-1995), an
    American physician, immunologist,
    and virologist
RENATO DULBECCO (1914-     ), an
    Italian-born American virologist who
    shared the 1975 Nobel Prize in
    Physiology or Medicine

## The Search for a Living Vaccine

A century ago, the first major poliomyelitis (polio) epidemic was recorded. Thereafter, epidemics of increasing frequency and severity struck the industrialized world. By the 1950's, as many as sixteen thousand individuals, most of them children, were being paralyzed by the disease each year.

The polio virus (poliovirus) enters the body through ingestion by the mouth. It replicates in the throat and the intestines and establishes an infection that normally is harmless. From there, the virus can enter the bloodstream. In some individuals it makes its way to the nervous system, where it attacks and destroys nerve cells crucial for muscle movement. The presence of antibodies in the bloodstream will prevent the virus from reaching the nervous system and causing paralysis. Thus, the goal of vaccination is to administer poliovirus that has been altered so that it cannot cause disease but nevertheless will stimulate the production of antibodies to fight the disease.

Albert Bruce Sabin received his medical degree from New York University College of Medicine in 1931. Polio was epidemic in 1931, and for

Sabin polio research became a lifelong interest. In 1936, while working at the Rockefeller Institute, Sabin and Peter Olinsky successfully grew poliovirus using tissues cultured in vitro. Tissue culture proved to be an excellent source of virus. Jonas Edward Salk soon developed an inactive polio vaccine consisting of virus grown from tissue culture that had been inactivated (killed) by chemical treatment. This vaccine became available for general use in 1955, almost fifty years after poliovirus had first been identified.

Sabin, however, was not convinced that an inactivated virus vaccine was adequate. He believed that it would provide only temporary protection and that individuals would have to be vaccinated repeatedly in order to maintain protective levels of antibodies. Knowing that natural infection with poliovirus induced lifelong immunity, Sabin believed that a vaccine consisting of a living virus was necessary to produce long-lasting immunity. Also, unlike the inactive vaccine, which is injected, a living virus (weakened so that it would not cause disease) could be taken orally and would invade the body and replicate of its own accord.

Sabin was not alone in his beliefs. Hilary Koprowski and Harold Cox also favored a living virus vaccine and had, in fact, begun searching for weakened strains of poliovirus as early as 1946 by repeatedly growing the virus in rodents. When Sabin began his search for weakened virus strains in 1953, a fiercely competitive contest ensued to achieve an acceptable live virus vaccine.

## Rare, Mutant Polioviruses

Sabin's approach was based on the principle that, as viruses acquire the ability to replicate in a foreign species or tissue (for example, in mice), they become less able to replicate in humans and thus less able to cause disease. Sabin used tissue culture techniques to isolate those polioviruses

**1257**

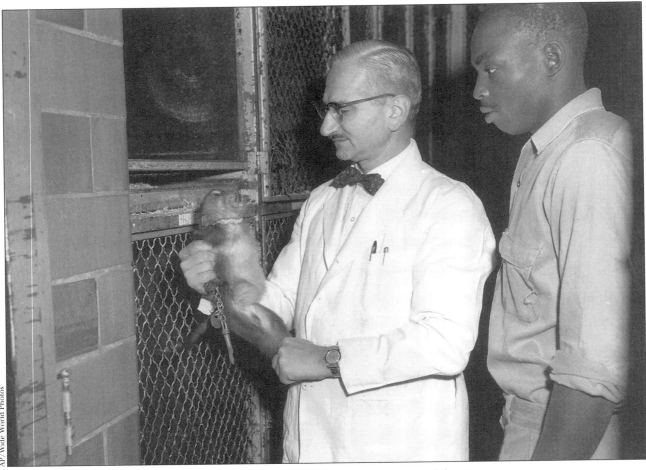

*Albert Bruce Sabin (left) examines a monkey in his laboratory.*

that grew most rapidly in monkey kidney cells. He then employed a technique developed by Renato Dulbecco that allowed him to recover individual virus particles. The recovered viruses were injected directly into the brains or spinal cords of monkeys in order to identify those viruses that did not damage the nervous system. These meticulously performed experiments, which involved approximately nine thousand monkeys and more than one hundred chimpanzees, finally enabled Sabin to isolate rare mutant polioviruses that would replicate in the intestinal tract but not in the nervous systems of chimpanzees or, it was hoped, of humans. In addition, the weakened virus strains were shown to stimulate antibodies when they were fed to chimpanzees; this was a critical attribute for a vaccine strain.

By 1957, Sabin had identified three strains of attenuated viruses that were ready for small experimental trials in humans. A small group of volunteers, including Sabin's own wife and children, were fed the vaccine with promising results. Sabin then gave his vaccine to virologists in the Soviet Union, Eastern Europe, Mexico, and Holland for further testing. Combined with smaller studies in the United States, these trials established the effectiveness and safety of his oral vaccine.

During this period, the strains developed by Cox and by Koprowski were being tested also in millions of persons in field trials around the world. In 1958, two laboratories independently compared the vaccine strains and concluded that the Sabin strains were superior. In 1962, after four years of deliberation by the U.S. Public Health Service, all three of Sabin's vaccine strains were licensed for general use.

## Consequences

The development of polio vaccines ranks as one of the triumphs of modern medicine. In the early 1950's, paralytic polio struck 13,500 out of every 100 million Americans. The use of the Salk vaccine greatly reduced the incidence of polio, but outbreaks of paralytic disease continued to occur: Fifty-seven hundred cases were reported in 1959 and twenty-five hundred cases in 1960. In 1962, the oral Sabin vaccine became the vaccine of choice in the United States. Since its widespread use, the number of paralytic cases in the United States has dropped precipitously, eventually averaging fewer than ten per year. Worldwide, the oral vaccine prevented an estimated 5 million cases of paralytic poliomyelitis between 1970 and 1990.

The oral vaccine is not without problems. Occasionally, the living virus mutates to a disease-causing (virulent) form as it multiplies in the vaccinated person. When this occurs, the person may develop paralytic poliomyelitis. The inactive vaccine, in contrast, cannot mutate to a virulent form. Ironically, nearly every incidence of polio in the United States is caused by the vaccine itself.

In the developing countries of the world, the issue of vaccination is more pressing. Millions receive neither form of polio vaccine; as a result, at least 250,000 individuals are paralyzed or die each year. The World Health Organization and other health providers continue to work toward the very practical goal of completely eradicating this disease.

*Robin S. Treichel*

# Esaki Demonstrates Electron Tunneling in Semiconductors

*A Japanese physicist demonstrated tunneling effects in electronic systems, a discovery that would help to revolutionize the field of electronics.*

**What:** Physics
**When:** 1957
**Where:** Tokyo, Japan
**Who:**
Leo Esaki (1925-     ), a Japanese physicist
Ivar Giaever (1929-     ), a Norwegian physicist
Brian Josephson (1940-     ), a British physicist
William Shockley (1910-1989), an American physicist

## The Semiconductor Revolution

Electrical and electronic devices both rely on electrons to perform work, such as generating heat, creating sound waves, or moving a machine part. In a simple electrical device (a light bulb, for example), the flow of electricity is controlled by a manual switch; in an electronic device, one electrical signal controls others. The advances that make up the electronic revolution may be thought of in simple terms as the creation of ever smaller, cheaper, and more sensitive switches for controlling electrical current.

In the period between the two world wars and before the invention of the transistor, electronic switching was done by using electron beams enclosed in vacuum tubes the size of small light bulbs; these devices needed to be heated and required substantial current to run. The earliest computers, which were built with this technology, occupied entire rooms and were less powerful than a modern hand-held calculator.

The invention of the transistor by William Shockley, John Bardeen, and Walter H. Brattain in 1948 profoundly changed electronics engi-

neering. A transistor uses the properties of a semiconductor, such as silicon, for switching. A semiconductor is a substance in which low-energy electrons are not free to move but higher-energy electrons are. Therefore, when voltage (a measure of electron energy) increases past a certain point, a semiconductor will conduct current. Thus, it acts as a voltage-sensitive switch, performing all the functions of a vacuum tube in much less space and with much less energy. Furthermore, the sensitivity of a silicon switch and the direction in which it passes current can be modified quite precisely by the addition of impurities (materials other than silicon). An entire new industry was created, in which the Japanese took the lead in applying the new technology to consumer products.

## Electron Tunneling and Tunnel Diodes

In 1956, Leo Esaki was a doctoral student at Tokyo University who also worked for the Sony Corporation. He was looking for ways of improving semiconductor diodes with p-n junctions. P-n stands for positive-negative. A p-n junction is formed when half of a piece of a semiconductor, such as silicon, is "doped" with impurities (that is, has impurities added to it) that have fewer electrons than the silicon does, creating a net positive charge, and the other half of the piece contains impurities with excess electrons, creating a net negative charge. P-n junctions are fundamental to modern electronics; much of the process of manufacturing a silicon chip consists of selectively introducing impurities into microscopic regions on the chip to create thousands of p-n junctions.

Esaki experimented with various levels of impurities in silicon and another semiconductor, germanium, taking advantage of improving technology for doping semiconductors to produce junctions that were far narrower and more clearly defined than had been possible a few years earlier. He then drew current across the junction as a function of increasing voltage. At the junction, a potential barrier exists that classical physics theory predicted electrons must surmount in order to flow through the semiconductor. Previous experiments had demonstrated current flow only at voltage levels greater than the potential barrier, but in heavily doped germanium crystals, current flow occurred at lower levels. This, Esaki showed, was the result of a tunneling effect.

The tunneling of electrons through a potential barrier, which is predicted by Albert Einstein's theory of general relativity, is related to the dual nature of an electron as both a particle and a wave. A relativistic particle (that is, a particle whose behavior is described by Einstein's theory) has a certain probability of being found in a region that is forbidden to it by classical physics. As long as the potential barrier region was broad (as it had been in earlier experiments), too few electrons tunneled through the barrier to be detected, but with the improved technology that made it possible to construct narrow p-n junctions, Esaki was able to demonstrate the tunneling phenomenon which had been predicted in mathematical models.

## Consequences

The techniques for the precise introduction of impurities into semiconductors developed by Esaki and his colleagues in the Sony laboratories in the 1950's are basic elements of the technology that produces the silicon chips that form the basis of modern electronics. Esaki "tunnel diodes" have both advantages and disadvantages when compared with nontunneling semiconductor transistors. Tunnel diodes are used together with transistors in computers, and they have been found to be particularly useful in sensitive microwave detectors, such as those used in radio telescopes.

The demonstration of electron tunneling across p-n junctions in semiconductors stimulated investigation in other areas—notably, in superconductivity. The work of Brian Josephson and Ivar Giaever, who shared the 1973 Nobel Prize in Physics with Esaki, also concerned tunneling in superconductors.

The silicon revolution has transformed manufacturing, communications, and information processing, and it has allowed virtually every task that can be done by electricity to be done with an efficiency and an economy that had not even been dreamed of before World War II. Its discoveries are not the product of a single person, laboratory, or nation; the silicon revolution is worldwide. The discoveries of Shockley, Esaki, Josephson, Giaver, and others who built upon their work helped to reshape the modern world.

*Martha Sherwood-Pike*

The Nobel Foundation

*Leo Esaki.*

**1261**

# Sony Develops Transistor Radio

*Sony introduced a transistorized pocket radio and opened up a mass market for electronics.*

**What:** Communications
**When:** 1957
**Where:** Tokyo, Japan
**Who:**
JOHN BARDEEN (1908-1991), an American physicist
WALTER H. BRATTAIN (1902-1987), an American physicist
WILLIAM SHOCKLEY (1910-1989), an American physicist
AKIO MORITA (1921-1999), a Japanese physicist and engineer
MASARU IBUKA (1908-1997), a Japanese electrical engineer and industrialist

## A Replacement for Vacuum Tubes

The invention of the first transistor by William Shockley, John Bardeen, and Walter H. Brattain of Bell Labs in 1947 was a scientific event of great importance. Its commercial importance at the time, however, was negligible. The commercial potential of the transistor lay in the possibility of using semiconductor materials to carry out the functions performed by vacuum tubes, the fragile and expensive tubes that were the electronic hearts of radios, sound amplifiers, and telephone systems. Transistors were smaller, more rugged, and less power-hungry than vacuum tubes. They did not suffer from overheating. They offered an alternative to the unreliability and short life of vacuum tubes.

Bell Labs had begun the semiconductor research project in an effort to find a better means of electronic amplification. This was needed to increase the strength of telephone signals over long distances. Therefore, the first commercial use of the transistor was sought in speech amplification, and the small size of the device made it a perfect component for hearing aids. Engineers from the Raytheon Company, the leading manufacturer of hearing aids, were invited to Bell Labs to view the new transistor and to help assess the commercial potential of the technology. The first transistorized consumer product, the hearing aid, was soon on the market. The early models built by Raytheon used three junction-type transistors and cost more than two hundred dollars. They were small enough to go directly into the ear or to be incorporated into eyeglasses.

The commercial application of semiconductors was aimed largely at replacing the control and amplification functions carried out by vacuum tubes. The perfect vehicle for this substitution was the radio set. Vacuum tubes were the most expensive part of a radio set and the most prone to break down. The early junction transistors operated best at low frequencies, and subsequently more research was needed to produce a commercial high-frequency transistor. Several of the licensees embarked on this quest, including the Radio Corporation of America (RCA), Texas Instruments, and the Tokyo Telecommunications Engineering Company of Japan.

## Perfecting the Transistor

The Tokyo Telecommunications Engineering Company of Japan, formed in 1946, had produced a line of instruments and consumer products based on vacuum-tube technology. Its most successful product was a magnetic tape recorder. In 1952, one of the founders of the company, Masaru Ibuka, visited the United States to learn more about the use of tape recorders in schools and found out that Western Electric was preparing to license the transistor patent. With only the slightest understanding of the workings of semiconductors, Tokyo Telecommunications purchased a license in 1954 with the intention of using transistors in a radio set.

The first task facing the Japanese was to increase the frequency response of the transistor to make it suitable for radio use. Then a method of manufacturing transistors cheaply had to be found. At the time, junction transistors were made from slices of germanium crystal. Growing the crystal was not an exact science, nor was the process of "doping" it with impurities to form the different layers of conductivity that made semiconductors useful. The Japanese engineers found that the failure rate for high-frequency transistors was extremely high. The yield of good transistors from one batch ran as low as 5 percent, which made them extremely expensive and put the whole project in doubt. The effort to replace vacuum tubes with components made of semiconductors was motivated by cost rather than performance; if transistors proved to be more expensive, then it was not worth using them.

Engineers from Tokyo Telecommunications again came to the United States to search for information about the production of transistors, visiting factories and laboratories to learn first-hand. In 1954, the first high-frequency transistor was produced in Japan. The success of Texas Instruments in producing the components for the first transistorized radio (introduced by the Regency Company in 1954) spurred the Japanese to greater efforts. Much of their engineering and research work was directed at the manufacture and quality control of transistors. In 1955, they introduced their transistor radio, the TR-55, which carried the brand name "Sony." The name was chosen because the executives of the company believed that the product would have an international appeal and therefore needed a brand name that could be recognized easily and remembered in many languages. In 1957, the name of the entire company was changed to Sony.

## Consequences

Although Sony's transistor radios were successful in the marketplace, they were still relatively large and cumbersome. Ibuka saw a consumer market for a miniature radio and gave his engineers the task of designing a radio small enough to fit into a shirt pocket. The realization of this design—given the trade name "Transistor Six"—was introduced in 1957. It was an immediate success. Sony sold the radios by the millions, and numerous imitations were also marketed under brand names such as "Somy" and "Sonny." The product became an indispensable part of the youth-oriented popular culture of the late 1950's and 1960's; its low cost enabled the masses to enjoy radio wherever there were broadcasts.

The pocket-sized radio was the first of a line of electronic consumer products that brought technology into personal contact with the user. Sony was convinced that miniaturization did more than make products more portable; it established a one-on-one relationship between people and machines. Sony produced the first all-transistor television in 1960. Two years later, it began to market a miniature television in the United States. The continual reduction in the size of Sony's tape recorders reached a climax with the portable tape player introduced in the 1980's. The Sony "Walkman" was a marketing triumph and a further reminder that Japanese companies led the way in the design and marketing of electronic products.

*Andre Millard*

# Velcro Is Patented

*Georges de Mestral invented Velcro, a material made of millions of tiny hooks and loops that could be used to make fasteners for a wide range of applications.*

**What:** Materials
**When:** 1957
**Where:** Switzerland
**Who:**
GEORGES DE MESTRAL (1904-1990), a
    Swiss engineer and inventor

## From Cockleburs to Fasteners

Since prehistoric times, people have walked through weedy fields and arrived at home with cockleburs all over their clothing. In 1948, a Swiss engineer and inventor, Georges de Mestral, found his clothing full of cockleburs after walking in the Swiss Alps near Geneva. Wondering why cockleburs stuck to clothing, he began to examine the cockleburs under a microscope. De Mestral's initial examination showed that each of the thousands of fibrous ends of the cockleburs was tipped with a tiny hook; it was the hooks that made the cockleburs stick to fabric.

This observation, combined with much subsequent work, led de Mestral to invent Velcro, which was patented in 1957 in the form of two strips of nylon material. One of the strips contained millions of tiny hooks, while the other contained a similar number of tiny loops. When the two strips were pushed together, the hooks were inserted into the loops, joining the two strips of nylon very firmly. This design makes Velcro extremely useful as a material for fasteners that is used in applications ranging from sneaker fasteners to fasteners used to join heart valves during surgery.

## Making Velcro Practical

Velcro is not the only invention credited to de Mestral, who also invented such items as a toy airplane and an asparagus peeler, but it was his greatest achievement. It is said that his idea for the material was partly the result of a problem his wife had with a jammed dress zipper just before an important social engagement. De Mestral's idea was to design a sort of locking tape that used the hook-and-loop principle that he had observed under the microscope. Such a tape, he believed, would never jam. He also believed that the tape would do away with such annoyances as buttons that popped open unexpectedly and knots in shoelaces that refused to be untied.

The design of the material envisioned by de Mestral took seven years of painstaking effort. When it was finished, de Mestral named it "Velcro" (a contraction of the French phrase *velvet crochet*, meaning velvet hook), patented it, and opened a factory to manufacture it. Velcro's design required that de Mestral identify the optimal number of hooks and loops to be used. Ultimately, he discovered that using approximately three hundred per square inch worked best. In addition, his studies showed that nylon was an excellent material for his purposes, although it had to be stiffened somewhat to work well. Much additional experimentation showed that the most effective way of producing the necessary stiffening was to subject the Velcro to infrared light after manufacturing it.

Other researchers have demonstrated that materials similar to Velcro need not be made of nylon. For example, a new micromechanical material (microvelcro) that medical researchers believed could be used to hold together blood vessels after surgery is made of minute silicon loops and hooks. This material is thought to be superior to other materials for such applications because it will not be redissolved prematurely by the body. Other uses for microvelcro may be to hold together tiny electronic components in miniaturized computers without the use of glue or other adhesives. A major advantage of the use of microvelcro in such situations is that

**1264**

it is resistant to changes of temperature as well as to most chemicals that destroy glue and other adhesives.

## Consequences

In 1957, when Velcro was patented, there were four main ways to hold things together. These involved the use of buttons, laces, snaps, and zippers (which had been invented by Chicagoan Whitcomb L. Judson in 1892). All these devices had drawbacks; zippers can jam, buttons and laces can come open at embarrassing times, and shoelaces can form knots that are difficult to unfasten. Almost immediately after Velcro was introduced, its use became widespread; Velcro fasteners can be found on or in clothing, shoes, watchbands, wallets, backpacks, bookbags, motor vehicles, space suits, blood-pressure cuffs, and in many other places. There is even a "wall jumping" game incorporating Velcro in which a wall is covered with a well-supported piece of Velcro. People who want to play put on jackets made of Velcro and jump as high as they can. Wherever they land on the wall, the Velcro will join together, making them stick.

Wall jumping, silly though it may be, demonstrates the tremendous holding power of Velcro; a Velcro jacket can keep a two-hundred-pound person suspended from a wall. This great strength is used in a more serious way in the design of the items used to anchor astronauts to space shuttles and to buckle on parachutes. In addition, Velcro is washable, comes in many colors, and will not jam. No doubt many more uses for this innovative product will be found.

*Sanford S. Singer*

# Eisenhower Articulates His Middle East Doctrine

*Concerned about growing Soviet power in the Middle East, U.S. president Dwight D. Eisenhower promised economic and military aid to Middle Eastern nations, especially those threatened by communist aggression.*

**What:** International relations
**When:** 1957-1958
**Where:** Washington, D.C., and Lebanon
**Who:**
DWIGHT D. EISENHOWER (1890-1969),
    president of the United States from
    1953 to 1961
GAMAL ABDEL NASSER (1918-1970),
    president of Egypt from 1956 to 1970
CAMILLE CHAMOUN (1900-1987), president
    of Lebanon from 1952 to 1958
FUAD SHEHAB (1902-1973), commander
    of the Lebanese army, and president
    of Lebanon from 1958 to 1964

## The Doctrine

After the Suez Canal crisis of October, 1956, the United States government began to reconsider its policies in the Middle East. Acting through the United Nations—and for once in agreement with the Soviet Union—the United States had persuaded Great Britain, France, and Israel to withdraw their troops from Egypt. Yet though the United States had succeeded, the incident had weakened unity among the Western Powers; at the same time, the Soviet Union had gained new influence in the Arab countries. Egyptian president Gamal Abdel Nasser had also become even more popular as a spokesman for Arab nationalism.

President Dwight D. Eisenhower and his advisers believed that there was a "power vacuum" in the Middle East—a gap that the Soviet Union would quickly try to fill. To address the situation, Eisenhower came up with a new policy that became known as the Eisenhower Doctrine.

After consulting with leaders in Congress and with Dag Hammarskjöld, secretary general of the United Nations, Eisenhower presented his doctrine in a message to Congress on January 5, 1957. The United States should fill the vacuum, he said, with economic and military aid. Eisenhower asked the new Eighty-fifth Congress to set aside $400 million for two years to help Middle Eastern nations. He also asked Congress for permission to send U.S. armed forces whenever a Middle Eastern government asked for help in defending itself against communist aggression.

In making his request, Eisenhower had several purposes in mind: He wanted Congress to supply the funds to aid the Middle East, but he also wanted to warn the Soviet Union that the United States intended to stop Soviet expansion in the Arab countries. Also, he hoped to show that the American government and people were united in this goal.

In some ways, the Eisenhower Doctrine was similar to the Truman Doctrine of 1947 and the Formosa Resolution passed by Congress in 1955. Yet it was different from the Truman Doctrine in that it was concerned with one particular geographic area (the Truman Doctrine had promised American support for any people resisting aggression). Another difference was that neither of the earlier policies had stated that armed forces would be sent only when the other nation requested them.

## Congress and the World React

The House Foreign Affairs Committee approved a resolution supporting the president's request on January 24, and the entire House followed suit on January 31 with a vote of 355-61. The Senate's response was slower. In debates in

**1266**

early March, Senator Richard Russell of Georgia proposed an amendment to remove the military and economic aid, but this amendment lost. Finally, the Senate passed the resolution, with a few changes, on March 5 by a vote of 72-19. On March 7, the House accepted the Senate version by 350-60, and the president signed the bill on March 9, 1957.

Eisenhower, a Republican president, had succeeded in gaining the approval of a Democratic Congress for his Middle Eastern policy. There are two likely reasons: Eisenhower had just been reelected, and the American people tended to trust him; also, because of the Cold War many Americans were very concerned about the Soviet Union.

Overseas, the reactions were less positive. As American leaders had expected, the Communist governments in Moscow and Beijing condemned the Eisenhower Doctrine. Prime Minister Jawarharlal Nehru of India thought that Eisenhower had exaggerated the dangers of Soviet aggression, and that the doctrine did not encourage peace. Led by Egypt, the Arab countries also criticized the new American policy.

## Consequences

One exception in the Middle East was Lebanon. All the other Arab countries had Muslim majorities, but Lebanon had a large Christian population. Lebanese Catholics had ties to Rome, while Lebanese Protestants had had contacts with American missionaries and teachers. Furthermore, Lebanon had been under French rule for a good part of its history. For all these reasons, the Lebanese people were more open to relations with the West than other Arab nations were.

After becoming independent, Lebanon had developed a political tradition to lessen religious tensions: The president had to be a Christian, the prime minister a Muslim. Yet there were still fierce rivalries within the country.

*Dwight D. Eisenhower.*

When it seemed that President Camille Chamoun intended to extend his term of office, contrary to the Lebanese constitution, strife broke out in Lebanon. Religious and political groups organized to oppose Chamoun—some of them with the help of Syria and Egypt.

This turmoil seemed to be subsiding when there was an unexpected crisis in Iraq. On July 14, 1958, a bloody revolution overthrew the pro-Western Iraqi government. President Chamoun was alarmed. Suspecting that the Iraqi coup had

**1267**

been plotted by the Soviet Union and followers of Egyptian president Nasser, he feared that a similar coup was about to take place in Lebanon. Chamoun asked the United States for help.

On July 15, on orders from President Eisenhower, units of the Sixth U.S. Marine Fleet landed in Lebanon to preserve order. With the help of U.S. ambassador Robert McClintock, the American troops were kept from influencing the political situation in Lebanon. Robert Murphy, an official in the U.S. State Department, worked with the rival forces in Lebanon to bring about a settlement.

Chamoun announced that he would not extend his term of office, and the negotiators per-suaded General Fuad Shehab to accept the Lebanese presidency. As Commander of the Lebanese army, Shehab had worked hard not to favor one group over another, and almost all factions in Lebanon were willing to support him. On July 31, he was elected president by the Lebanese parliament, and the U.S. Marines left Lebanon on October 25, 1958.

Eisenhower had been able to exercise his doctrine to promote peace in Lebanon. For the next few decades, however, the Middle East would continue to be a shifting ground of Arab nationalism and competition between the United States and the Soviet Union.

*George J. Fleming*

# Senning Invents Implanted Heart Pacemaker

*Ake Senning's development of transistor circuitry in 1958 made possible the development of permanently implanted heart pacemakers.*

**What:** Medicine
**When:** 1957-1972
**Where:** Stockholm, Sweden
**Who:**

AKE SENNING (1915-        ), a Swedish physician

RUNE ELMQUIST, co-inventor of the first pacemaker

PAUL MAURICE ZOLL (1911-1999), an American cardiologist

## Cardiac Pacing

The fundamentals of cardiac electrophysiology (the electrical activity of the heart) were determined during the eighteenth century; the first successful cardiac resuscitation by electrical stimulation occurred in 1774. The use of artificial pacemakers for resuscitation was demonstrated in 1929 by Mark Lidwell. Lidwell and his coworkers developed a portable apparatus that could be connected to a power source. The pacemaker was used successfully on several stillborn infants after other methods of resuscitation failed. Nevertheless, these early machines were unreliable.

Ake Senning's first experience with the effect of electrical stimulation on cardiac physiology was memorable; grasping a radio ground wire, Senning felt a brief episode of ventricular arrhythmia (irregular heartbeat). Later, he was able to apply a similar electrical stimulation to control a heartbeat during surgery.

The principle of electrical regulation of the heart was valid. It was shown that pacemakers introduced intravenously into the sinus node area of a dog's heart could be used to control the heartbeat rate. Although Paul Maurice Zoll utilized a similar apparatus in several patients with cardiac arrhythmia, it was not appropriate for extensive clinical use; it was large and often caused unpleasant sensations or burns. In 1957, however, Ake Senning observed that attaching stainless steel electrodes to a child's heart made it possible to regulate the heart's rate of contraction. Senning considered this to represent the beginning of the era of clinical pacing.

## Development of Cardiac Pacemakers

Senning's observations of the successful use of the cardiac pacemaker had allowed him to identify the problems inherent in the device. He realized that the attachment of the device to the lower, ventricular region of the heart made possible more reliable control, but other problems remained unsolved. It was inconvenient, for example, to carry the machine externally; a cord was wrapped around the patient that allowed the pacemaker to be recharged, which had to be done frequently. Also, for unknown reasons, heart resistance would increase with use of the pacemaker, which meant that increasingly large voltages had to be used to stimulate the heart. Levels as high as 20 volts could cause quite a "start" in the patient. Furthermore, there was a continuous threat of infection.

In 1957, Senning and his colleague Rune Elmquist developed a pacemaker that was powered by rechargeable nickel-cadmium batteries, which had to be recharged once a month. Although Senning and Elmquist did not yet consider the pacemaker ready for human testing, fate intervened. A forty-three-year-old man was admitted to the hospital suffering from an atrioventricular block, an inability of the electrical stimulus to travel along the conductive fibers of the "bundle of His" (a band of cardiac muscle fibers). As a result of this condition, the patient required repeated cardiac resuscitation. Similar

**1269**

types of heart block were associated with a mortality rate higher than 50 percent per year and nearly 95 percent over five years.

Senning implanted two pacemakers (one failed) into the myocardium of the patient's heart, one of which provided a regulatory rate of 64 beats per minute. Although the pacemakers required periodic replacement, the patient remained alive and active for twenty years. (He later became president of the Swedish Association for Heart and Lung Disease.)

During the next five years, the development of more reliable and more complex pacemakers continued, and implanting the pacemaker through the vein rather than through the thorax made it simpler to use the procedure. The first pacemakers were of the "asynchronous" type, which generated a regular charge that overrode the natural pacemaker in the heart. The rate could be set by the physician but could not be altered if the need arose. In 1963, an atrial-triggered synchronous pacemaker was installed by a Swedish team. The advantage of this apparatus lay in its ability to trigger a heart contraction only when the normal heart rhythm was interrupted. Most of these pacemakers contained a sensing device that detected the atrial impulse and generated an electrical discharge only when the heart rate fell below 68 to 72 beats per minute.

The biggest problems during this period lay in the size of the pacemaker and the short life of the battery. The expiration of the electrical impulse sometimes caused the death of the patient. In addition, the most reliable method of checking the energy level of the battery was to watch for a decreased pulse rate. As improvements were made in electronics, the pacemaker became smaller, and in 1972, the more reliable lithium-iodine batteries were introduced. These batteries made it possible to store more energy and to monitor the energy level more effectively. The use of this type of power source essentially eliminated the battery as the limiting factor in the longevity of the pacemaker. The period of time that a pacemaker could operate continuously in the body increased from a period of days in 1958 to five to ten years by the 1970's.

## Consequences

The development of electronic heart pacemakers revolutionized cardiology. Although the initial machines were used primarily to control cardiac bradycardia, the often life-threatening slowing of the heartbeat, a wide variety of arrhythmias and problems with cardiac output can now be controlled through the use of these devices. The success associated with the surgical implantation of pacemakers is attested by the frequency of its use. Prior to 1960, only three pacemakers had been implanted. During the 1990's, however, some 300,000 were implanted each year throughout the world. In the United States, the prevalence of implants is on the order of 1 per 1,000 persons in the population.

Pacemaker technology continues to improve. Newer models can sense pH and oxygen levels in the blood, as well as respiratory rate. They have become further sensitized to minor electrical disturbances and can adjust accordingly. The use of easily sterilized circuitry has eliminated the danger of infection. Once the pacemaker has been installed in the patient, the basic electronics require no additional attention. With the use of modern pacemakers, many forms of electrical arrhythmias need no longer be life-threatening.

*Richard Adler*

# Bardeen, Cooper, and Schrieffer Explain Superconductivity

> *John Bardeen, Leon N. Cooper, and John Robert Schrieffer were the first physicists to explain how some metals, as they approach absolute zero (−237.59 degrees Celsius), lose their resistance to electricity.*

**What:** Physics
**When:** February-July, 1957
**Where:** Urbana-Champaign, Illinois
**Who:**

JOHN BARDEEN (1908-1991), an American physicist

LEON N. COOPER (1930-        ), an American physicist

FRITZ WOLFGANG LONDON (1900-1954), an American physicist

HEINZ LONDON (1907-1970), an American physicist

HEIKE KAMERLINGH ONNES (1853-1926), a Dutch physicist

JOHN ROBERT SCHRIEFFER (1931-        ), a graduate student in physics

## A Scientific Miracle

When an electric current is run through a piece of metal, a considerable amount of the energy is lost to what is called "electrical resistance." Different metals have different resistances. In 1911, Heike Kamerlingh Onnes, a Dutch physicist who later won a Nobel Prize in Physics, made a startling discovery. He found that when the temperature of mercury was lowered almost to absolute zero (−273.15 degrees Celsius), it seemed to lose all of its electrical resistance. Onnes had discovered "superconductivity." For the next fifty years, scientists would repeat this experiment with other metals with the same results, but no one was able to explain it.

John Bardeen first became interested in superconductivity in 1938, when he read David Shoenberg's new book, *Super-Conductivity.* He had already heard of the report given by Fritz London and Heinz London to a meeting of the

Royal Society in London that established a link between superconductivity and quantum mechanics, something else that physicists were only beginning to understand. By 1940, Bardeen had begun to formulate his own explanation for superconductivity. Unfortunately, his thinking was sidetracked by World War II. Between 1941 and 1945, Bardeen worked at the Naval Ordnance Laboratory, doing research on the transistor. He was awarded his first Nobel Prize for this research in 1956.

Bardeen resumed his study of superconductivity in 1950, after some new discoveries helped to explain how electricity works. These discoveries turned out to be some of the missing pieces in the puzzle of superconductivity, though the puzzle still seemed almost impossible to solve. Bardeen decided that the only way to solve the puzzle was to break it up into smaller, less difficult pieces. In 1951, having become a professor of physics and engineering at the University of Illinois at Urbana-Champaign, he asked Leon N. Cooper, who had a doctorate in physics from Columbia University, to help him with some of the problems. In 1956, they were joined by John Robert Schrieffer, one of Bardeen's graduate students.

It was already known that twenty-six metals and ten alloys (combinations of metals) are superconductors at different temperatures. The highest temperature at which any of them became superconductors was −214.44 degrees Celsius. As yet, there was no real evidence that superconductivity had to take place at such extremely low temperatures. Although using liquid nitrogen is a fairly cheap way to bring temperatures down close to absolute zero, Bardeen and his co-workers knew that superconductivity would never be truly practical unless it could be achieved at higher temperatures.

AP/Wide World Photos

*John Bardeen*

*John Robert Schrieffer.*

## Cooper Pairs

As early as 1950, Bardeen had understood that the key to understanding superconductivity lay in the way in which electrons move and interact with one another. Earlier theories had supposed that superconductivity involved atomic vibration. Bardeen suggested possible ways of measuring the change in electrical resistance when superconducting temperatures had been reached. Until this time, there had been no instruments sensitive enough to measure this resistance.

By 1956, Cooper, Bardeen's assistant, had finally taken the first step toward solving the puzzle. He discovered that free electrons (electrons that are not bound to any one molecule) are attracted to each other in pairs during superconductivity. These pairs are often called "Cooper

pairs." After a year of working with this discovery, Bardeen and his coworkers finally came up with a successful model for understanding superconductivity. They published this model in February, 1957, and followed it with supporting evidence for the next six months.

One way to view their model is to think of the electrons as people in a crowded railroad station. The people are squeezed together so tightly that they bump into anything that gets in their way—mostly other people. This is how electrons move when there is little or no current running through a metal.

When a current is introduced, it is as though the people on one side of the station were suddenly pushed very hard. The people in their immediate vicinity are pushed very hard as well, but the people who are farther away feel the push only slightly, and the people on the other side of the station may not feel anything. If electrons are weakly paired, however, as the Cooper pairs are, the resistance to any one pair from all the electrons that are not a part of that pair goes down about one hundred times. It is this state that comes close to superconductivity. Bardeen and his coworkers also discovered that besides lacking resistance, superconductors can also prevent magnetic fields from entering them.

## Consequences

For centuries, people have been seeking a way to use perpetual motion to create energy without using fuel. When Bardeen proposed his theoretical explanation of superconductivity, scientists began to think that the search was almost over. When the theory was announced, scientists immediately started to devise ways to put superconductivity to use. They thought of electrical power lines made of superconductive wire. Without any electrical resistance to stop much of the electricity from reaching the end of the line, electrical power plants would be much cheaper and more efficient. Scientists also suggested superfast trains that would be built on top of powerful magnets and hover over superconductive "rails." They imagined computers running hundreds of times faster than had been possible before the development of superconductors. These ideas and many more were made possible by the theory set forth by Bardeen, Cooper, and Schrieffer. Many ideas have already been put to work, and others, still unthought of, will be developed in the years to come.

*R. Baird Shuman*

# Ghana Is First Black African Colony to Win Independence

*Led by the Convention People's Party and Kwame Nkrumah, the British-ruled Gold Coast was the first sub-Saharan African colony to gain political independence, under the new name of Ghana.*

**What:** Political independence

**When:** March 6, 1957

**Where:** Accra, Ghana

**Who:**

KWAME NKRUMAH (1909-1972), leader of the Convention People's Party, and prime minister of Ghana from 1957 to 1966

## Colonial History

Portuguese explorers along the coast of West Africa in the fifteenth century were soon followed by English, Dutch, Swedish, Danish, and German traders. They named areas of the coast for the products they found there: the Grain Coast (modern Liberia), the Ivory Coast (Côte d'Ivoire), the Slave Coast (between the Volta River and the Niger Delta), and the Gold Coast (modern Ghana). After 1650, however, the slave trade became more important than trade in gold and ivory, and European competition along the Gold Coast decreased.

By 1824, the Asante Confederacy, a government led by the Akan ethnic group, was able to conquer most of the Gold Coast. The Fante state, along the southern coast, rebelled many times against Asante (also known as Ashanti) rule; there were nine wars between the Asante and Fante during the nineteenth century.

To weaken the Asante and help the British gain control over trade routes, Great Britain helped the Fante and other southern groups wage war against the Asante in 1824, 1826, and 1863. In 1874, the Asante were finally defeated by this coalition.

Meanwhile, the Danes and the Dutch had left

the Gold Coast area. The Council of Merchants, which was headed by George Maclean from 1829 to 1843, helped the British establish colonial rule in the area. Maclean was not a governor general and did not officially represent the British government, yet he acted as a judge in the southern Gold Coast and encouraged missionary societies to become active there.

After Maclean died in 1847, the British government decided to take direct control of the southern part of the Gold Coast. A legislative council (with no African members) was set up, and a poll tax was established in 1852. Once the Asante were defeated, the British created the "Gold Coast Colony" in the area south of the Asante. Colonial rule became official in the Asante region itself in 1901.

Many European nations launched new efforts to set up colonies in Africa in the decades after 1880. The Germans and French moved into the areas east of the Volta and north of the Asante, and in response the British began moving beyond the Gold Coast Colony. Between 1890 and 1900, there were several wars and minor conflicts between the British and the Africans, French, and Germans. The British succeeded in taking control over all of what is now Ghana.

## The Africans' Resistance

One of the first to challenge British colonialism was Chief John Aggrey, who was king of the Cape Coast and had been educated in mission schools. He said that the British had no legal right to rule and complained that they were imposing martial law.

The major protest organization of the nineteenth century was the Aborigines Rights Protection Society (ARPS), which was formed in 1897.

The ARPS, made up of Ghanaians who had attended mission schools, called for constitutional reforms and protested when the British took land from Africans. This organization was active until about 1930.

The Pan-Africanist National Congress of British West Africa was influential between 1920 and 1930. This group demanded reforms in all areas of colonial life: education, sanitation, politics, medicine, and agriculture.

A variety of new organizations arose in the 1930's; together they were known as the "young men," because they were not led by the traditional chiefs. One of these groups, the West African Youth League, began to demand independence rather than colonial reforms. It was criticized by many chiefs, and its leader was forced to leave the colony.

After World War II, people all over Asia and Africa began demanding independence. The British responded with the Burns Constitution, which united the Gold Coast Colony and Asante and gave Africans representation in the legislative council. The Africans, however, did not

think that these reforms went far enough.

In 1947 the United Gold Coast Convention (UGCC) was formed to push for greater reforms. The UGCC leaders did not support the traditional chiefs, but they were reformers rather than radicals. Only reluctantly did they agree to make Kwame Nkrumah the secretary of the UGCC.

Nkrumah, who was thirty-nine years old at the time, had been studying in the United States and Great Britain for twelve years. When he returned to his homeland, he found a changed country. Large numbers of lower-class Ghanaians were ready to join the nationalist cause.

There were riots in early 1948, and soon Nkrumah and others broke away from the UGCC to form the Convention People's Party (CPP), whose slogan was "Self-Government Now." When Nkrumah called the people to resist the government, he and other CPP leaders were arrested. But this only made the CPP more popular.

Elections were held in February, 1951, and the CPP won thirty-four of thirty-eight seats. Between 1951 and 1954, the CPP was able to begin important reforms to benefit the lower classes. New opposition parties were formed, but the CPP was able to win the elections of 1954 and 1956.

On March 6, 1957, the CPP became the government of Ghana, the first African country to gain independence. The country's new name was taken from an inland trading state that had been powerful in the eleventh century.

**Consequences**

Ghana immediately began trying to help other African states become independent. Nkrumah, the first prime minister of Ghana, declared that "the independence of Ghana is meaningless unless it is linked up with the total liberation of the African continent." Conferences in Accra, Ghana's capital, were attended by nationalist leaders from all over Africa.

**1275**

*A crowd at Accra celebrates the birth of independent Ghana after the stroke of midnight on March 6.*

AP/Wide World Photos

Nkrumah also called for African states to work toward political unity. In 1958, Ghana and Guinea formed a political union, and Mali joined the association in 1961. Nkrumah helped to form the Organization of African Unity (OAU) in 1963.

*Catherine Scott*

# Western European Nations Form Common Market

*After the success of the Marshall Plan, European leaders designed the European Common Market to bring about economic cooperation and growth.*

**What:** Economics; International relations
**When:** March 25, 1957
**Where:** Rome
**Who:**

JEAN MONNET (1888-1979), head of French National Planning Council from 1945 to 1947, and president of the European Coal and Steel Community from 1952 to 1955

ROBERT SCHUMAN (1886-1963), French minister of foreign affairs from 1948 to 1952

KONRAD ADENAUER (1876-1967), chancellor of West Germany from 1949 to 1963

ALCIDE DE GASPERI (1881-1954), prime minister of Italy from 1945 to 1953

PAUL-HENRI SPAAK (1899-1972), premier of Belgium from 1947 to 1950, and secretary general of NATO from 1957 to 1961

## Steps Toward Unity

After World War II, the nations of Europe found themselves devastated: politically weak and with their economy in ruins. As they looked about, they saw that their former glory and power had been captured by the United States and the Soviet Union. Leaders such as Konrad Adenauer of Germany, Alcide De Gasperi of Italy, and Jean Monnet of France believed that if Europe was to regain its strength, it would have to be unified. The European nations would need to begin cooperating rather than competing with one another.

The United States provided Europe with a gift, the Marshall Plan, to help rebuild industry and trade. In 1948, the Organization of European Economic Cooperation was formed, with fifteen nations participating, to supervise the use of Marshall Plan funds. The Netherlands, Belgium, and Luxembourg had already created "Benelux," an economic union, and had abolished all duties charged on goods crossing their borders.

In 1951, Robert Schuman led in the formation of the European Coal and Steel Community (ECSC); its purpose was to abolish trade barriers, coordinate industry, restrict trusts, and control the production of weapons. Another step toward unity was the European Defense Community, but this plan was rejected in the summer of 1954 by the French Assembly. Still, working together had made Europe economically healthy and prosperous, and many European leaders wished for even more economic cooperation.

In 1955, the foreign ministers of the ECSC nations met in Messina, Italy, to decide what would be next. The British walked out of this meeting, thinking that it was of no importance. Under the leadership of Paul-Henri Spaak of Belgium, however, the Messina talks continued in a chateau outside Brussels. In May, 1956, the Spaak Committee came up with a proposal that became the basis for the European Common Market.

## The Market Is Created

The Suez Canal crisis and the Hungarian uprising in 1956 made it clear that Europe was still weak and lacked direction. In response, the foreign ministers of Belgium, France, Germany, Italy, Luxembourg, and the Netherlands met and signed the Treaty of Rome on March 25, 1957, to create the Common Market. At the same time they formed the European Atomic Energy Community (Euratom), which was to coordinate the development of nuclear power as a source of energy.

The Treaty of Rome consists of 248 articles setting up a process of economic partnership. According to the treaty, economic union was to be a first step in "an even closer union among European peoples." The ideals of Adenauer, De Gasperi, and Schuman were finally being given a chance.

Over a period of twelve to fifteen years, each nation's customs barriers were to be eliminated. As a result, a free trade area would be established that would include 160 million people. Laws concerning social welfare and taxes would be harmonized, and eventually the Common Market nations would work to set up shared agriculture and transportation policies.

Action would be taken through several new institutions. The Council of Ministers was composed of ministers of member states; they would make major decisions by coming to a unanimous vote. The European Commission was an executive body with nine members; its decisions could be vetoed by the Council of Ministers. There was also an Assembly of 132 delegates appointed by the parliaments of member nations; its task would be to discuss proposals from the council and the commission. The Court of Justice, with seven judges, would deal with cases where the treaty had been violated. Brussels, Belgium, was chosen as the Common Market's headquarters.

**Consequences**

The Common Market began to operate in January, 1958, and by the end of 1961 it had pro-

duced results. Tariffs had been cut by 40 percent on industrial products and 30 percent on agricultural products. Twelve former African colonies were given certain privileges in the Common Market: tariff advantages, funds to help in their development, and protection for their new industries. By the 1970's, the Common Market had brought about strong economic growth in Europe.

In 1961 the Common Market nations considered admitting Great Britain to membership, though the British had not been interested earlier. After two years of studying the issue, French president Charles de Gaulle vetoed British membership. He did not consider Great Britain a part of the European Continent and did not like the fact that the British had close economic ties with the United States.

After de Gaulle died in 1970, France took a new position. In May, 1971, British prime minister Edward Heath and French president Georges Pompidou met in Paris. Their talks were positive, setting the stage for more negotiations between Great Britain and the Common Market.

Finally, on January 1, 1973, not only Great Britain but also Ireland and Denmark officially joined their European neighbors and became participating members of the Common Market. At the beginning of the twenty-first century, with more members than ever before, the Common Market—which had come to be called the European Economic Community, or EEC—was living up to the ideals of its founders, creating a Europe with close economic and political ties.

*Thomas P. Wolf*

**1278**

# Jodrell Bank Radio Telescope Is Completed

> *The Big Dish at Jodrell Bank, the world's largest fully steerable radio telescope dish for nearly twenty years, greatly advanced radio astronomy.*

**What:** Astronomy
**When:** August, 1957
**Where:** Jodrell Bank, near Manchester, England
**Who:**
SIR BERNARD LOVELL (1913-     ), an English radio astronomer
KARL JANSKY (1905-1950), an American radio engineer
GROTE REBER (1911-     ), an American radio engineer

## Widening the Spectrum

Until about the 1930's, astronomy relied almost exclusively on visible light. This changed in 1931, amid New Jersey potato fields. While searching for the source of static in ship-to-shore radio communications, Karl Jansky, an engineer for Bell Laboratories, had built a large radio antenna. What he discovered was that radio waves were coming from space, most strongly from just above the stars that formed the "stinger" of the scorpion in the constellation Scorpius. This was soon recognized as the direction of the center of the Milky Way galaxy.

Radio waves and visible light are members of a family called "electromagnetic waves," to which X rays, infrared waves, ultraviolet waves, and microwaves belong as well. Electromagnetic waves differ in frequency—that is, in the number of waves that pass a certain point per second. Borrowing a musical term, astronomers sometimes refer to a certain set of frequencies as an "octave." While visible light contains only two octaves, the entire electromagnetic spectrum contains about fifty. Jansky's detection of radio waves allowed astronomers to "hear" many new octaves. The explosion of new astronomical infor-

mation during the twentieth century was a direct result of Jansky's discovery.

Astronomers recognized the potential of Jansky's discovery only after Grote Reber created a radio map of the galaxy using a radio telescope he had built in his garden in Wheaton, Illinois. He reported his findings to the scientific community in the early 1940's, during World War II.

Many historians attribute the success of the Royal Air Force during the Battle of Britain to the invention and refinement of "radar," which stands for *r*adio *d*etecting *a*nd *r*anging. The English led the field in radar and radio astronomy because, unlike optical signals, radio waves are unaffected by clouds, of which England has many. When the war in Europe was over, surplus equipment and veteran experimenters, including Sir Bernard Lovell, the eventual creator of the Big Dish, became available for general scientific research.

After completing his doctorate in physics at Bristol University in 1936, Lovell had become assistant lecturer in physics at the University of Manchester. Returning to the University of Manchester after the war with two trailers of radar equipment borrowed from the army, Lovell followed up on radar studies of meteor showers. To avoid radio interference from the city of Manchester, Lovell conducted his research at the university's Jodrell Bank Experimental Station in Cheshire. At the time, Jodrell Bank was a botanical research station; no one suspected it would become a global center for a new branch of astronomical research.

## Bigger and Better

Soon afterward, Lovell erected a stationary radio aerial antenna almost 67 meters across. Although this large dish soon afforded fascinating glimpses of the radio universe, its immobility frustrated Lovell and his colleagues. Carried by the rotation of the earth, it collected radio waves from only a small swath of the sky. Named in 1951 as director of the Jodrell Bank Laboratory, Lovell convinced the Nuffield Foundation and the English government to share the cost of building a giant, fully steerable radio telescope, which would span about as many octaves of radio waves as a piano keyboard spans in sound.

Sir H. Charles Husband, the engineer in charge of constructing the dish, faced two major difficulties. The first—how delicately to align something so large—was solved by Patrick M. S. Blackett, who had won the 1948 Nobel Prize in Physics. Blackett proposed that the telescope be moved by the gear-and-rack mechanisms from the dismantled battleships *Royal Sovereign* and *Revenge*. These mechanisms had previously turned the turrets of their ships' 38-centimeter guns.

The second and greater difficulty concerned the rigidity of the huge reflector: It must retain its precise shape when moved by the gear-and-rack mechanisms as well as when subjected to wind. In the preceding decade, a suspension bridge over the Tacoma Narrows in Washington had been shaken apart by vibrations from a mild wind. To avoid such a problem, engineers studied scale models of the Big Dish in wind tunnels under gale conditions and then refined its design. Husband used some 300,000 kilograms of steel in the 76-meter-wide bowl, a diameter that is more than two and one-half times larger than the dome of the Grand Rotunda in the United States Capitol Building.

Construction of the facility was authorized in 1952. The Big Dish cost almost $2 million, about $800,000 more than was expected. Nevertheless, the scientific return was worth the money: Dur-

ing the evening of August 2, 1957, the Milky Way wrote its radio signature on the detectors of the Big Dish for the first time.

## Consequences

During its long and distinguished lifetime, the Big Dish would hear many signals of great scientific interest. Few, if any, would create the public sensation occasioned by Lovell's announcement, on October 13, 1957, that he had tracked the carrier rocket of Sputnik 1, the first artificial satellite, as it had flown over England the previous night. Two years later, on September 13, 1959, Lovell verified that Luna 2, the first spacecraft to reach the lunar surface, had crashed on the Moon. He also tracked the first lunar orbital mission by Luna 10 in 1966 and the history-making manned landing on the Moon by Apollo 11 on July 20, 1969.

The Big Dish is not the only instrument that Lovell contributed to astronomy. The Multi-Element Radio-Linked Interferometer (MERLIN), named for a famous English scholar, used two separate receivers that were connected to achieve superior resolution. With a resolution thousands of times finer than what the unaided human eye can achieve, MERLIN has disclosed the cores of radio galaxies and quasars in unexpected detail.

On September 30, 1980, Lovell retired as director of Jodrell Bank (now more properly called the Nuffield Radio Astronomy Laboratories), his career marked by spectacular triumphs and swirling controversy. In his 1968 book *The Story of Jodrell Bank*, Lovell wrote, "It is unlikely that a scientific project has ever survived so many crises to attain ultimate success."

*Clyde Smith*

## 1280

# GREAT EVENTS

## 1900-2001

# CATEGORY INDEX

## LIST OF CATEGORIES

## AGRICULTURE

Congress Reduces Federal Farm Subsidies and Price Supports, **7**-2760

Gericke Reveals Significance of Hydroponics, **2**-569

Insecticide Use Increases in American South, **1**-359

Ivanov Develops Artificial Insemination, **1**-44

Morel Multiplies Plants in Vitro, Revolutionizing Agriculture, **3**-1114

Müller Develops Potent Insecticide DDT, **2**-824

## ANTHROPOLOGY

Anthropologists Find Earliest Evidence of Modern Humans, **6**-2220

Benedict Publishes *Patterns of Culture*, **2**-698

Boas Lays Foundations of Cultural Anthropology, **1**-216

Boule Reconstructs Neanderthal Man Skeleton, **1**-183

Dart Finds Fossil Linking Apes and Humans, **2**-484

Humans and Chimpanzees Are Found to Be Genetically Linked, **5**-2129

Johanson Discovers "Lucy," Three-Million-Year-Old Hominid Skeleton, **5**-1808

Last Common Ancestor of Humans and Neanderthals Found in Spain, **7**-2841

Leakeys Find 1.75-Million-Year-Old Hominid Fossil, **4**-1328

Mead Publishes *Coming of Age in Samoa*, **2**-556

117,000-Year-Old Human Footprints Are Found Near South African Lagoon, **7**-2860

Pottery Suggests Early Contact Between Asia and South America, **3**-1230

Simons Identifies Thirty-Million-Year-Old Primate Skull, **4**-1553

Weidenreich Reconstructs Face of Peking Man, **2**-786

Zdansky Discovers Peking Man, **2**-476

IV

**XIV**

**XVI**

## LABOR

## XXI

## PHOTOGRAPHY

## PHYSICS

**XXII**

**XXIII**

## POLITICS

**XXV**